HANDBOOK OF NANOSAFETY

HANDBOOK OF NANOSAFETY

MEASUREMENT, EXPOSURE AND TOXICOLOGY

ULLA VOGEL, MSc, PhD
*Professor and Head of Nanotoxicology, Danish Nanosafety Centre,
National Research Centre for the
Working Environment (NRCWE), Copenhagen, Denmark*

KAI SAVOLAINEN, PhD
*Professor and Head, Department of Industrial Hygiene and Toxicology,
Finnish Institute of Occupational Health (FIOH), Helsinki, Finland*

QINGLAN WU, PhD
*Coordinator of the EU project NANOTRANSPORT,
Principal Researcher at Det Norske Veritas As., Oslo, Norway*

MARTIE VAN TONGEREN, PhD
*Director, Centre for Human Exposure Science (CHES),
Institute of Occupational Medicine (IOM), Edinburgh, UK*

DERK BROUWER, PhD
*Senior Scientist, Netherlands Organisation for
Applied Scientific Research, Zeist, The Netherlands*

MARKUS BERGES, PhD
*Head of Exposure Assessment division of Berufsgenossenschaftliches
Institut für Arbeitsschutz, Sankt Augustin, Germany*

AMSTERDAM • BOSTON • HEIDELBERG • LONDON
NEW YORK • OXFORD • PARIS • SAN DIEGO
SAN FRANCISCO • SINGAPORE • SYDNEY • TOKYO
Academic Press is an imprint of Elsevier

ELSEVIER

Academic Press is an imprint of Elsevier
32 Jamestown Road, London NW1 7BY, UK
225 Wyman Street, Waltham, MA 02451, USA
525 B Street, Suite 1800, San Diego, CA 92101-4495, USA

Notice

Knowledge and best practice in this field are constantly changing. As new research and experience broaden our understanding, changes in research methods, professional practices, or medical treatment may become necessary. Practitioners and researchers must always rely on their own experience and knowledge in evaluating and using any information, methods, compounds, or experiments described herein. In using such information or methods they should be mindful of their own safety and the safety of others, including parties for whom they have a professional responsibility. To the fullest extent of the law, neither the Publisher nor the authors, contributors, or editors, assume any liability for any injury and/or damage to persons or property as a matter of products liability, negligence or otherwise, or from any use or operation of any methods, products, instructions, or ideas contained in the material herein.

British Library Cataloguing-in-Publication Data
A catalogue record for this book is available from the British Library

Library of Congress Cataloging-in-Publication Data
A catalog record for this book is available from the Library of Congress

ISBN: 978-0-12-416604-2

For information on all Academic Press publications visit our website at elsevierdirect.com

Typeset by TNQ Books and Journals
www.tnq.co.in

Printed and bound in the United States

Transferred to Digital Printing, 2013

Full Contents

Foreword

Looking back at previous accidents and occupational diseases one very quickly realizes that most of them, if not all, could have been prevented by better management. Sometimes with better knowledge and design, sometimes with better rules and enforcement and sometimes with better availability of human capacities to face adverse work environments or to run ahead of events.

Looking back at technological progress it is clear that risks evolve; they emerge, get mature and die-out. The emerging phase is very knowledge heavy. Existing knowledge is inadequate to thoroughly understand processes, there is a general tendency to apply old trusted assumptions, technical approaches and mind-sets that worked successfully in the past are, somehow blindly, copied. Uncertainties, conflicting research results and inadequacy of equipment and of skills play their role. Public perceptions do not escape this overall obscure reality. Old technologies that have been successful in ensuring better living conditions are the hardest to replace precisely for their same success.

Risk in itself is only an expectation, an analyst's assessment of possible events with possible adverse effects. As such it does not "exist" outside the analysts' minds. Even a zero value does not mean that the event is impossible; it only means that the analyst (or several analysts) expects that the event is impossible. Risk analysis is, nevertheless, a powerful tool, since it allows scientific insight into the future.

Risk is defined as the product of two probabilities: That the event/hazard occurs and that it has adverse effects. In the case of occupational risks the risk is the product of the probability that a material has a toxic behavior and of the probability that exposure occurs above thresholds that trigger adverse effects.

The hazard quantification depends on the material properties (physical, chemical, biological, etc.), the environment in which the material will be found in the one or the other phase of its life and its behavior with regards to the toxicity end-points considered.

Exposure quantification depends on the material behavior with regard to the possible uptake by organisms, either directly from the original source or indirectly.

Different risks related to the same material and the same exposure scenario do not simply add up. They should come together combined in a risk assessment, which completes the risk analysis.

What comes next is the question of risk acceptance. Acceptance or non-acceptance depends on the value the society as final risk bearer reserves to human health and integrity and to environmental protection and the amount of resources the society accepts to commit. Both vary with space and time. This question, although supported by science, is basically societal in nature, and is outside the scope of this book.

Unacceptable risk needs to be managed by eliminating/reducing the hazard or the exposure or both. It usually requires a decision on the amount of today's resources that can be committed for the benefit of tomorrow. An over-managed risk commits present resources unnecessarily, while an under-managed risk compromises tomorrow's well-being.

Risk management is a different story for which some remarks on regulation need to be recalled.

Regulation essentially consists in formalization of behavior. It is either agreed or imposed. Its driving force is the perception on the side of one societal group that another group does not behave as it would be legitimately expected to by the first. The formalization of behavior is directed to jobs (e.g., building regulation) or to skills (e.g., architect's degree) or both. Violation of regulation is punishable according to pre-set standards and procedures. It is interesting to note that regulation implementation can take an a-posteriori form, punishing for damages that have occurred, or an a-priori form, punishing for behavior that could cause, but has not actually caused, damage.

In this second case the burden of proof falls with the enforcement authority which faces the challenge of objectively measuring both hazard and exposure to demonstrate beyond doubt that risks values, again the value of something that does not physically exist, are in the forbidden area and as a consequence corrective action must be taken. This challenge necessitates scientific knowledge and skills universally recognized as adequate, or simply imposed, for that purpose.

After these words, which simply try to cut short a very long story, the structure of this book is more than evident. After the General Introduction, exposure scenarios are discussed, and the possible Effects of Nanomaterials on Human Health are analyzed. From Source to Dose: Emission, Transport, Aerosol Dynamics and Dose Assessment for Workplace Aerosol Exposure comes next, logically leading to the technologies for Measurement of Aerosol Nanomaterials and the associated Quality Control of Measurement Devices. Exposure Assessment Strategies come next, and the book concludes its main part with the Risk Assessment and Risk Management chapter. Examples and Case Studies are given as guidance, followed by some information on the Safe Use of Nanomaterials and the Future, though this is not the main scope of this book.

Most of the information presented is the result of the hard work mainly undertaken inside the EU-RTD project, NANODEVICE, partly financed by the NMP programme itself as part of the larger Framework Programme 7 of the European Union. It integrates knowledge from several other projects and sources, and the NANODEVICE team is thankful to all those who one way or another have contributed and continue to do so. It does not claim to provide definitive answers, as this would be unreasonable for any emerging risk, but I hope it will provide the foundation for these answers.

Finally, I wish to thank all the researchers for their groundbreaking work, for their collaboration spirit along many European countries and many scientific cultures, and for the diligence and openness they demonstrated in the very often tough scientific discussions. Many thanks are extended to the project managers and the several chapter editors for their efforts in giving this information the shape of a book that I believe will come out to be a reference for future work.

Georgios Katalagarianakis

Disciaimer

The author expresses his own thoughts not necessarily reflecting those of the EC.

List of Contributors

Olivier Aguerre Institut National de l'Environnement industriel et des RISques (INERIS), Verneuil-en-Halatte, France

Robert J. Aitken Institute of Occupational Medicine (IOM), Edinburgh, United Kingdom; Institute of Occupational Medicine (IOM), Singapore

Harri Alenius Systems Toxicology, Finnish Institute of Occupational Health, Helsinki, Finland

Christof Asbach Institute of Energy and Environmental Technology (IUTA), Air Quality & Sustainable Nanotechnology, Duisburg, Germany

Markus G.M. Berges Institute of Occupational Safety and Health of the German Social Accident Insurance (DGUV-IFA), Sankt Augustin, Germany

Christophe Bressot Institut National de l'Environnement industriel et des RISques (INERIS), Verneuil-en-Halatte, France

Thomas Brock Berufsgenossenschaft Rohstoffe und chemische Industrie (BG RCI), Heidelberg, Germany

Derk H. Brouwer TNO, Quality of Life, Research & Development, Zeist, The Netherlands

Julia Catalán Systems Toxicology, Finnish Institute of Occupational Health, Helsinki, Finland

Dirk Dahmann IGF Institut für Gefahrstoff-Forschung, Institut an der Ruhr-Universität Bochum, Bochum, Germany

Udo Gommel Fraunhofer Institute for Manufacturing Engineering and Automation IPA, Department Ultraclean Technology and Micromanufacturing, Stuttgart, Germany

Boris Gorbunov Naneum Ltd., Canterbury Innovation Center, Canterbury, United Kingdom

Hans-Georg Horn TSI GmbH, Aachen, Germany

Keld Alstrup Jensen Danish Nanosafety Centre, National Research Centre for the Working Environment, Copenhagen, Denmark

Heinz Kaminski Institute of Energy and Environmental Technology (IUTA), Air Quality & Sustainable Nanotechnology, Duisburg, Germany

Markus Keller Fraunhofer Institute for Manufacturing Engineering and Automation IPA, Department Ultraclean Technology and Micromanufacturing, Stuttgart, Germany

Ismo Kalevi Koponen Danish Nanosafety Centre, National Research Centre for the Working Environment, Copenhagen, Denmark

Thomas A.J. Kuhlbusch   Institute of Energy and Environmental Technology (IUTA), Air Quality & Sustainable Nanotechnology, Duisburg, Germany; Center for NanoIntegration Duisburg-Essen (CENIDE), Duisburg, Germany

Olivier Le Bihan   Institut National de l'Environnement industriel et des RISques (INERIS), Verneuil-en-Halatte, France

Andre Lecloux   Belgium Nanocyl s.a., Sambreville, Belgium

Göran Lidén   Stockholm University, Stockholm, Sweden

Hanna Lindberg   Systems Toxicology, Finnish Institute of Occupational Health, Helsinki, Finland

Marita Luotamo   European Chemicals Agency (ECHA), Helsinki, Finland

Martin Morgeneyer   Université de Technologie de Compiègne (UTC), Compiègne, France

Robert Muir   Naneum Ltd., Canterbury Innovation Center, Canterbury, United Kingdom

Hannu Norppa   Systems Toxicology, Finnish Institute of Occupational Health, Helsinki, Finland

Jaana Palomäki   Systems Toxicology, Finnish Institute of Occupational Health, Helsinki, Finland

Uwe Rating   Institute of Energy and Environmental Technology (IUTA), Air Quality & Sustainable Nanotechnology, Duisburg, Germany

Sheona A.K. Read   Institute of Occupational Medicine (IOM), Edinburgh, United Kingdom

Bryony L. Ross   Institute of Occupational Medicine (IOM), Edinburgh, United Kingdom

Araceli Sánchez Jiménez   Institute of Occupational Medicine (IOM), Edinburgh, United Kingdom

Kai Savolainen   Nanosafety Research Centre, Finnish Institute of Occupational Health, Helsinki, Finland

Martin Seipenbusch   Karlsruhe Institute of Technology KIT, Karlsruhe, Germany

Neeraj Shandilya   Institut National de l'Environnement industriel et des RISques (INERIS), Verneuil-en-Halatte, France; Université de Technologie de Compiègne (UTC), Compiègne, France

Burkhard Stahlmecke   Institute of Energy and Environmental Technology (IUTA), Air Quality & Sustainable Nanotechnology, Duisburg, Germany

Ana María Todea   Institute of Energy and Environmental Technology (IUTA), Air Quality & Sustainable Nanotechnology, Duisburg, Germany

Martie van Tongeren   Centre for Human Exposure Science (CHES), Institute of Occupational Medicine (IOM), Edinburgh, United Kingdom

Mingzhou Yu   Institute of Earth Environment, Chinese Academy of Sciences, Xi'an, China; China Jiliang University, Hangzhou, China

Acknowledgements

The editors wish to thank Kirsten Rydahl and Ms. Leila Ahlström for their immense help in the final editing of the handbook. Dr. Ewen McDonald's remarkable efforts to improve the quality of the language of many of the chapters of the book are also greatly acknowledged. The handbook is prepared as part of the NANODEVICE project and was supported by EU FP7 CP-IP 211464 (NANODEVICE).

1

General Introduction

Kai Savolainen

Nanosafety Research Centre, Finnish Institute of Occupational Health, Helsinki, Finland

The potential of engineered nanomaterials (ENM) and nanotechnologies to improve the quality of life, to contribute to economic growth and to sharpen the competitiveness of industry is now widely recognized, not only in Europe, but globally. Nanotechnologies can permit remarkable technological advances and innovations in many industrial sectors including the chemical, pharmaceutical, pulp and paper, food and information industries, as well as energy production and consumer items. However, there is an ongoing scientific and political debate about the potential risks of ENM and nanotechnologies [1–6]. One must not simply praise the technological benefits; the concepts of safety and health have to be incorporated into all thinking related to the production of engineered nanomaterials and emerging nanomaterials, especially the future generations of engineered nanomaterials [7], the so-called second- (active nanomaterials), third- (self-assembling nanomaterials) and fourth-generation nanomaterials (nano-robots, countless new nanomaterial innovations), which will briefly be mentioned here, although they are largely outside the scope of this discussion.

The EU 2020 strategy defines smart, sustainable and inclusive growth as the principal European 2020 objective. Research and innovation have been identified as the twin drivers of European social and economic prosperity, i.e., capable of generating growth combined with environmental sustainability [8]. Recently, the same issues have been highlighted in several American documents on environmental health and safety strategies [9–11].

The competitiveness of the European industry is the crucial factor in achieving these challenging goals, i.e., the role of innovations and an accelerated pace of the commercialization of innovations have been

Handbook of Nanosafety
http://dx.doi.org/10.1016/B978-0-12-416604-2.00001-9

1

recognized as being the foundation stones of growth. The recent Communication from the Commission on Horizon 2020 — The Programme for Research and Innovation [12] emphasizes the importance of research and innovation for society at large. The same issues have been stressed in the EU Nanotechnology Action Plan 2005—2009 [1].

These considerations all mean that in the future there will be remarkable changes in the way that the European Commission and the EU Member States evaluate these emerging materials and technologies. The proposal for establishing the new Programme for Research and Innovation, Horizon 2020 [13] places the major emphasis on securing a strong position in key enabling technologies (KET) such as information and communication technologies (ICT), nanotechnology, advanced materials, space technology and biotechnology. The document underlines their significance to Europe's competitiveness and its ability to provide the new products and innovative services essential for meeting global challenges. In particular, nanotechnology offers substantial possibilities for improving the competitive position of the EU and for responding to key societal challenges. The need to ensure the safe development and application of nanotechnologies has been included in the broad line of activities of the Horizon 2020 proposal [14]. Over the years (see 15—18) it has become increasingly clear that one cannot have successful nanotechnologies without the parallel assurance that these novel materials are safe and pose no threat to human health or the environment. Hence, both engineered nanomaterials and the nanotechnologies utilizing these materials have to prove their trustworthiness [19,20].

In fact, it is not enough that the new technology applications should be safe in themselves, they should also confer substantial improvements on human health and offer environmental protection. Due to the rapidly increasing production and use of ENM and utilization of nanotechnologies, these safety aspects must be fully understood and addressed. Even though it is unlikely that the size of nanomaterials *per se* can be viewed as a hazard or pose a threat to human health or the environment [3], their small size does mean that they gain ready access to living organisms, and this raises the question of their biocompatibility [21]. At present, there is scientific uncertainty about the safety of several of these materials. Emphasizing the importance of the safety assessment of the nanosized substances, a facet raised not only by the European Commission and several EU Member States, but also by authorities outside Europe, e.g., in the US [10,11,22]. It is recognized that the safety of the manufacturing processes, as well as the technologies and products utilizing engineered nanomaterials, will be crucial if these materials, technologies and products are to be successful.

1.1 USE AND APPLICATIONS OF ENGINEERED NANOMATERIALS

If one wishes to assess the usefulness of the potential benefits of nanotechnologies in the future, then it is important to have a reasonably good understanding of the complexities associated with these novel technologies and materials, especially those related to safety and health. During its exploration and assessment of the protection of human health in the context of engineered nanomaterial and nanotechnologies, this book also considers their impact on the environment. There is an urgent need for finding reliable ways to predict the potential health and environmental hazards of these technologies and materials; only in this way can the safety of engineered nanomaterials and their nanotechnology applications be assured. Engineered nanomaterials and their nanotechnology applications offer huge benefits and potential for the future both in terms of scientific and technological progress and economic expectations, but these materials and technologies have to be proved to be safe; there has to be a guarantee that the various stakeholders can rely on the safety of these materials and technologies, especially when one considers the wide range of applications where they will be used, e.g., consumer products, manufacturing and industrial processes [23]. Assuring the safe use of engineered nanomaterials and nanotechnologies is an integral part, not only of risk management, but also of the risk governance, of these materials and technologies. This needs to be done in such a way that the legitimate interests of various stakeholders in society are taken into account, i.e., from the manufacturer on to the consumer and ultimately to the environment [24].

To date, the number of consumer products containing engineered nanotechnologies according to producers exceeds several thousands [23], and the annual turnover of goods incorporating engineered nanomaterials has been predicted to exceed three trillion dollars by 2020. These numbers highlight the predicted economic value of these technologies in the future; see Figure 1.1 for the economic expectations [25].

When one attempts to sketch the value chain of nano-enabled products, as well as the safety and health issues related to the various steps of this chain, then all steps in the life-cycle of the nanotechnologies have to be evaluated. These steps, which will be discussed later in detail, include 1) production, transportation and storage; 2) incorporating engineered nanomaterials into primary products such as powders or polymers; 3) generating secondary products in which the primary products are used; and 4) processes leading to side-products and waste, and incorporation of recycled engineered nanomaterials

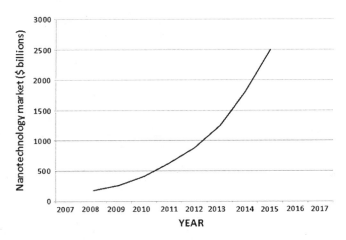

FIGURE 1.1 **The predicted size of the global market of products incorporating engineered nanomaterials.** *Adapted from Lux Research, 2009 [25].*

back into the value chain, thus prolonging the life-cycle of these materials. In all these stages releases may take place and lead to the exposure of humans or the environment, with workers handling these materials being the most likely population to be exposed. The value chain of engineered nanomaterials and the steps at which releases into the environment possibly can take place are depicted in Figure 1.2.

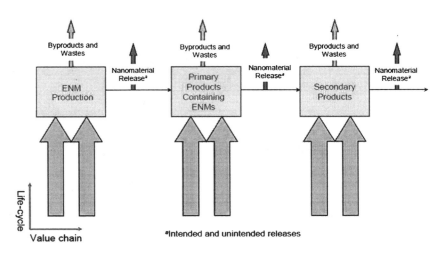

FIGURE 1.2 **Potential human and ecosystem exposure through the value chain and life-cycle of engineered nanomaterial production, use and disposal.** *Source: National Academy of Sciences: EHS Strategy of Engineered Nanomaterials, 2012 [11].*

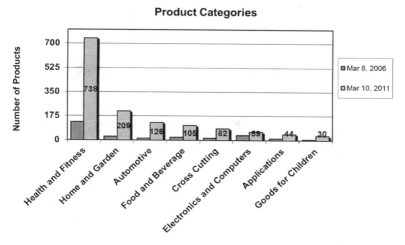

FIGURE 1.3 Different consumer product categories incorporating engineered nanomaterials. The figure clearly shows that the healthcare sector has become the leader in using these materials in healthcare products. *Adapted from PEN; www.nanotechproject.org/ [23,26].*

The reader also needs to have a realistic appreciation of what may be expected in the near future with respect to major scientific observations and breakthroughs in those areas in which nanotechnologies and engineered nanomaterials may play a significant role, and in which safety and health are important issues. The range of products and activities that encompasses these materials and technologies is rapidly growing as man's abilities to manipulate materials at the atomic or near atomic scales improve. At the same time, the use of engineered nanomaterials continues to expand. Clearly, when the production of engineered nanomaterials changes from small scale to mass production, the likelihood of leaks into the occupational environment increases [3,11]. Figure 1.3 depicts different groups of consumer products incorporating engineered nanomaterials and their relative importance in terms of numbers of products in several product categories [26].

1.2 WHAT IS A NANOMATERIAL?

Recently, one of the important discussion topics has been the definition of what constitutes an engineered nanomaterial. Some experts have strongly opposed any definition, as they claim that the definition will soon become obsolete and it even may hinder the progress of research [4], whereas other experts strongly favor a clear definition, as this can assist in developing regulations and legislation of engineered nanomaterials [27]. The European Commission has recently adopted a recommendation [28] about the definition of nanomaterial, according to which

> 'nanomaterial' means a natural, incidental or manufactured material containing particles, in an unbound state or as an aggregate or as an agglomerate and where, for 50% or more of the particles in the number size distribution, one or more external dimensions is in the size range 1 nm—100 nm. In specific cases and where warranted by concerns for the environment, health, safety or competitiveness the number size distribution threshold of 50% may be replaced by a threshold between 1 and 50%.

This definition is currently being incorporated into different pieces of horizontal EU legislation. Indeed, the potential importance of this definition lies mainly in the implementation of the current EU chemical legislation, REACH [29], and it will provide a definition that can be used in various items of EU horizontal chemical legislation e.g., cosmetics [30], food and other consumer products. However, when exploring the many unknown characteristics of engineered nanomaterials, as well as the development of methods for measuring exposure to these materials or identifying possible hazards of a range of engineered nanomaterials, then the definition is likely to be of lesser importance. This is because our understanding of the various physical chemical characteristics of engineered nanomaterials continues to increase, and this new information will be incorporated into all research endeavors [21,31].

1.3 EXPOSURE TO ENGINEERED NANOMATERIALS MERITS ATTENTION

The rapid growth of the utilization of engineered nanomaterials in consumer and industrial products and processes has led to an increased likelihood of human exposure as well as elevated concentrations of engineered nanomaterials in occupational environments. This has also triggered concerns about the exposure levels to these materials, an issue that has been largely overlooked so far [33—35].

Kuhlbusch et al. [36] have carefully evaluated the currently available data on exposure to different engineered nanomaterials and exposure situations. They emphasized that different approaches can be used to assess exposure in workplaces and these include 1) studies based on real workplaces and 2) process-based studies in simulated workplaces and of simulated work processes. The major benefits of the first approach are that the data originate from a workplace with real work conditions. However, due to the presence of background aerosols in the general work environment or produced during the process itself, extensive and time-consuming measurement campaigns have to be conducted. The latter approach, which is based on work done in laboratories, allows the clear differentiation between a release from the investigated process and that of background aerosols emerging from other sources than the process itself. Both protocols may be valuable when assessing the potential exposure to engineered nanomaterials in workplaces, which will be the primary contact point between humans and the materials. Both approaches also need to be used because an increasing number of workers are becoming exposed to elevated levels of these materials in their workplaces (see 36).

In global terms, the increased probability of exposure to these materials (see 34,36) has been recognized as an emerging risk associated with these materials and technologies. This is due to the potential health risks of engineered nanomaterials exposure, both to human health and the environment [3]. The consequences of this development are also the raised public awareness and concerns about the safety of engineered nanomaterials and their associated technologies. This is why it is so important to set realistic priorities and goals for evaluating the safety of these materials; in fact the safety and health issues surrounding nanomaterials and nanotechnologies are surrounded by uncertainties [17]. The monitoring of exposure levels at workplaces requires affordable, easy-to-use, preferably portable measuring devices that can be used by the companies to assess the true exposure to the engineered nanomaterials they produce or utilize in their production processes. Previously, these kinds of monitoring devices were unavailable, but now they are emerging onto the market for use by all who are interested in assessing workplace exposure to these materials. Even though it will be some time before comprehensive data sets on exposure at workplaces are available for experts and regulators, the recent developments have made it possible to devise regulations for monitoring exposure levels to engineered nanomaterials with an agreed set of metrics. This development will also enable the regulators to promote safety in workplaces where engineered nanomaterials are being used, e.g., it will allow them to impose occupational exposure levels as more information on the workplace exposure and potential hazards associated with the exposure becomes available. Nonetheless, it needs to be emphasized that before one can predict hazards in workplaces by assessing exposure in occupational environments, major breakthroughs in understanding the associations between the characteristics of materials and hazard will be required (see 37–39).

1.4 HOW TO MEASURE EXPOSURE TO ENGINEERED NANOMATERIALS

It is important to emphasize that measurement of exposure to engineered nanomaterials is not simply concerned with measuring the concentration of these materials. If one wishes to assess exposure to these materials, it is important to appreciate the magnitude of the particles' number, concentration, their size (size distribution), their physical form (powder or slurry), route of absorption, frequency and duration of exposure and distribution (spatial, within-worker, between-workers, possible control measures). It is also important to have practical insights at the workplace into the level of occupational hygiene at a given site so that necessary precautions can be initiated [36].

Engineered nanoparticles express several characteristics that can be measured and that reflect different features of these materials. These characteristics are called metrics, and the most frequently measured metrics include mass concentration, number concentration and surface concentration. All these metrics contain weaknesses, and hence in many cases all of them need to be measured. For example, the mass of engineered nanomaterials is typically so small that in many instances accurate measurement of mass may be inaccurate and require a long sampling time. Hence, detection of high peak exposures might be impossible. On the other hand, the measurement of number concentrations may encounter huge spatial distributions and temporal variations. Furthermore, the distinction between process-derived engineered nanoparticles and the concentration of ubiquitous ultrafine background particle concentrations, which are also in the nano-size, may be very difficult. This is further complicated by the fact that currently there are no available technologies that can make a chemical distinction between process-derived engineered nanoparticles and ubiquitous background ultrafine particles. Finally, surface concentrations are difficult to use alone because the rapid agglomeration and aggregation of nano-sized particles complicates the reliable assessment of the presence of nano-sized particles.

Currently, there is a consensus that the best approach to assess the exposure to engineered nanomaterials, especially in the workplace, is represented by the nanoparticle emission assessment technique (NEAT). This strategy is based on the detection of airborne nanomaterials by using portable instruments measuring the number concentrations of the nanomaterials together with filters that allow offline analysis of the samples on the filters for particle morphology, size, count and elemental composition.

Nonetheless, implementation of the measurement of metrics described above and using the NEAT strategy continues to be a challenge. One of the reasons has been that affordable, easy-to-use and also portable instruments have been largely unavailable. This situation is now rapidly changing as new instruments are entering the market (www.nano-device.eu). However, even if the measurement of engineered nanomaterials in workplaces and elsewhere, especially in aerosols, will become easier than before, the incomplete understanding of the biological significance of the many measures of hazard will remain a challenge until the association between material characteristics and their health consequences is better understood (for this general discussion see [36,40]). Hence, in addition to knowledge on exposure and the measurement of exposure to engineered nanomaterials, it is also important to obtain an understanding of the potential hazards, i.e., toxicity, of engineered nanomaterials, preferably at reliable exposure concentrations or doses (see, e.g., 41).

1.5 HOW ABOUT THE HAZARDS?

In experimental animal studies, engineered nanomaterials have been associated with a number of hazards. Several organ systems have been affected in one way or another, displaying some harmful effects of engineered nanomaterials. These include, for example, 1) the lungs; 2) the cardiovascular system; 3) the central and peripheral nervous systems; 4) the immunological system; 5) the skin; 6) the gastrointestinal tract; 7) the reproductive system; and 8) as a more general toxic form of effects, the induction of genotoxicity and increased probability of cancer induction (see 42). Several engineered nanomaterials such as metal oxides (titanium dioxide) and single- and multi-walled carbon nanotubes (SWCNT; MWCNT) have induced pulmonary fibrosis and granuloma formation as well as subpleural fibrosis in experimental studies [18,37,43–45]. Similarly, engineered nanomaterials such as titanium dioxide and iron oxide have been shown to impair the microcirculation in various organs [46] and to disturb both cardiac functions and effects of neurotransmitters on the cardiovascular system [47]. Manganese and iron oxide nanoparticles have been demonstrated to enter the olfactory bulb through the axons of the olfactory nerve in the nasal olfactory epithelium [15,48], and Ceccatelli and Bardi [49] have shown that these materials can be transported to other areas in the brain after inhalation through the nose. In more recent studies, several engineered nanomaterials have been observed to disturb the immunological system by increasing the production of pro-inflammatory mediators [18,37,38,45,47,50,51]. There are also a few studies that show that when metal oxide nanoparticles are applied on the skin, they can penetrate through the skin to some extent, although usually without reaching the systemic circulation [52]. There is even a study that revealed that when sunblock creams containing a natural isotope of zinc oxide were applied to human skin, the isotope could be reliably measured in both blood and urine [53]. It is not thought likely that engineered nanomaterials have any major effect on the reproductive system, but this cannot be fully excluded either [54]. The effects of these materials on the gastrointestinal tract have not been reliably demonstrated in experimental studies so far, but in the future the use of these materials or their applications will increase in food items. Hence, when the exposure through this route increases, understanding of the potential gastrointestinal toxicity will become much more important [55].

Recently, possible genotoxic effects of metal and metal oxide nanoparticles have also received some attention. Titanium dioxide nanoparticles have been shown to be slightly genotoxic [56], and silver nanoparticles coated with polyvinyl-pyrrolidine can evoke dose-dependent DNA damage *in vitro* [57]. There are studies indicating that long, multi-walled carbon nanotubes can cause an asbestos-like effect after a low single intraperitoneal dose and one week follow-up [58]. In another study, it has been shown that an identical material, also given into the abdominal cavity of a sensitive mouse strain, could induce mesotheliomas in a dose-dependent manner in the cavity after one single low dose with a one year follow-up [59]. Hence, it is not surprising that even though most nanomaterials seem to have low biological activities, there are some that may have the potential of harming human health.

In order to be able to respond to the societal needs for safe nanomaterials and nanotechnologies it is necessary to complement valuable mechanistic studies with systematic short-term and long-term animal studies to achieve a reliable estimate of the possible hazards and risks of engineered nanomaterials. A substantial majority of the studies have investigated only a relatively limited number of nanomaterials, and they cannot be used for risk assessment and safety assessment of ENM unless they have been adequately validated in appropriate animal models to demonstrate that they have true predictive power. So far, these kinds of validated *in vitro* studies do not exist, and this has greatly hindered the development of novel intelligent testing strategies of these materials [60–62].

In a recent paper, Nel et al. [38] called for totally novel testing strategies for engineered nanomaterials to enable the available human and other resources to be able to cope with this ever-increasing challenge, thus echoing the report of Hartung [63]. It is evident that the development of quick, affordable and reliable methods for the assessment of safety of engineered nanomaterials is more important than ever.

1.6 REQUIREMENTS FOR THE ASSESSMENT AND MANAGEMENT OF RISKS OF ENGINEERED NANOMATERIALS

Currently, the concern about the safety of engineered nanomaterials and nanotechnologies is overshadowed by the uncertainty of the potential associated harms and risks; this may prevent investment into the entire technological field. Hence, the major challenge for nanosafety research today and into the foreseeable future will be to reduce the uncertainty around the safety of these materials and technologies. The uncertainty arises from our incomplete ability to assess reliably exposure to these materials and hazards associated with the exposure. It is clear that no matter how appropriate the current legislative frameworks, for example, in the United States [64] or the European Union Chemicals legislation [29] or elsewhere seem to be, as long as their effective implementation has to be based on unreliable assessment methods, the reliable risk assessment, management and governance of the engineered nanoparticles will be fraught with difficulties (see 62).

It will be important in the future that reliable assessment and management of risks of engineered nanomaterials can become possible for regulators. The manufacturing industries would benefit from adopting a way of working that would guarantee the incorporation of safety and health into the novel engineered nanomaterials and nano-based products and processes (see 7,11,62). This would mean adoption of a safe-by-design principle in the design of novel engineered nanomaterials and in the development of novel industrial processes involving these materials.

The life-cycle of engineered nanomaterials refers to the use, incorporation into products, use of the products and their final disposal or recycling. This kind of approach requires the establishment a new safety culture developing and implementing a complete system of methods, techniques and technologies, a competent scientific and technical community and a portfolio of measures to ensure a total safety paradigm and to inform the general public about the safety management and governance of the technologies utilizing engineered nanomaterials [65].

The current situation is especially challenging for the risk assessment of engineered nanomaterials since novel materials incorporating engineered nanomaterials are continuously being developed, and the current testing abilities and resources do not guarantee an adequate assessment of safety and risks of these materials [66]. Hartung [63] called for abandoning the existing experimental animal-testing paradigms and replacing them with high-throughput testing approaches. Attempts to develop such approaches are ongoing, but still require a large amount of work [37,38]. As long as there is no clear evidence of the predictive power of these approaches for *in vivo* systems, it is highly unlikely that they will be approved as a part of the regulatory framework of nano-materials or other chemicals, since these new approaches are associated with too many uncertainties.

1.7 TO BE EXPECTED IN THE FUTURE

At the same time as more data are generated, expanding our un-derstanding of the potential hazards associated with exposure to the first-generation nanomaterials, there is an urgent need to develop totally novel paradigms for the testing of the existing and novel nano-materials, whether second-, third- or fourth-generation materials. This will require a systematic initiative to investigate the causalities and associations between nanomaterial characteristics and the interactions of nanomaterials with biological organisms. These studies should aim at identifying the material features and characteristics that are associ-ated with a harmful endpoint and at which level so that they can be used to assess exposure and simultaneously to predict the existing risks online.

1.8 CONCLUSIONS

Challenges associated with the safety of engineered nanomaterials and nanotechnologies are global in their scale. The significant obvious benefits of engineered nanomaterials and nanotechnologies are over-shadowed by the potentially harmful characteristics of these materials, and these are often intimately associated with the technological benefits. Since these emerging materials and technologies have been identified as the cutting-edge technologies that are predicted to hold the keys to economic prosperity, their success also requires an assurance of their safety. The current challenges are that the available tools that can be used for assessing their safety are often inappropriate or so laborious that adequate assessment of safety of these materials and technologies remains highly problematic, and the resources or testing tools are not likely to enable safety assessment of the novel nanomaterials that are flooding onto the market. This means that totally new safety assessment paradigms will need to be developed during the coming years to resolve this problem. What is needed is a rapid identification of research priorities and development of a roadmap that will cut a clear pathway to assure the safety of nanomaterials and nanotechnologies in the

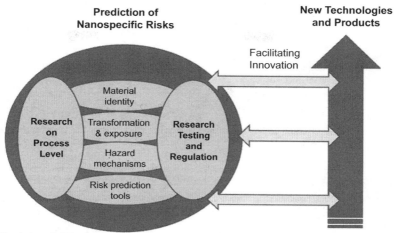

Nanosafety for Innovation and Sustainability

All subaims shall feed to the overall aims of predicting and controlling possible nanospecific risks.

FIGURE 1.4 **Key challenges in promoting the safety of engineered nanomaterials and nanotechnologies.**

future. The close correlation between basic and applied research and innovations is depicted in Figure 1.4. This concept has been developed by the NanoSafety Cluster (www.nanosafetycluster.eu/), which incorporated all nanosafety research projects and their scientists funded by the European Commission Seventh Framework Programme (FP7) (http://cordis.europa.eu/fp7/home_en.html).

References

[1] European Commission COM. 543. Nanosciences and Nanotechnologies: An Action Plan for Europe 2006-2009. Available at: http://ec.europa.eu/nanotechnology/pdf/nano_action_plan2005_en.pdf; 2005.

[2] Maynard AD, Aitken RJ, Butz T, et al. Safe handling of nanotechnology. Nature 2006; 444:267—9.

[3] European Commission COM. 572 final. Second regulatory review on nanomaterials. Available at: http://eur-lex.europa.eu/LexUriServ/LexUriServ.do?uri=COM:2012: 0572:FIN:EN:PDFM; 2012.

[4] Maynard AD. Don't define nanomaterials. Nature 2011;475:31.

[5] Nature Nanotechnology Editorial. Join the dialogue. Nature Nanotech 2012;7:545.

[6] Nature Nanotechnology Editorial. The dialogue continues. Nature Nanotech 2013;8:69.

[7] Roco MC, Mirkin CA, Hersam MC. Nanotechnology research directions for societal needs in 2020: retrospective and outlook summary. NSF/WTEC (National Science Foundation/World Technology Evaluation Center) report (summary of the full report published by Springer); 2010.

[8] European Commission COM. 48. Green Paper From Challenges to Opportunities: Towards a Common Strategic Framework for EU Research and Innovation Funding. Available at: http://ec.europa.eu/research/horizon2020/pdf/com_2011_0048_csf_green_paper_en.pdf; 2011.

[9] NIOSH. Occupational Exposure to Carbon Nanotubes and Nanofibers. Current Intelligence Bulletin 161-A 2010:1–149.

[10] National Nanotechnology Initiative (NNI). Environmental, Health, and Safety (EHS) Research Strategy. Available at: http://www.nano.gov/sites/default/files/pub_resource/nni_2011_ehs_research_strategy.pdf; 2011.

[11] National Academy of Sciences (NAS). A Research Strategy for Environmental, Health, and Safety Aspects of Engineered Nanomaterials. National Research Council of the National Academies; 2012. P. 1–211.

[12] European Commission COM. 808 final. Communication from the Commission to the European Parliament, the Council, the European Economic and Social Committee and the Committee of the Regions: Horizon 2020 – The Framework Programme for Research and Innovation. Available at: http://eur-lex.europa.eu/LexUriServ/LexUriServ.do?uri=COM:2011:0808:FIN:en:PDF; 2011.

[13] European Commission COM. 809 final. Proposal for a Regulation of the European Parliament and of the Council: Establishing Horizon 2020 – The Framework Programme for Research and Innovation (2014-2020). Available at: http://eur-lex.europa.eu/LexUriServ/LexUriServ.do?uri=COM:2011:0809:FIN:en:PDF; 2011.

[14] Horizon 2020. http://ec.europa.eu/research/horizon2020/index_en.cfm; 2013.

[15] Oberdörster G, Sharp Z, Atudorei V, et al. Translocation of inhaled ultrafine particles to the brain. Inhal Toxicol 2004;16:437–45.

[16] Borm PJA, Robbins D, Haubold S, et al. The potential risks of nanomaterials: a review carried out for ECETOC. Part Fibre Toxicol 2006;3:11.

[17] Savolainen K, Alenius H, Norppa H, et al. Risk assessment of engineered nanomaterials and nanotechnologies-A review. Toxicology 2010;269:92–104.

[18] Donaldson K, Schinwald A, Murphy F, et al. The Biologically Effective Dose in Inhalation Nanotoxicology. Acc Chem Res 2012 [Epub ahead of print].

[19] Fadeel B, Savolainen K. Broaden the discussion. Nature Nanotech 2013;8:71.

[20] Savolainen K, Alenius H. Disseminating widely. Nature Nanotech 2013;8:72.

[21] Tenzer S, Docter D, Rosfa S, et al. Nanoparticle size is a critical physicochemical determinant of the human blood plasma corona: a comprehensive quantitative proteomic analysis. ACS Nano 2011;5:7155–67.

[22] National Research Council (NRC). A Research Strategy for Environmental, Health and Safety Aspects of Engineered Nanomaterials. Available at: https://download.nap.edu/catalog.php?record_id=13347#toc; 2012.

[23] PEN. The Project of Emerging Nanotechnologies. http://www.nanotechproject.org/.

[24] International Risk Governance Council (IRGC). White paper on nanotechnology risk governance. Geneva: IRGC; 2006. Available at: http://www.irgc.org/IMG/pdf/IRGC_white_paper_2_PDF_final_version-2.pdf.

[25] Lux Research Inc. Nanomaterials of the Market Q1 2009. Cleantech's Dollar Investments, Penny Returns; 2009.

[26] Savolainen K, Vainio H. Health risks of engineered nanomaterials and nanotechnologies (in Finnish). Duodecim 2011;127(11):1097–104.

[27] Stamm H. Risk factors: Nanomaterials should be defined. Nature 2011;476:399.

[28] European Commission 2011/696/EU. Recommendation of 18 October 2011 on the definition of nanomaterial. Available at: http://eur-lex.europa.eu/LexUriServ/LexUriServ.do?uri=OJ:L:2011:275:0038:0040:EN:PDF; October 2011.

[29] European Commission Regulation no 1907/2006 of the European Parliament and of the Council of 18 December 2006 concerning the Registration, Evaluation, Authorisation

and Restriction of Chemicals (REACH). Available at: http://eur-lex.europa.eu/ LexUriServ/LexUriServ.do?uri=oj:l:2006:396:0001:0849:en:pdf.

[30] European Commission Regulation (EC) No 1223/2009 of the European Parliament and of the Council of 30 November 2009 on cosmetic products. Available at: http://eur-lex. europa.eu/LexUriServ/LexUriServ.do?uri=OJ:L:2009:342:0059:0209:en:PDF.

[31] Auffan M, Rose J, Bottero JY, et al. Towards a definition of inorganic nanoparticles from an environmental, health and safety perspective. Nature Nanotech 2009;4:634−41.

[32] Monopoli MP, Åberg C, Salvati A, Dawson KA. Biomolecular coronas provide the biological identity of nanosized materials. Nature Nanotech 2012;7:779−86.

[33] Peters TM, Elzey S, Johnson R, et al. Airborne monitoring to distinguish engineered nanomaterials from incidental particles for environmental health and safety. J Occup Environ Hyg 2009;6:73−8.

[34] Brouwer D. Exposure to manufactured nanoparticles in different workplaces. Toxicology 2010;269:120−7.

[35] van Broekhuizen P, van Broekhuizen F, Cornelissen R, Reijnders L. Workplace exposure to nanoparticles and the application of provisional nano reference values in times of uncertain risks. J Nanopart Res 2012;14:770.

[36] Kuhlbusch TA, Asbach C, Fissan H, et al. Nanoparticle exposure at nanotechnology workplaces: a review. Part Fibre Toxicol 2011;8:22. http://dx.doi.org/10.1186/1743-8977-8-22.

[37] Palomäki J, Välimäki E, Sund J, et al. Long, needle-like carbon nanotubes and asbestos activate the NLRP3 inflammasome through a similar mechanism. ACS Nano 2011; 5:6861−70.

[38] Nel A, Xia T, Meng H, et al. Nanomaterial Toxicity Testing in the 21st Century: Use of a Predictive Toxicological Approach and High-Throughput Screening. Acc Chem Res 2012 [Epub ahead of print].

[39] Fadeel B, Feliu N, Vogt C, et al. Bridge over troubled waters: understanding the synthetic and biological identities of engineered nanomaterials. WIREs Nanomed Nanobiotechnol 2013;5:111−29.

[40] Bergamaschi A, Iavicoli I, Savolainen K. Exposure assessment. In: Fadeel B, Pietroiusti A, Shvedova AA, editors. Adverse effects of engineered nanomaterials: Exposure, Toxicology, and Impact on Human Health. San Diego: Academic Press; 2012. p. 25−43.

[41] Hunt SR, Wan D, Khalap VR, et al. Scanning Gate Spectroscopy and Its Application to Carbon Nanotube Defects. Nano Lett 2011;11:1055−60.

[42] Fadeel B, Pietroiusti A, Shvedova AA. Adverse effects of engineered nanomaterials: Exposure, Toxicology, and Impact on Human Health. San Diego: Academic Press; 2012.

[43] Shvedova AA, Sager T, Murray AR, et al. Critical Issues in the Evaluation of Possible Adverse Pulmonary Effects Resulting from Airborne Nanoparticles. In: Monteiro-Riviere NA, Tran CL, editors. Nanotoxicology - Characterization, Dosing and Health Effects. New York: Informa Healthcare; 2007. p. 225−36.

[44] Ryman-Rasmussen JP, Cesta MF, Brody AR, et al. Inhaled carbon nanotubes reach the subpleural tissue in mice. Nature Nanotech 2009;4:747−51.

[45] Rossi EM, Pylkkänen L, Koivisto AJ, et al. Airway exposure to silica-coated TiO_2 nanoparticles induces pulmonary neutrophilia in mice. Toxicol Sci 2010;113:422−33.

[46] Bihari P, Holzer M, Praetner M, et al. Single-walled carbon nanotubes activate platelets and accelerate thrombus formation in the microcirculation. Toxicology 2010;269: 148−54.

[47] Castranova V, Schulte PA, Zumwalde RD. Occupational Nanosafety Considerations for Carbon Nanotubes and Carbon Nanofibers. Acc Chem Res 2012 [Epub ahead of print].

[48] Elder A, Gelein R, Silva V, et al. Translocation of Inhaled Ultrafine Manganese Oxide Particles to the Central Nervous System. Environ Health Perspect 2006;114:1172–8.

[49] Ceccatelli S, Bardi G. Neurological System. In: Fadeel B, Pietroiusti A, Shvedova AA, editors. Adverse effects of engineered nanomaterials: Exposure, Toxicology, and Impact on Human Health. San Diego: Academic Press; 2012. p. 157–68.

[50] Nel A, Xia T, Mädler L, Li N. Toxic potential of materials at the nanolevel. Science 2006;311:622–7.

[51] Nel AE, Mädler L, Velegol D, et al. Understanding biophysicochemical interactions at the nano–bio interface. Nat Mater 2009;8:543–57.

[52] Monteiro-Riviere NA, Inman AO, Zhang LW. Limitations and relative utility of screening assays to assess engineered nanoparticle toxicity in a human cell line. Toxicol Appl Pharmacol 2009;234:222–35.

[53] Gulson B, McCall M, Korsch M, et al. Small Amounts of Zinc from Zinc Oxide Particles in Sunscreens Applied Outdoors Are Absorbed through Human Skin. Toxicol Sci 2010;118:140–9.

[54] Hougaard KS, Campagnolo L. Reproductive toxicity. In: Fadeel B, Pietroiusti A, Shvedova AA, editors. Adverse effects of engineered nanomaterials: Exposure, Toxicology, and Impact on Human Health. San Diego: Academic Press; 2012. p. 225–42.

[55] Jepson MA. Gastrointestinal tract. In: Fadeel B, Pietroiusti A, Shvedova AA, editors. Adverse effects of engineered nanomaterials: Exposure, Toxicology, and Impact on Human Health. San Diego: Academic Press; 2012. p. 209–24.

[56] Lindberg HK, Falck GC-M, Catalán J, et al. Micronucleus assay for mouse alveolar type II and Clara cells. Environ Molec Mutag 2010;51:164–72.

[57] Nymark P, Catalán J, Suhonen S, et al. Genotoxicity of Polyvinylpyrrolidone-coated silver nanoparticles in BEAS 2B cells. Toxicology 2012 pii: S0300–483X(12)00369-1. doi: 10.1016/j.tox.2012.09.014. [Epub ahead of print].

[58] Poland CA, Duffin R, Kinloch I, et al. Carbon nanotubes introduced into the abdominal cavity of mice show asbestos-like pathogenicity in a pilot study. Nature Nanotech 2008;3:423–8.

[59] Takagi A, Hirose A, Futakuchi M, et al. Dose-dependent mesothelioma induction by intraperitoneal administration of multi-wall carbon nanotubes in p53 heterozygous mice. Cancer Sci 2012;103:1440–4.

[60] Morris J, Willis J, De Martinis D, et al. Science policy considerations for responsible nanotechnology decisions. Nature Nanotech 2011;6:73–7.

[61] Gulumian M, Kuempel ED, Savolainen K. Global challenges in the risk assessment of nanomaterials: Relevance to South Africa. S Afr J Sci 2012;108. Art. #922, 9 pages, http://dx.doi.org/10.4102/sajs.v108i9/10.922.

[62] Savolainen K, Gulumian M. Nanotechnology — Emerging research and implications for health and policy. In: Fan AM, Alexeef GV, Khan EM, editors. Toxicology and Risk Assessment: Principles and Applications. Singapore: Pan Stanford Publishing; 2013.

[63] Hartung T. Toxicology for the twenty-first century. Nature 2009;460(7252):208–12. http://dx.doi.org/10.1038/460208a.

[64] Toxic Substances Control Act (TSCA). Chemical Substance inventory. United States Environmental Protection Agency, Office of Toxic Substances; 1979.

[65] Kahru A, Dubourguier HC. From ecotoxicology to nanoecotoxicology. Toxicology 2010; 269:105–19.

[66] Choi J-Y, Ramachandran G, Kandlikar M. The impact of toxicity testing costs on nanomaterial regulation. Environ Sci Toxicol 2009;43:3030–4.

2

Nanotechnology and Exposure Scenarios

*Sheona A.K. Read[1], Araceli Sánchez Jiménez[1],
Bryony L. Ross[1], Robert J. Aitken[1,2],
Martie van Tongeren[1]*

[1] Institute of Occupational Medicine (IOM),
Edinburgh, United Kingdom
[2] Institute of Occupational Medicine (IOM), Singapore

2.1 INTRODUCTION

Exposure scenarios (ES) are an essential tool for estimating exposure and risk. Exposure scenario descriptions generally include information on the substance, the process and activities, the presence of any risk management measures (RMMs) and estimates of exposure that can be expected under the conditions described. Within guidance for the Regulation on Registration, Evaluation, Authorisation and Restriction of Chemicals (REACH), ES are defined as sets of information describing the conditions under which the risk associated with the identified use(s) of a substance can be controlled, including operational conditions (OCs) and RMMs. Harmonization of the description of exposure scenarios for nanomaterials together with standardization of the actual exposure assessment is needed allows proper interpretation of the information contained within the scenario ensures pragmatic implementation of safe practices and effectively communicates information on safe use and practices up and down the supply chain.

This chapter firstly describes the development of the nanotechnology industrial sector, including the manufacturing processes and applications of engineered nanomaterials (ENMs). Secondly, the way in which ES can be developed and some initial experiences from Framework Programme (FP)

7 project 'NANEX' are discussed. Finally, a number of ES case studies are presented, based on information collected during the FP7 project 'NANODEVICE'. Throughout this chapter, we will only focus on occupational exposure, although exposure scenarios for consumer use and environmental release can and should also be developed to ensure that exposure is assessed and managed across the life-cycle of the nanomaterials.

2.2 DEVELOPMENT OF NANOTECHNOLOGY

Although nanotechnology is a relatively recent development in scientific research, nanoparticles are not a new phenomenon. Scientists have been aware of colloids and sols for over a century [1,2], and many other well-known industrial processes produce materials with nano-scale dimensions [3]. An example of this is the synthesis of carbon black by flame pyrolysis, which produces powdered carbon with a very high surface-to-mass ratio. Although this carbon powder is typically highly agglomerated, it has a primary particle size in the range 20–100 nm. In 2005, worldwide capacity for carbon black production was estimated to be 10 million tons [4]. Silicon dioxide (SiO_2), ultrafine titanium dioxide (TiO_2) and ultrafine metals such as nickel are among the other common materials produced by flame pyrolysis or similar thermal processes. In addition, several other industrial processes create and use nanoparticles as part of the process, such as thermal spraying and coating.

Interest in nanotechnology began to emerge in the 1980s, stimulated by experimental advances such as the invention of the Scanning Tunneling Microscope (STM) for imaging surfaces at the atomic level [5], the discovery of fullerenes [6], and later carbon nanotubes [7] with their unique properties at the nano-scale. By the early 2000s, nanotechnology had become a rapidly evolving and expanding discipline, with a vast array of new nanomaterials being discovered or produced, accompanied by heightened claims concerning their properties, behaviors and potential applications. According to analyses by the National Institute for Public Health and the Environment (RIVM), the number of consumer products with a *claim* to contain nanomaterials on the European market showed a six-fold increase between 2007 and 2010, from 143 products to 858 products in 2010 [8].

In 2004, Roco [9] anticipated the development of nanotechnology through a series of four generations:

- First generation ($\sim$2004–2012): Passive nanostructures, e.g., nanoparticles, nanotubes, nanocomposites, nanocoatings, nanostructured materials moving from passive nanomaterials and nanostructures;
- Second generation ($\sim$2008–2016): Active nanostructures, e.g., electronics, sensors, targeted drugs, adaptive structures;

- Third generation ($\sim$2018–2026): Systems of nanosystems, e.g., guided molecular assembly, 3D networking, robotics, supra-molecules;
- Fourth generation ($\sim$2022–2040): Molecules 'by design', hierarchical functions, evolutionary systems.

Production is now well into the first-generation stage, with research and development driving towards the commercialization of second-generation nanostructures.

2.3 PRODUCTION OF ENGINEERED NANOMATERIALS

Engineered nanomaterials and their products can be produced by a variety of methods, with different methods also being used to synthesize the same type of material. An overview of common production methods for ENMs has been published by Sengül et al. [10]. Generally, nano-materials are produced in two ways:

i. A 'top-down' approach, in which larger bulk materials are 'broken' into smaller, nano-scale materials by mechanical, chemical or other energy-requiring means; and

ii. A 'bottom-up' approach, whereby nano-scale materials are synthesized from atomic or molecular subcomponents through precisely controlled chemical reactions.

Several of the specific methods involved in both approaches are summarized in Table 2.1. Top-down manufacturing is currently the more

TABLE 2.1   Common Nanomaterial Manufacturing Methods

Top-Down Techniques	Bottom-Up Techniques
Electrospinning	Vapor-phase techniques
Milling	Deposition techniques, e.g., CVD, sputtering, evaporation
Mechanical milling	Nanoparticle/nanostructured materials synthesis techniques,
Cryo-milling	e.g., evaporation, laser ablation, flame synthesis, arc discharge
Mechanochemical	Liquid-phase techniques
bonding	Precipitation
Etching	Sol-gel
Wet/dry etching	Solvothermal- and sonochemical-synthesis
Lithography	Microwave irradiation
	Reverse micelle
	Electrodeposition
	Self-assembly techniques
	Electrostatic self-assembly
	Self-assembled monolayers
	Langmuir-Blodgett formation

(Adapted from [11])

common approach used to produce nanomaterials, although bottom-up methods allow for customized design of reactions and processes at the molecular level [11].

A recently published International Organization for Standardization (ISO) technical specification document [12] provides a useful categorization of eight common types of engineered nanomaterials, which differ according to their morphologies and material composition. The characteristics of these materials and their production processes are summarized in Table 2.2.

2.4 APPLICATIONS OF NANOTECHNOLOGY

Progress towards harnessing and utilizing the unique properties of current generation nanomaterials in new commercial and industrial applications is developing across a broad range of areas. Nowack et al. [13] identified ten core technology sectors that are developing nanotechnology applications. A summary of the nanoproducts and applications for the ten industrial sectors is given below.

2.4.1 Automotive and Aeronautics

Activities in this sector are focused on using lighter materials with improved performance for producing lower-weight vehicle structural parts [14]. In particular, efficient use of energy resources, reduction in CO_2 emissions and improved safety and comfort are major drivers. The sector can be divided into the development of three key nano-enabled product types for automotive and aeronautics applications: nanostructured metals and alloys, polymer nanocomposites and tribological nanocoatings. The properties, production techniques, applications and state of development of these materials are outlined in Table 2.3.

2.4.2 Agrifood

Nanotechnology applications are expected to bring a range of benefits to the food sector, including new tastes, textures and sensations, a decrease in the use of fat, enhanced absorption of nutrients and improved packaging, traceability and security of food products [15]. The agrifood sector can be divided into three main subsectors:

i) agricultural production (production of materials from raising domesticated animals and cultivation of plants);

ii) food processing and functional food (turning agricultural produce into consumer products); and

TABLE 2.2 Description of Eight Typical Forms of Engineered Nanomaterials

Types of ENMs	Description	Manufacturing Process	Production
Carbon black	Virtually pure form of elemental carbon. Physical appearance is that of a very fine, black powder. Primary particle size is typically less than 100 nm	Thermal decomposition of gaseous or liquid hydrocarbons under controlled conditions	Manufactured since start of 20th century. In the top five industrial chemicals produced worldwide
Metals and metal oxides, ceramics	Wide range of compact forms of nanoparticles, including ultrafine titanium dioxide and fumed silica. Often available only in agglomerated or aggregated form. Composites having, for example, a metal core with an oxide shell or alloys	Flame pyrolysis methods	Produced in relatively high bulk and incorporated into several commercial products (e.g., TiO_2 in sunscreens)
Fullerenes	Hollow spheres that are composed entirely of carbon. Most common type is fullerene (C_{60}), also known as a Buckminster fullerene or a 'buckyball'	Production based on carbon electric arc	Relatively small scale, although commercial production of fullerenes using this method began in 1991
Carbon nanotubes	Allotropes of carbon with a cylindrical structure. Carbon nanotubes (CNT) usually consist of curved graphene layers. Can be either single-wall carbon nanotubes (SWCNT) with one graphene layer or multi-wall carbon nanotubes (MWCNT) with several graphene layers	Produced using a variety of methods. The main methods are chemical vapor deposition (CVD), arc-discharge and laser ablation. Typically requires use of metal catalysts (e.g., iron, nickel) and includes a purification step	Over 100 companies worldwide producing CNT, with this number expected to double in the next 5 years. Several CNT-enabled products on the marketplace
Nanowires	Small conducting or semiconducting nanofibers with a single crystal structure, typical diameter of a few 10s of nm and a large aspect ratio. Various metals have been used to fabricate nanowires, including cobalt, gold, copper and silicon	Typically involves manufacture of a template followed by the deposition of a vapor to fill the template and grow the nanowire. Deposition processes include electrochemical deposition	

(Continued)

TABLE 2.2 Description of Eight Typical Forms of Engineered Nanomaterials—cont'd

Types of ENMs	Description	Manufacturing Process	Production
Quantum dots	Small (2–10 nm) assemblies of semiconductor materials with novel electronic, optical, magnetic and/or catalytic properties. Quantum dots (QD) are considered to be between an extended solid structure and a single molecular entity. Semiconductor QD exhibit distinct photoelectronic properties related directly to their size	Grown by advanced epitaxial techniques in nanocrystals. Produced by wet chemical colloidal processes or by ion implantation, or in nanodevices made by state-of-the-art lithographic techniques	
Dendrimers	Polymer particles in which the atoms are arranged in a branching structure, usually symmetrically around a core. Typically monodisperse with a large number of peripheral groups that can be functionalized	Either a divergent method where the dendron originates from a central core and branches out; or a convergent method where the dendrimers grow inwards to a focal point	Synthesis usually involves an iterative sequence of two straightforward chemical reactions, and therefore large-scale production is technically and economically viable. Several types of dendrimers are produced commercially on a large scale
Nanoclays	Ceramic nanoparticles of layered mineral silicates. Depending on chemical composition and nanoparticle morphology, nanoclays are organized into several classes such as montmorillonite, bentonite, kaolinite, hectorite and halloysite. Nanoclays are typically modified with an organic surface treatment to allow for dispersion into low-polarity (polymeric) systems. The organic surface treatments are based on quaternary amine salt chemistry	Naturally occurring and engineered. Organically modified nanoclays are manufactured by a wet chemical process, whereby the positively charged nitrogen of a quaternary amine ion exchanges into the clay platelet for either sodium or calcium interstitial cations. Further size classification is typically carried out using traditional ball milling techniques	

(Based on [12])

TABLE 2.3 Overview of Nano-Enabled Products Used for Automotive and Aeronautics Applications

	Nanostructured Metals & Alloys	Polymer Nanocomposites (PNCs)	Tribological Nano-Coatings
Description/ key properties	Solids having structural features in the range of 1–100 nm in at least one dimension; Exhibit high strength and high corrosion resistance compared to bulk counterparts.	Two-phase material where one phase has at least one dimension in the nanometer range; Enhanced mechanical and electrical properties in terms of strength, weight, flame retardancy and electrical conductivity.	Coatings that are applied to the surface of a component in order to control its friction and wear; Can be single-phase, mutli-phase or composite coatings.
Common types	Most relevant nanostructured metals applied for transport applications are aluminium, magnesium, titanium and their alloys.	Most commonly applied PNCs are nanoclay incorporated polymers. Most common 'fillers' investigated for application in PNCs are nanoclays, CNTs, carbon nanofibers and graphene.	Wide variety of materials can be used in tribological coatings including carbides, cermets, metal ceramic oxides, metals, nitrides and diamond.
Main production techniques	Severe plastic deformation (SPD), nanopowder sintering, melt spinning and electrodeposition.	In-situ polymerization, solution exfoliation and melt intercalation.	Thermal spraying and vapor deposition processes.
State of development	Most of these techniques are currently used mainly in lab-scale production. Few exceptions where commercial products exist include small parts e.g., bolts, fasteners or screws.	In commercial use Nanocomposite usage by this sector represents almost 80% of global consumption.	In commercial use

(Based on [14])

iii) food packaging and distribution (materials designed to preserve or protect fresh or processed foods) [16].

Nanoclays are the most commonly utilized type of nanomaterial in agricultural production, with gold nanoparticles and carbon nanotubes being used predominantly in food processing and titanium and silver used in particular in food packaging [17]. Currently, only a limited number of nanotechnology applications within the agrifood sector are on the market, for example nanoemulsions in pesticides, equipment coatings for disinfection and delivery systems for nutrients.

2.4.3 Construction

Material innovations are continuously finding their way into modern building structure, including innovations based on nanotechnology. Applications of nanotechnology in the construction industry include cement-based materials (e.g., cement additives and ultra-high-performance concrete), coatings (e.g., self-cleaning glass, antireflective coatings), living comfort and building safety applications (e.g., flame-retardant coatings and antibacterial surfaces) and sustainability and environmental applications (e.g., thermal insulation) [18].

Some applications have been brought to the market, such as self-cleaning windows and roof tiles that utilize titanium dioxide (TiO_2) or nano-enhanced cement that uses silica or carbon nanotubes (CNT), but use is not widespread mainly due to expense [19]. Titanium dioxide and silicon dioxide (SiO_2) are currently the most commonly used nano-materials in the construction industry. Other nanomaterials such as CNTs are being discussed intensively as potential candidates for future construction materials but have no significant influence at present. It is likely that the potential for exposure will be high in the construction industry if the use of nanomaterials becomes widespread, in particular for those applications where nanomaterials are likely to be incorporated into production in powder or slurry form [20].

2.4.4 Energy

Nanotechnology has the potential to provide a multitude of improvements and new solutions in the energy sector, enabling a reduction in the amount of material required, the introduction of new properties to existing materials, and thus the development of alternative and more efficient energy solutions. Potential nanotechnology-based applications within this sector include photovoltaics, thermoelectricity, fuel cells, batteries and super-capacitors, fossil fuel, nuclear energy and energy harvesting [21].

Types of nanomaterials that are being investigated for use within these applications include CNTs and other carbon-based nanomaterials (e.g., fullerenes, graphene, diamond), metal oxide nanoparticles and nanowires (e.g., oxides of zinc, titanium, yttrium, silicon, etc.) and, primarily for batteries, lithium-based nanoparticles (e.g., $LiFePO_4$). In general, it is considered that as nanomaterials are bound to a substrate and fully confined, human or environmental exposure is unlikely during normal use of the final application [19]. However, many nanomaterial-based applications within the energy sector are still at an early stage of development, and in some cases there is still a wide gap between industrial specifications and the state-of-the-art. One of the main nanotechnology-enabled products to have emerged onto the marketplace in this area is batteries, using nano-structured or nanoparticle-coated anodes or cathodes to increase battery capacity. In some areas the application of nanotechnology is anticipated to be of little value for the foreseeable future, for example in nuclear energy, fossil fuel or high-temperature fuel cells, mainly because the high temperature destroys the nano-scale structure of materials.

2.4.5 Environment

Maintaining and restoring the quality of air, water and soil is one of the great challenges of our time [22]. Conventional remediation technologies have so far shown only limited effectiveness in reducing the levels of pollutants, particularly in soil and water. Nanomaterials may significantly improve the performance of remediation approaches due to their greater surface area [23].

Mueller and Nowack [22] have provided an overview of the potential applications for nanomaterials in the environmental sector, with applications across the areas of air purification (e.g., catalytic traps, photo-catalysis, static filters), wastewater purification, drinking water treatment and ground water remediation (e.g., photocatalysis, nanofiltration, metals and metal oxide nanoparticles, CNTs), as well as soil remediation (e.g., nano zero valent iron, iron compounds and carbon-based nanomaterials). The two most commonly used nanoparticles within the sector are TiO_2 and iron/iron oxides [24]. Various applications have been successfully demonstrated at the laboratory scale, and some are now in market production. For example, within Europe, nanostructured membranes are in market use for drinking water production and wastewater treatment. In addition, in the US 10% of all soil/groundwater remediation projects involve the use of nano zero valent iron, although this is considerably lower in Europe [19].

2.4.6 Health, Medicine and Nanobio

The potentially huge impact of nanotechnology in the area of medicine has been recognized, and a significant amount of funding has been provided to this sector. Key developments and technologies within this sector have been reviewed [25], highlighting potential applications within the following subsectors: therapeutics (e.g., drug delivery systems); sensors and diagnostics (e.g., *in vitro* and *in vivo* nanosensors); regenerative medicine (tissue and cell engineering); surgery, implants and coatings; novel bionanostructures (e.g., synthetic cells); and cosmetics (e.g., sunscreens). A wide variety of nanomaterials are used in nanomedicine, including [26]:

- structures based on lipids, proteins, DNA/RNA or other naturally occurring materials and substances;
- many different nanomaterials based on polymers, both degradable and nondegradable;
- carbon-based nanomaterials (e.g., CNT, diamond, carbon black, fibers, nanowires);
- metals and metal oxides;
- silica; and
- quantum dots.

The majority of applications are still at the R&D stage or entering into early clinical trials. However, there are products at all stages of development, including some therapeutic drug delivery tools that are already on the market. Further research into the potential hazardous side effects of various nanomedicine applications that are applied or administered into the body is required.

2.4.7 Information and Communications Technology (ICT)

ICT is a technology sector upon which nanotechnologies will have a significant impact, as much of the technology development in this area is already at the nanometer scale. Technological developments can be identified in four major subsectors [27]:

- integrated circuits (e.g., high-k dielectrics, transistors);
- memory (e.g., MRAM, Phase Change RAM);
- photonics (e.g., optical networking, photonic/electronic integration); and
- displays (e.g., organic LEDs, field emission displays).

It is important to note that while most current nanotechnology-enabled ICT applications are based on technological advances developed at the nano-scale, including materials nanostructured in the bulk or on surface (e.g., layers and films), relatively few applications involve the use of engineered nanomaterials. However, of those used, the most common are CNT, nanoparticles of zinc oxide, quantum dots (QDs) and nanowires of semiconductors. Some ICT applications utilizing nanomaterials are already available on the market including, for example, CNT sensors, probe tips and transparent conductive films. Research is still very active in this area, however, and a significant number of new applications are expected to emerge over the coming years.

2.4.8 Security

Nanotechnologies hold great potential for security capabilities, and the applications can be divided into four subsectors [28]:

i. detection (e.g., chemical and biological devices, explosives weapons);
ii. protection (e.g., personal protection, infrastructure and equipment);
iii. incident support (e.g., decontamination, forensics); and
iv. anticounterfeiting, authentication and positioning and localization (AAPL).

A wide variety of nanoparticles are being investigated for security applications, including CNTs and other carbon-based nanomaterials, oxides and metal nanoparticles and quantum dots. The vast majority of applications in this sector are at the applied research stage of development, although some are undergoing field trials.

2.4.9 Textiles

Nanotechnology has the potential to be applied across the entire pipeline of the development of textiles, as indicated in Figure 2.1.

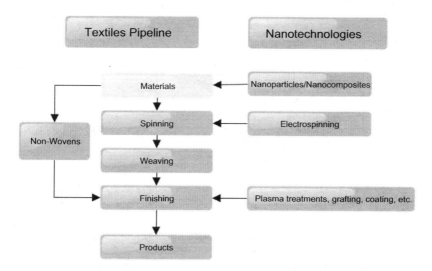

FIGURE 2.1   **Application of nanotechnology in the textiles pipeline.** *Adapted from [29].*

The key aims of applying nanotechnologies within the textiles sector are to i) upgrade present functions and performance of textile materials and ii) develop innovative products, in particular smart/functional textiles with new features and functions (e.g., stain-proof, self-cleaning, conductive). Mantovani and Zappelli [29] reviewed the potential applications for nanomaterials across four key subsectors of textile production: nanomaterials and nanocomposites (currently used in textiles); fiber production (innovative technologies for obtaining nanofibers, e.g., electrospinning); finishing treatments (using both wet and gas-phase processes); and nano-related textile products. A range of nanomaterial types are applied within the textile sector, with the most commonly used being CNTs, zinc oxide nanoparticles and montmorillonite [30]. Textile products incorporating nanotechnology are already on the market, primarily related to the conventional garment and furnishing industry. Looking to the medium—long term future, nanotechnology is expected to find application in areas where performances outweigh costs, such as sport/outdoor textiles, advanced/nonconventional textiles, medical textiles and military textiles.

2.4.10 Chemistry and Materials

This sector spans all major industry sectors, covering processing technologies and applications for carbon-based nanomaterials, nanocomposites, nanostructured metals and alloys, nanopolymers and nanoceramics.

2.5 EXPOSURE SCENARIOS FOR NANOMATERIALS

Table 2.4 shows the potential for occupational exposure to engineered nanomaterials in the ten core industrial sectors involved in developing application of nanomaterials identified by Nowack et al. [13]. The highest potential for exposure occurs when handling powder materials, for example during packing of powders following nanomaterial manufacture.

Human exposure to engineered nanomaterials can occur during all life-cycle stages. The primary route of occupational exposure to ENMs is through inhalation of airborne nanoparticles or agglomerates of nanoparticles. In addition, dermal exposure will also be possible in many exposure scenarios. Exposure may also occur through inadvertent ingestion, depending on the hygienic behavior of workers (or lack thereof).

TABLE 2.4 Core Technology Sectors Involved in the Development of Nanotechnology Applications and Potential for Exposure. Exposure Potential Ranges from 'Unlikely' (−) to 'Low' (+), 'Medium' (++) and 'High' (+++)

Technology Sector	Subsector	Potential Occupational Exposure
Agrifood	Agricultural production	++
	Food processing	++
	Food packaging	++
Automotive	Bulk nanostructured metals	++
	Polymer nanocomposites	++
	Tribiological nanocoatings	++
Environment	Air purification	−
	Wastewater treatment	−
	Drinking water treatment	−
	Groundwater remediation	−
	Soil remediation	+
Security	Detection	+
	Protection	+
	Incident support	+
	Anti-counterfeiting, authentication, positioning and localization	+
Chemistry/ Materials		+++
Construction	Cement additives	++
	Coatings	++
	Flame retardants	++
	Insulation	+
Energy	All subsectors exhibit the same exposure potential	+
ICT	All subsectors exhibit the same exposure potential	−
Textiles	Sports & outdoor clothing (based on textile treatments)	++
	Sports & outdoor clothing (NM are incorporated into fibers)	++
	Medical textiles	++
Health	Therapeutics	+
	Sensors and diagnostics *in vitro*	−
	Sensors and diagnostics *in vivo*	+
	Regenerative medicine	+
	Implants, surgery and coatings	+
	Novel bionanostructures	+
	Cosmetics	++

(Adapted from [13])

In an occupational setting, exposure to nanomaterials can occur for workers at all phases of production [31]. During the development of a new material, the quantities are generally small and materials will usually be produced under highly controlled conditions. Hence, exposures should generally be fairly low, although high exposures are possible following accidental releases (e.g., due to spills and accidents).

In the commercial production phase, exposures can potentially occur during synthesis or in downstream activities such as recovery, packaging, transport and storage. Under these circumstances, much larger quantities of materials are likely to be handled. In such production facilities exposure can occur during activities such as harvesting, post-treatment, packing, cleaning and maintenance. Some nanomaterials have been produced in high volumes for a number of decades (e.g., carbon black), and relatively high exposures have been observed among workers involved in packing of carbon black and cleaning of the plants [32].

Some materials will be subsequently incorporated in a range of other products or may be used in other processes as a feedstock material. Again these scenarios of product manufacturing have the potential to result in exposure to workers involved in them. Nanomaterials may also be incorporated, for example, into composites, which may be subsequently reengineered by cutting, sawing or finishing. Again, in these circumstances, the potential for exposure exists. In addition to this, there would also be the potential for exposure to both professional and consumer users, as well as the environment, during the use phase of the product. Finally, we could consider end-of-life scenarios where the material is disposed of, perhaps by incineration or some other process such as shredding or grinding.

Hence, it may be easily seen that for a single material, there are multiple exposure scenarios that may or may not occur depending on the details of manufacture, use and disposal of that material. Throughout these scenarios, the population exposed, the levels of exposure, the duration of exposure and the nature of the material to which people are exposed (e.g., pristine material, composite, agglomerate) are all different.

Under the EU REACH Regulation [33], concerning the *Registration, Evaluation, Authorisation* and *Restriction* of *Chemicals*, exposure scenarios must be prepared when a substance is manufactured or imported in quantities of 10 tons per year and above and classified as dangerous or as PBT (persistent, bioaccumulative, toxic) or vPvB (very persistent, very bioaccumulative). Exposure scenarios within REACH are defined as sets of information describing the conditions under which the risk associated with the identified use(s) of a substance can be controlled, including operational conditions (amounts used, frequency and duration of use, etc.) and risk management measures (measures to limit/prevent release, personal protection, etc.) [34]. Exposure scenarios are the basis for quantitative exposure assessment but are also used as a communication tool in the supply chain.

Under REACH, occupational ES should be developed for:

- the manufacturing process;
- identified uses, including uses by the manufacturer/importer of the substance concerned, uses further down the chemical supply chain and consumer uses; and
- life-cycle stages resulting from manufacture and identified uses (e.g., article service life and waste life stages).

Although there are currently no provisions in REACH referring specifically to nanomaterials, it is considered that REACH deals with substances in whatever size, shape or physical state. It follows that nanomaterials are implicitly covered by REACH, and its provisions apply [35]. There are currently no well-established occupational exposure limits (OELs) or derived no-effect levels (DNELs) for engineered nanomaterials, although the National Institute for Occupational Safety and Health (NIOSH) has proposed some standards for carbon nanotubes [36] and nano-scale titanium dioxide [37]. Therefore, it may not always be feasible to develop safe exposure scenarios in the context of REACH based on existing exposure and hazard information. However, it may be feasible to develop 'exposure scenarios' for specific data-rich applications of certain ENMs. Although such ES may not be integrated into a quantitative risk assessment, they can be used to benchmark different exposure scenarios with respect to state-of-the-art process operations and control measures and to provide guidance to reduce exposure.

Growth in the manufacturing and use of engineered nanomaterials will undoubtedly result in an increased potential for occupational exposure. A study funded by the National Science Foundation projects that 6 million nanotechnology workers will be required worldwide by 2020, with 2 million of those jobs in the United States [38].

The main life-cycle stages for ENMs are shown in Figure 2.2 and can be summarized as follows [39]:

i. manufacturing of nanoparticles;
ii. formulation of nanomaterials and nanoproducts;
iii. industrial use of nanomaterials or products;
iv. professional and consumer uses of nanoproducts;
v. service life of nanoproducts; and
vi. waste life stage of nanoproducts.

For each material and product type, it is possible to envisage a theoretical map of potential exposure scenarios across its entire life-cycle. An example of such a life-cycle map for carbon nanotubes is depicted in Figure 2.3. The map covers exposure scenarios during the main stages of

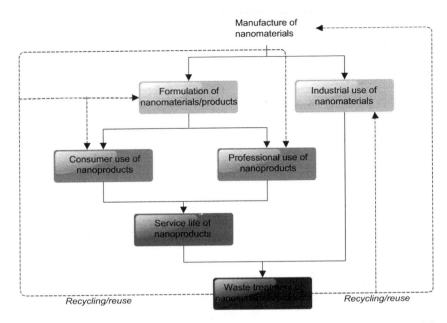

FIGURE 2.2 **Simplified overview of life-cycle stages of manufactured nanomaterials.** *Adapted from [39].*

the life-cycle of carbon nanotubes, from the synthesis of the primary materials to the manufacture of intermediate and final products containing these carbon nanotubes — in this case the focus was on composite materials, textile fibers and concrete products — as well as the professional and consumer use of these products and their eventual recycling and waste disposal.

For each ES, a number of Contributing Exposure Scenarios (CES) may be identified. For example, during the manufacturing of textile fibers containing carbon nanotubes, different CES may be identified for weighing, mixing and loading of the raw materials, the spinning of carbon nanotube fibers, the weaving of these fibers into the textile product, as well as the finishing and packing of the product, any testing of the products (e.g., durability, strength, function, etc.), cleaning and maintenance of the room and equipment and storage and distribution of the final product.

Such life-cycle maps allow a useful visual representation of potential exposure scenarios throughout the life-cycle of an ENM and can help to identify where critical exposures may occur. However, due to a lack of available data for many of the exposure scenarios, particularly in the use and service life and waste treatment stages, it is currently not feasible to

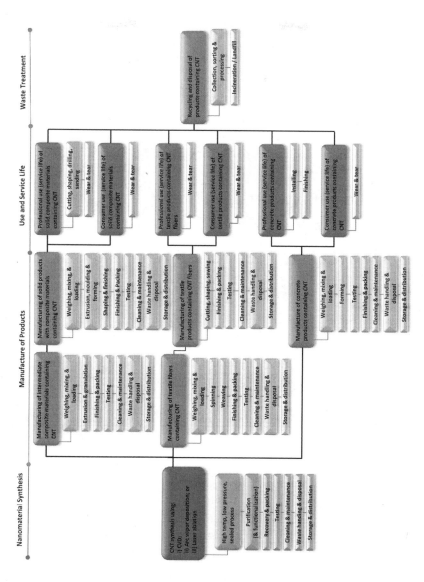

FIGURE 2.3   Exposure scenario lifecycle map for carbon nanotubes in three products: composite materials, textiles and concrete.

obtain detailed descriptions and representative exposure estimates for all currently existing exposure scenarios across the life-cycle, as well as those that will become reality in the near future.

In 2011, Kuhlbusch et al. [40] reviewed the peer-reviewed literature for available evidence of release and exposure to ENMs in the workplace. This information included workplace measurements in industrial production and processing plants, as well as laboratory simulations. Activities were reviewed along the production pathway, including production, handling and refinement of the raw material, bagging and shipping of the nanomaterial, processing of the nanomaterial and work processes with the nanomaterial product. Comparisons across the different studies were complicated because of the different measurement strategies and techniques, metrics and size ranges used as well as the different data analysis procedures employed.

All studies highlighted that the concentration of nanoparticles released into the workplace was influenced by the ENM employed and the process treatment. The physical state of the ENM, in suspension or powder, was an important determinant of exposure, with the prior significantly decreasing the potential for release. The characteristics of the process (e.g., open or closed production; collection by filtration or opening the reactor) also contributed significantly to the release. During processing, release of free ENM from treatment processes on coatings and composites was not observed.

Overall, 13 of the 25 reviewed workplace exposure studies reported release of ENM of particle size less than 100 nm. However, only three studies were able to clearly identify the ENM: during a leak in the pelletizing area of a carbon black industrial production plant [41]; during bagging and vacuum cleaning in a plant producing fullerenes [42]; and in the production area of SWCNT and MWCNT by chemical vapor deposition [43].

The authors acknowledged that a coherent approach for the likelihood of release and workplace exposure assessment is needed. They highlighted that exposure assessment should include a detailed description of the ENM and the processes, including contextual information, to enable the observed concentrations to be correlated with the ENM and process under investigation.

In order to build further knowledge regarding exposure scenarios for ENMs, the FP7 project 'NANEX' — a 12-month project that commenced in December 2009 — aimed to develop a catalog of occupational ES for engineered nanomaterials, taking account of the entire life-cycle of these materials, focusing in particular on carbon nanotubes (CNTs), nano-sized titanium dioxide (nano-TiO_2) and nano-sized silver (nano-Ag). In total, information on 57 occupational exposure scenarios were summarized using publicly available data and data collected in several large-scale sampling campaigns (e.g., European projects 'NANOSH' and 'Nano-INNOV'). In parallel, NANEX carried out case studies in partnership with several companies that manufacture ENMs. During these case studies, exposure scenarios were developed for CNTs, nano-Ag and nano-TiO_2, mainly for manufacture but also for some downstream use. Compared to the ES that were based on data from the open literature and measurement surveys, more detailed information on operating conditions and risk management measures were available for the development of the case study scenarios.

Most of the exposure scenarios related to potential exposure to CNTs, although they predominantly pertained to research-scale activities of production/synthesis or handling of the CNTs (weighing, removing, sonication, etc.). There were a small number of scenarios related to machining of composites containing CNTs. In contrast, for TiO_2 most scenarios were described based on information from commercial-scale manufacturing and formulation. Only a few occupational exposure scenarios could be developed for nano-Ag, either describing manufacturing in a wet chemistry process or handling of nano-Ag.

From the process of developing these exposure scenarios, several main conclusions could be drawn. Most studies reported in the literature or as part of the measurement campaigns had an explorative character and were focused on concentration/emission analysis. Therefore, these studies did not include most of the information that is necessary to build exposure scenarios (e.g., amount used, frequency of activities). Basic characterization of the products used and operational conditions were often not described. The most important observation was the lack of harmonization of the measurement strategy and the analysis and reporting of measurement data. Brouwer et al. [44] provided recommendations for harmonization for measurement strategies for exposure to manufactured nano-objects, which also included a list of core information to be collected during measurement surveys. This includes an adequate description of the process and facility, the activities that are performed during the measurement and exposure mitigation (exhaust ventilation, PPE, etc.) and climate conditions (e.g., temperature and relative humidity).

Based on the information collected by NANEX, a comprehensive analysis of key data gaps and research needs was carried out. A list of minimum data needs was established outlining categories of information that should be included when describing results from exposure studies of ENMs, including both nano-specific and generic data needs. Short-term research priorities identified included i) establishing harmonized methods for data collection/reporting (e.g., exposure metrics); ii) verification of effectiveness of risk management measures; and iii) development of risk management strategies to be applied while waiting for development of more detailed exposure and risk assessment methodologies. Longer-term research priorities were identified as i) increased collection of high-quality data over the life-cycle of ENMs; ii) the development of more advanced understanding of ENM exposure and the key exposure determinants; and iii) development, calibration and validation of ENM-specific exposure estimation models.

2.6　EXPOSURE SCENARIO CASE EXAMPLES

As part of FP7 project 'NANODEVICE', exposure scenario information was gathered from four European research institutes. In order to facilitate the data gathering, a standardized exposure scenario format was employed, detailed in Table 2.5, which was adapted from the ES format template published by the European Chemicals Agency (ECHA) as part of the guidance supporting the REACH Regulation [45]. Information was primarily available from workplace surveys and thus corresponded to the earlier stages of the nanomaterial life-cycle (primary synthesis, manufacture of products). A selection of case studies, based on interviews with the research institutes and published information, is summarized below. Summary measurement data described in this section were obtained by reading approximate measurement points from the graphs as published in the relevant reports/posters. More detailed data are available in the referenced documents.

Case Study ES1: Production of Nanocomposites Containing Silica Nanospheres

This exposure scenario focuses on the potential exposure of laboratory workers during the production of nanocomposites containing silica nanospheres, based on a study undertaken as part of FP7 project 'NANOSH'. The exposure scenario consists of five contributing exposure scenarios as detailed below:

CES 1.1 Grinding and packing of silica nanospheres (70 nm)
CES 1.2 Grinding and packing of silica nanospheres (110 nm)
CES 1.3 Weighing of silica nanospheres and other chemicals
CES 1.4 Transfer of silica nanospheres into plastic bags and mixing with other chemicals
CES 1.5 Loading the reactor with powder containing silica nanospheres

TABLE 2.5 Exposure Scenario Format Used to Develop NANODEVICE Case Examples

---

Exposure Scenario Format Addressing Uses Carried Out by Workers

---

EXPOSURE SCENARIO (1)

Title of exposure scenario

List of names of contributing worker scenarios (1-n) and corresponding Process Category (PROC - from ECHA Guidance R12); include market sector Product Category (PC), if relevant

Name of contributing environmental scenario (n) and corresponding Environmental Release Category (ERC - from ECHA Guidance R12)

Further explanations - include short description of the process

---

CONTRIBUTING EXPOSURE SCENARIO (1) CONTROLLING WORKER EXPOSURE FOR ...

Name of contributing exposure scenario 1

Further specification

OPERATIONAL CONDITIONS AND RISK MANAGEMENT MEASURES

Product characteristic

What was the physical state of the substance the company was using - powder (agglomerated or fine), liquid or solid (granular or pellet-type)? If it was a solid, was the presence/creation of dust observed during this task? What was the level of dustiness (propensity of the material to generate airborne dust during handling)?
Choose from: Very high (> 500 mg dust/kg test sample); High (> 150-500 mg/kg); Medium (50-150 mg/kg)
If it was a liquid, what is the vapor pressure?

What were the dimensions of the primary particles or agglomerates?

What shape and form were the particles - were they spherical, regular (e.g., cubic), irregular (e.g., bundle), platelet or a fiber-type?

If a mixture was being used, what concentration of nanomaterial was present in the mixture? (e.g., for liquid state % in volume [ml NM/ml product]; for solid state g NM/g product)

Amounts used

What quantity of the substance was used at the workplace (per task)?

Frequency and duration of use/exposure

How long was the substance used for and how often [per worker]? State duration per task/ activity and frequency (e.g., single events or repeated) of exposure. In cases where specific information is not known, duration and frequency for a worker should be stated as 220 days per year, each with 8 hrs work (worst case).

Human factors not influenced by risk management

What were the primary and secondary routes of exposure - inhalation, dermal, oral?

(Continued)

TABLE 2.5   Exposure Scenario Format Used to Develop NANODEVICE
                Case Examples—cont'd

Exposure Scenario Format Addressing Uses Carried Out by Workers

Were there any other particular conditions of use not influenced by risk management (e.g., body parts potentially exposed as a result of the nature of the activity)?

Other given operational conditions affecting worker exposure

What was the volume of the room that the task was carried out in? If not known exactly, provide an estimate based on following options: < 100 m^3 (e.g., meeting room); 100-1000 m^3 (e.g., open-plan office); > 1000 m^3 (e.g., warehouse)

What was the energy level of the process? Choose from Zero (e.g., removal and handling of clean barrels and plastic containers); Very low (e.g., weighing small amounts of material); Low (e.g., < 5 cm drop height); Moderate (e.g., drop between 5-30 cm, mixing nanomaterials in a liquid medium); High (e.g., drop between 30-100 cm); Very high (e.g., drop from > 100 cm, dry mixing, dry cleaning with a brush)

Was the process manual, automatic, semiautomatic?

What was the temperature and pressure of the process?

Technical conditions and measures at process level (source) to prevent release

What was the level of containment of the process - closed, semi-enclosed (e.g., fumehood) or open-air process?

What was the level of effectiveness of the containment (% reduction)? If this is not known for nanomaterials, what is it for other substances?

Technical conditions and measures to control dispersion from source towards the worker

What engineering controls were in place to control exposure? e.g., LEV, general mechanical ventilation or natural ventilation (e.g., through windows and doors)

What kind of filter was used in the ventilation system?

What is the level of effectiveness of this control measure (% reduction)? If this is not known for nanomaterials, what is it for other substances?

Were any other technical measures taken to control dispersion from the source to the worker (e.g., separation & segregation; cabin)?

Organizational measures to prevent /limit releases, dispersion and exposure

Were the engineering controls properly maintained?

Was cleaning & maintenance of the room and equipment undertaken? How often?

Were the employees trained in handling a) nanomaterials in general; b) the specific task that he/she carried out; c) good housekeeping procedures; d) emergency response procedures?

Was routine monitoring of exposure/exposure measurements undertaken by the company?

Were the nanomaterials used properly disposed of (e.g., solid, liquid waste)?

TABLE 2.5   Exposure Scenario Format Used to Develop NANODEVICE
Case Examples—cont'd

Exposure Scenario Format Addressing Uses Carried Out by Workers

Were any other measures to prevent or limit substance release, dispersion and exposure in place (e.g., organizational measures to limit the no. of exposed workers or duration of exposure; segregation of dirty departments, safe storage, fire/explosion protection & prevention, eyebaths/showers)?

Conditions and measures related to personal protection, hygiene and health evaluation

Was personal protective equipment (PPE) provided to the workers? If so, what type (e.g., wearing of gloves, face protection, full body dermal protection, goggles, respirator, etc.)?

What is the level of effectiveness of the PPE provided (% reduction)? If not known for nanomaterials, what is it for other substances?

Were the workers using the PPE properly? How long was the PPE used by the workers before replacement?

Was PPE (including lab coats) removed before leaving the task area?

Was eating/drinking allowed in the area where the task was being carried out?

Were other personal hygiene procedures observed (e.g., washing hands after handling of substances, changing contaminated clothes, etc.)?

EXPOSURE ESTIMATION AND REFERENCE TO ITS SOURCE

Is the measurement data you gathered for this task available (e.g., in a paper or report)? Provide reference.

Particle mass concentration

Did you gather data for particle mass concentration over the duration of this task? If so, what method(s)/instruments were used?

If multiple methods were used, in your opinion which method provided the most reliable data for this substance and task? Explain reasoning (e.g., sensitivity of the analysis, direct measure, etc.).

What was the mean, median and geometric mean range of the measurements (mg/m^{-3}) over the duration of this task?

What was the fraction of the total airborne particles that were monitored (e.g., inhalable, respirable, <100 nm)?

Were measurements collected for the entire duration of the task? If it was for a shorter or longer period, state how long the measurements were taken for.

Where were the measurements taken - within the breathing zone of the worker, at the source point, at a distance from the source? If at a distance, state how far from the source.

Was the data corrected for the presence of background aerosols (e.g., comparison between near- and far-field, comparison between periods of activity and non-activity, etc.)?
If no, what was the background, how was it calculated and applied to the measurements?

(Continued)

TABLE 2.5   Exposure Scenario Format Used to Develop NANODEVICE
Case Examples—cont'd

Exposure Scenario Format Addressing Uses Carried Out by Workers

---

Particle number concentration

Did you gather data for particle number concentration over the duration of this task? If so, what method(s)/instruments were used?

If multiple methods were used, in your opinion which method provided the most reliable data for this substance and task? Explain reasoning (e.g., sensitivity of the analysis, direct measure, etc.)

What was the mean and geometric mean range of the measurements (no. particles per m^3) over the duration of this task?

What was the fraction of the total airborne particles that were monitored (e.g., inhalable, respirable, <100 nm)?

Were measurements collected for the entire duration of the task? If it was for a shorter or longer period, state how long the measurements were taken for.

Where were the measurements taken - within the breathing zone of the worker, at the source point, at a distance from the source? If at a distance, state how far from the source.

Was the data corrected for the presence of background aerosols (e.g., comparison between near- and far-field, comparison between periods of activity and non-activity, etc.)?
If no, what was the background, how was it calculated and applied to the measurements?

Surface area concentration

Did you gather data for surface area concentration over the duration of this task? If so, what method(s)/instruments were used?

If multiple methods were used, in your opinion which method provided the most reliable data for this substance and task? Explain reasoning (e.g., sensitivity of the analysis, direct measure, etc.).

What was the mean and geometric mean range of measurements ($m^2\,m^{-3}$) over the duration of this task?

What was the fraction of the total airborne particles that were monitored (e.g., inhalable, respirable, <100 nm)?

Were measurements collected for the entire duration of the task? If it was for a shorter or longer period, state how long the measurements were taken for.

Where were the measurements taken - within the breathing zone of the worker, at the source point, at a distance from the source? If at a distance, state how far from the source.

Was the data corrected for the presence of background aerosols (e.g., comparison between near- and far-field, comparison between periods of activity and non-activity, etc.)?
If no, what was the background, how was it calculated and applied to the measurements?

Particle size distribution

Did you gather data for particle size distribution over the duration of this task? If so, what method(s)/instruments were used?

TABLE 2.5   Exposure Scenario Format Used to Develop NANODEVICE
Case Examples—cont'd

Exposure Scenario Format Addressing Uses Carried Out by Workers

If multiple methods were used, in your opinion which method provided the most reliable data for this substance and task? Explain reasoning (e.g., sensitivity of the analysis, direct measure, etc.)

What was the mass or number median aerodynamic diameter and standard deviation (μm) over the duration of this task? Are PSD plots available?

What was the fraction of the total airborne particles that were monitored (e.g., inhalable, respirable, <100 nm)?

Were measurements collected for the entire duration of the task? If it was for a shorter or longer period, state how long the measurements were taken for.

Where were the measurements taken - within the breathing zone of the worker, at the source point, at a distance from the source? If at a distance, state how far from the source.

Was the data corrected for the presence of background aerosols (e.g., comparison between near- and far-field, comparison between periods of activity and non-activity, etc.)?
If no, what was the background, how was it calculated and applied to the measurements?

Physicochemical characterization

Did you undertake any physicochemical characterization of the airborne particles? If so, provide a description

CONTRIBUTING EXPOSURE SCENARIO (N) CONTROLLING WORKER
EXPOSURE FOR ...

Name of contributing scenario (n)

CES 1.1 and 1.2: Grinding and Packing of Silica Nanospheres

The grinding and packing of the silica nanospheres (CES 1.1 and CES 1.2) was undertaken in a fumehood within a laboratory room. Results of this part of the study have been published in a report for the European Agency for Safety and Health at Work [46].

OPERATIONAL CONDITIONS AND RISK MANAGEMENT MEASURES

In both CES 1.1 and 1.2, solid, raw granules of silica nanospheres were ground manually using a pestle and mortar in a fumehood to produce spherical silica nanoparticles suitable for use in nanocomposite production. The granules were pure (with minor impurities) and of almost spherical shape prior to grinding (as confirmed by TEM), with a high viscosity (resin/paste-like). In CES 1.1, the nanospheres had a particle size of 70 nm. Nine samples of approximately 10—15 g each (122 g in total) were ground and packed into a glass tube over a period of approximately 60 minutes. Following this, in CES 1.2, silica nanospheres with a particle size of 110 nm were ground using the same procedure. In this case, two samples of approximately 66 g each (132 g total) were ground and packed into a glass tube over a period of approximately 25 minutes.

The laboratory room in which the tasks were undertaken had a volume of 89 m^3 and was at room temperature (22–25°C) with a relative air humidity of 24–26% and negligible air velocity. General ventilation was operational in the laboratory room, and the door between the laboratory and the corridor was kept open. The fumehood had operational local exhaust ventilation (LEV) and the sash was kept up for the duration of the activity. During the grinding activity the workers wore personal protective equipment (PPE), specifically laboratory coats, half-face masks and gloves, which were removed and disposed of after the task. In addition, good housekeeping procedures were in place at this site, with daily cleaning of the room undertaken.

EXPOSURE ESTIMATION

The following suite of instruments was used to measure airborne exposure concentrations in a range of locations:

- Particle number concentration was measured using an ultrafine particle counter (P-TRAK, TSI model 8525, range 20–1000 nm) and a Scanning Mobility Particle Sizer (SMPS, CPC with long DMA, range 15–661 nm);
- Surface area concentration was measured using a Handheld Particle Counter (AERO-TRAK 9000, TSI, range 10–1000 nm: TB fraction);
- Particle size distribution was measured using a Scanning Mobility Particle Sizer (SMPS, CPC with long DMA, range 15–661 nm).

Particle number and surface area concentration measurements are presented in Table 2.6. Particle size distribution data remain company-confidential.

Noticeable peaks in the total number concentration and surface area concentration were observed during grinding of 70 nm nanospheres (CES 1.1), as compared to a background concentration prior to any activity. However, during grinding of 110 nm nanospheres (CES 1.2), the total number and surface area concentrations were found to be lower than the background concentration measured in the fumehood between grinding activities. During this time in the laboratory a perceptible draft was observed (air velocity increased to 1.3 m/s) as a result of the operation of the general ventilation, which could explain this observation. After both grinding activities, number and surface concentrations of particles as measured on the desk were similar or lower than that determined as 'background' before grinding and were comparable to concentrations obtained in the corridor.

TABLE 2.6   Measurement Data for the Grinding and Packing of Silica Nanospheres

Monitoring Location	Total Particle Number Concentration (particles/cm³)	Surface Area Concentration ($\mu m^2/cm^3$)
On the desk as 'background' in the laboratory room prior to any activity	P-TRAK: 12,000-19,000 SMPS: 12,000-17,000	16-20
In the fumehood as 'background' before the activity	P-TRAK: ~14,000 SMPS: ~15,000	~19
In the fumehood during grinding of 70-nm nanospheres (CES 1.1)	P-TRAK: Peak of 25,000, with a sub-peak of 16,000 SMPS: Peak of 30,000, with a sub-peak of 19,000	Peak of 27, with a sub-peak of 21
In the fumehood between grinding of 70- and 110-nm nanospheres	P-TRAK: ~15,000 SMPS: ~17,000	~20
In the fumehood during grinding of the nanospheres (110 nm)	P-TRAK: ~12,000 SMPS: ~14,000	17
On the desk in the laboratory room after grinding	P-TRAK: ~13,000 SMPS: ~14,000	15-18
In the corridor close to the door of the laboratory room	P-TRAK: ~13,000 SMPS: ~14,000	17-19

CES 1.3-1.5: Weighing, Mixing and Loading of Powder Containing Silica Nanospheres

The other activities of weighing, transferring, mixing and loading of powder containing silica nanospheres (CES 1.3−1.5) were undertaken in a workroom in a separate building. Results of this part of the study have been presented as conference posters [47,48].

OPERATIONAL CONDITIONS AND RISK MANAGEMENT MEASURES

The workroom had a volume of 95 m³ and was at room temperature (24−26°C) with a relative air humidity of 24−25% and negligible air velocity. General ventilation was operational in the laboratory room, and the door between the laboratory and the corridor was kept closed during the tasks. Good housekeeping procedures were in place at this site, with daily cleaning of the room undertaken.

During CES 1.3, a powder of ground silica nanospheres was weighed into small batches on an open workbench in the workroom. The powder consisted of pure silica nanospheres (with minor impurities), 110 nm in particle size. In this scenario, five batches of varying weights from 7.5−60 g (ca. 120 g in total) were manually weighed during a period of 24 minutes. Other chemicals for the production of nanocomposites were also weighed during the measurement period, the exact composition and quantities of which were confidential. During this activity, the workers wore PPE, specifically a laboratory coat and nitrile gloves, which were removed and disposed of after the task.

Following this, during CES 1.4, the five batches of silica nanospheres (110 nm) powder were transferred to separate clear plastic bags along with other chemicals. The bags were then sealed and manually shaken to mix the materials. The exact composition and quantities of the other chemicals used during this task are confidential. In total, approximately 120 g of nanospheres were transferred and mixed during a period of 12 minutes. This task was undertaken on the open workbench. It is not known what PPE the workers were wearing during this period.

CES 1.5 describes the manual loading of bags with silica nanospheres powder (110 nm) and chemicals into a reactor for the production of polymer nanocomposites. Six batches containing silica nanospheres powders (over 120 g in total) were poured into the reactor over a period of approximately 90 minutes. The sixth batch was prepared during the loading period using an unknown quantity of silica nanospheres. Thus, only an estimate of over 120 g nanospheres in total for loading the six batches can be made. This task was undertaken as an open task in the workroom; however an LEV extraction system was in place near the reactor opening. It is not known what PPE the workers were wearing during this period.

EXPOSURE ESTIMATION

Monitoring of airborne particle levels was undertaken using the same suite of instruments as described for CES 1.1—1.2. Results of measurements taken in a range of locations during CES 1.3—1.4 are summarized in Table 2.7.

TABLE 2.7   Summary of Measurement Data for the Weighing and Mixing of Silica Nanospheres

Monitoring Location	Particle Number Concentration (particles/cm^3)	Surface Area Concentration ($\mu m^2/cm^3$)
On the workbench as background for approximately 1 hour prior to weighing	P-TRAK: 12,000-17,500 SMPS: ~12,000-18,000	~16-22
On the workbench during the weighing task (CES 1.3)	P-TRAK: Peak of 22,500 SMPS: Peak of 24,000	Peak of ~26
On the workbench between the weighing and mixing tasks	P-TRAK: ~17,500-20,000 SMPS: ~21,000-22,000	~24-26
On the workbench during the mixing task (CES 1.4)	P-TRAK: Peak of 19,500 SMPS: Peak of 24,000	Peak of ~26
On the workbench for approximately 30 mins after the mixing task	P-TRAK: 17,500-19,000 SMPS: ~22,000-24,000	~26

During both the weighing and mixing tasks, an increase in the total number concentration was observed using both P-TRAK and SMPS. Similarly, the surface area concentration was also observed to increase during the weighing task, and remained approximately at this level during the period between task and during the mixing task itself.

During CES 1.5, measurements were collected before, during and after the loading period, as summarized in Table 2.8.

Total number concentrations determined with the use of SMPS were higher than those determined with P-TRAK. During loading of the first and second batch of components, total number concentrations were observed to increase, particularly in case of fine particles. When loading the next batches of components noticeable peaks of concentrations were not observed. The concentrations decreased until they were lower than the background concentrations. A perceptible draft (air velocity 1.3 m/s) was observed in the room during these latter loadings, which could explain this observation. A similar pattern was seen in the surface area concentration measurements.

Case Study ES2: Production of Silicon Nanoparticles in a Pilot Plant Facility

This exposure scenario focuses on the emission of nanoparticles into the workplace during the production of silicon nanoparticles in a pilot plant facility, based on a study undertaken by Wang et al. [49]. The exposure scenario consists of three CES as detailed below:

CES 2.1 Synthesis of silicon nanoparticles in a pilot plant facility using a hot wall reactor
CES 2.2 Bagging and packaging of silicon nanoparticles in a pilot plant facility
CES 2.3 Cleaning of the tubing system within a pilot plant facility used to synthesize silicon nanoparticles.

TABLE 2.8 Summary of Measurement Data for the Loading of Silica Nanospheres

Monitoring Location	Particle Number Concentration (Particles/cm^3)	Surface Area Concentration (μm^2/cm^3)
At the loading point as background for 1 hour prior to loading	P-TRAK: 17,000-19,000 SMPS: 21,000-24,000	~25-26
At the loading point during the loading task (CES 1.5)	P-TRAK: 20,000-29,000 SMPS: 24,000-36,000	28-37
At the loading point for approximately 45 mins after the final loading	P-TRAK: 14,000-19,000 SMPS: 15,000-24,000	19-28

CES 2.1: Synthesis of Silicon Nanoparticles in a Pilot Plant Facility Using a Hot Wall Reactor

OPERATIONAL CONDITIONS AND RISK MANAGEMENT MEASURES

CES 2.1 describes the production of silicon nanoparticles in a hot wall reactor housed within the pilot plant facility, in which the precursor gas was thermally decomposed to form silicon nanoparticles as a powder.

The pilot plant facility, designed to produce nanoparticles in the kg/day range, was a three-story building ($\sim$150–200 m^3) within a large hall (>1000 m^3). The plant was separated from the rest of the hall by acrylic sheets at the upper levels and sliding doors at the ground level. Air exchange in and out of the plant could still take place through gaps around the sliding doors. A general mechanical ventilation system was in place at the top of the plant, and good housekeeping procedures were in place with daily cleaning and maintenance of the hall and equipment undertaken.

The reactor was located on the third floor in the plant. Carrier gas from the reactor flowed through tubing to a filter housing on the second floor, where the particles were collected on filters. The use of metal grids for the floors meant that air exchange could take place between the levels. Production was carried out at high temperature ($\sim$1000°C) and pressure (100–150 mbar) in a closed system. A double-filter system was installed to allow continuous production. Particles were filtered alternately by one of the filter cartridges, while reverse pulsing was applied to the other filter to blow the particles off into a plastic bag underneath for bagging.

Approximately 750 g of silicon nanoparticles were produced over the duration of 1 hour. The size of the primary particles was approximately 60 nm, however the powder was typically present as agglomerates of $\sim$200–300 nm. The generated particles consisted of pure silicon core, covered by a naturally grown oxide layer on the surface. A limited number of people (approximately four) were allowed into the pilot plant facility while production was taking place. No PPE was used by workers during the synthesis stage.

EXPOSURE ESTIMATION

Monitoring of airborne exposure concentrations was undertaken as follows:

- Particle number concentration was measured at the source point using an ultrafine water-based condensation particle counter (UWCPC, TSI model 3786, particle size range >2.5 nm, 1 s time resolution) and a handheld CPC (TSI model 3007, particle size range >10 nm, 1 s time resolution);
- Surface area concentration was measured at the source point using a Nanoparticle Surface Area Monitor (NSAM, TSI model 3550, particle size range <1 µm, 1 s time resolution);

- Particle size distribution was measured at the source point using a Fast Mobility Particle Sizer (FMPS, TSI model 3091, particle size range 5.6—560 nm, 1 s time resolution) and an SMPS (TSI model 3936 with long DMA, size range 15—750 nm, 5 min time resolution).

A second set of instruments was placed outside the enclosure of the pilot plant facility to monitor the particle level at the background (far-field), both prior to the activity (overnight) and during activities (daytime).

According to the UWCPC, the total particle number concentration was in the range 10,000—12,000 particles/cm^3 during production. On average, the surface area concentration during production as measured by NSAM was determined to be approximately 60 $\mu m^2/cm^3$. According to the FMPS, the electrical mobility diameter ranged from 10—300 nm during production. All of these measurements are comparable with background levels measured in the hall area. It is important to note that welding activity, as well as other activities, was being undertaken within the hall at the same as these measurements were taken.

CES 2.2: Bagging and Packaging of Silicon Nanoparticles in a Pilot Plant Facility

OPERATIONAL CONDITIONS AND RISK MANAGEMENT MEASURES

CES 2.2 describes the semiautomatic process of bagging and packing of the silicon nanospheres (primary particles: 60 nm; agglomerates: 200—300 nm) following collection of the produced nanoparticles on the bag house filters. This task was undertaken in the same pilot plant facility and building as described in CES 2.1. In this task, approximately 750 g of silicon nanopowder was filled into two big plastic bags over a period of approximately 10 minutes.

The filter housing on the second floor was connected by a vertical pipe, which extended to a height of approximately 1 m above the ground. The pipe was comprised of two stages of valves and a lock. A plastic bag was attached to the end of the pipe with double clamps to collect the nanoparticles. A reverse pulsing procedure was repeated several times to transport the maximum amount of particles into the bag. The bag was sealed using a hot bar sealer, before being cut from the pipe with scissors. A limited number of people (approximately 1—2) were allowed into the pilot plant enclosure while bagging was taking place. Workers wore PPE, specifically laboratory coats, filtering half-masks and gloves.

EXPOSURE ESTIMATION

Monitoring of airborne particle levels was undertaken using the same suite of instruments and procedures as described for CES 2.1.

According to FMPS measurements, total particle number concentrations were approximately 10,000 particles/cm^3 before and after the thermal seal process, comparable to the background level measured in the hall. Again, a jump up to ~58,500 particles/cm^3 was observed during the thermal seal process itself where the hot clamp was used. Particle size concentrations in the size range 20–60 nm increased significantly during the thermal seal process. This is thought to be due to the formation of organic particles via condensation of evaporated plastic material from the bag at high temperatures. On average, the surface area concentration as measured by NSAM was determined to be approximately 45 μm^2/cm^3, comparable to the background level measured in the hall. A jump up to ~79 μm^2/cm^3 was observed during the thermal seal process. According to the FMPS, the electrical mobility diameter ranged from 10–300 nm during production, again comparable to the background level measured in the hall. Thus, overall no increase in airborne concentrations was seen above background during the bagging activity, except during thermal seal activity.

CES 2.3: Cleaning of the Tubing System within a Pilot Plant Facility Used to Synthesize Silicon Nanoparticles

OPERATIONAL CONDITIONS AND RISK MANAGEMENT MEASURES

CES 2.3 describes cleaning of the synthesis system, including the reactor, the tubing system and the filters, of the pilot plant facility to avoid contamination of the products. The cleaning procedure consisted of two major steps. In stage 1 (semiautomatic, closed process), a vacuum pump was connected to the synthesis system and the pressure was lowered to 50 mbar. Nitrogen was injected into the system, through the opening and closing of valves every few seconds, in order to remove residual particles on the walls of the reactor and tubing. During nitrogen injection, the system pressure increased to about 300 mbar. The residual particles were collected on a filter installed between the pump and the synthesis system. After approximately 25 minutes, the vacuum pump was switched off. During this stage, workers wore gloves and laboratory coats. In stage 2 (manual, open process), different tubing sections were disconnected and manually vacuum-cleaned. Respirators were worn by the technicians and researchers prior to opening the system. Substantial particle deposition on the inside walls of the tubing was visible. The vacuum pump was then switched on and its connecting pipe used by the technicians to draw the particles off the tubing wall. A wrench was used to knock on the tubing in order to dislodge particles from the wall. Following cleaning of the tubing, both the vacuum pump and its pipe were wiped with isopropyl alcohol-saturated wipes. During this stage, workers wore half-face masks, as well as gloves and laboratory coats. The total duration of both stages of the task was approximately 3 hours.

EXPOSURE ESTIMATION

Monitoring of airborne particle levels was undertaken using the same suite of instruments and procedures as described for CES 2.1.

According to FMPS measurements, the total particle number concentration was in the range of 7000–8000 particles/cm^3 before cleaning started. This increased gradually to 8000–9000 particles/cm^3 during stage 1 (cleaning with nitrogen injection). After opening the system and commencing stage 2 (manual cleaning), the particle concentrations continued to gradually increase up to approximately 10,000 particles/cm^3. Peaks in the particle concentrations were observed when the system tubing was knocked on with a wrench, with the total number concentration reaching 17,000 particles/cm^3, more than twice that before cleaning.

The NSAM reading increased from approximately 38 $\mu m^2/cm^3$ before cleaning to 49 $\mu m^2/cm^3$ during stage 1 (cleaning with nitrogen injection). After opening the system and commencing stage 2 (manual cleaning), the particle concentrations continued to gradually increase up to approximately 60 $\mu m^2/cm^3$. Peaks of the particle concentrations were observed when the system tubing was knocked on with a wrench, with the surface area concentration reaching 174 $\mu m^2/cm^3$, more than four times that before cleaning.

The electrical mobility diameter, as measured by FMPS, ranged from 10–300 nm. The particle concentration was observed to increase after the cleaning started, and further so after opening of the tubing system. The shape of the distributions did not change significantly, except when the tubing system was knocked on by a wrench.

Case Study ES3: Production of MWCNT at Pilot Plant Scale

This exposure scenario focuses on the production of multi-walled carbon nanotubes (MWCNT) in a commercial pilot scale facility. Details of this study have recently been presented at a conference [50] and will be published in the near future as part of a report for FP6 project 'NANOSH'. The exposure scenario consists of three CES:

CES 3.1 Continuous weighing and pouring of MWCNT
CES 3.2 Vacuum cleaning of the reactor used to produce MWCNT
CES 3.3 Semiautomatic filling of bins with MWCNT

CES 3.1: Continuous Weighing and Pouring of MWCNT
OPERATIONAL CONDITIONS AND RISK MANAGEMENT MEASURES

CES 3.1 describes the manual weighing of quantities of dry MWCNT powder, which were then poured into cans for further use, shipping, etc.

This task was undertaken on an open-air workbench in a dedicated laboratory room, 52 m^3 in volume, with general mechanical ventilation and unfiltered external supply air.

The powder was pure MWCNT (with minor impurities due to the iron catalyst used in production) and consisted mainly of agglomerates. The agglomerates were approximately 50–900 μm in size, while the primary fibers were between 10–15 nm in diameter and 1–10 μm in length. During the task, 1 kg of MWCNT powder was manually weighed and poured continuously over a period of 40 minutes (the same 1 kg of material was repetitively weighed and poured approximately 50 times). This was performed to simulate a worst-case scenario and does not represent normal operation.

Good housekeeping practices were in place at the facility, with the room and equipment cleaned after each operation and waste nanomaterials disposed of in a special waste. The number of workers in the room was limited to one person during the task, and the worker wore a whole-body chemical safety suit, under positive air pressure, and nitrile gloves, which were removed and disposed of at the end of each operation before leaving the room.

EXPOSURE ESTIMATION

Particle number concentration was measured at the source point using a CPC (Grimm 5400, range 10–2000 nm), with measurements taken for approximately 40 minutes before the task, during the task and for approximately 40 minutes after the task. The arithmetic mean total particle number concentration during the task was 94,000 particles/cm^3, a three-fold increase above the background level of 31,000 particles/cm^3.

Surface area concentration was measured at the source point using the same protocol with a nanoparticle surface area monitor (NSAM 3550, TSI, range 10–400 nm in the alveolar deposition fraction). The arithmetic mean surface area concentration during the task was determined to be 63 μm^2/cm^3, compared to a background of 34 μm^2/cm^3.

Particle size distribution was measured at the source point using an electrical low-pressure impactor (ELPI, range 30 nm–10 μm) using the same protocol. The modal (maximum) value of the size distribution was 43 nm during activity, comparable to the background measurement. An increase was noticed in the range 1–10 μm.

CES 3.2: Vacuum Cleaning of the Reactor Used to Produce MWCNT

OPERATIONAL CONDITIONS AND RISK MANAGEMENT MEASURES

CES 3.2 describes manual cleaning of a reactor used to produce MWCNT using a vacuum cleaner (suction tube) fed through a small opening at the top of the reactor. This task was undertaken in the upper level of a plant room of volume 300 m^3. General mechanical ventilation was present in the room, and there was also a supply of fresh air. The room was at room temperature (24—25°C) and atmospheric pressure throughout the task. The MWCNT produced in the reactor consisted of a dry agglomerated powder. The agglomerates were approximately 50—900 μm in size, while the primary fibers were between 10—15 nm in diameter and 1—10 μm in length. Residual MWCNT powder was present on the inside of the reactor following production, which was removed during the cleaning activity over a period of 20 minutes. It is not known how often the area surrounding the reactor was cleaned. The number of workers in the area was limited to two persons, both of whom wore a full-body chemical safety suit, a disposable half-face mask and nitrile gloves, which were removed before leaving the task area and disposed of in a special waste. The filters of the vacuum cleaner were disposed of in a special waste.

EXPOSURE ESTIMATION

The arithmetic mean total particle number concentration during the task, as measured by CPC, was 72,000 particles/cm^3, lower than the background level before the task of 94,000 particles/cm^3 due to influences in the unfiltered external air due to other sources (e.g., motor engines, incineration, etc.). The arithmetic mean surface area concentration during the task, measured by NSAM, was determined to be 150 μm^2/cm^3 (alveolar deposited fraction), again lower than the background level before the task of 212 μm^2/cm^3. The modal (maximum) value of the size distribution, determined using ELPI, was 43 nm during activity, comparable to the background measurement.

CES 3.3: *Semiautomatic Filling of Bins With MWCNT*

OPERATIONAL CONDITIONS AND RISK MANAGEMENT MEASURES

CES.3.3 describes a scenario in which an automatic machine was used to fill bins (approximately ~200 L in volume) with MWCNT from a storage unit, operated by a worker. The task was undertaken in a large hall of approximately 600 m^3. However, the bins were housed in a closed glove box (approximately 5 m^3 in size), such that the worker was separated from the source during the task. A general mechanical ventilation system was in place, sucking air from inside the filling station. Workers wore goggles and gloves while performing this task, the latter of which were disposed of after one operation.

EXPOSURE ESTIMATION

Measurements in this scenario were taken outside of the glove box, up to 1 m from the process. The arithmetic mean total particle number concentration during the task, as measured by CPC, was 6700 particles/cm^3, again lower than the background level before the task of 10,000 particles/cm^3 due to other external sources (e.g., motor engines, incineration, etc.). The modal (maximum) value of the size distribution, determined using ELPI, was 74 nm during activity; no background measurements were taken in this case. Surface area concentration was not measured during this task.

Case Study ES4: Production of Paints Containing TiO$_2$ Nanoparticles

This exposure scenario describes the pouring and mixing of various nanopowders (primarily nano-TiO$_2$) in two paint production facilities. Details of this study have been published in a Danish report [51], and a manuscript for publication is in preparation. The exposure scenario consists of two CES:

CES 4.1 Pouring and blending powders in a mixer with external LEV (around the rim of the mixer)
CES 4.2 Pouring and blending powders in a mixer with internal LEV (in the mixer).

Although these CES are developed from measurements at two different production facilities, in this instance they have been included in one overall ES due to the similarities in terms of material types and processes used.

CES 4.1: Pouring and Blending Powders in a Mixer with External LEV (Around the Rim of the Mixer)
OPERATIONAL CONDITIONS AND RISK MANAGEMENT MEASURES

In CES 4.1, batches of 500 and 25 kg of powder, including nanopowders, were manually poured sequentially into an automatic mixer. This task was undertaken as an open-air process in a work room approximately 1500 m^3 in volume. The work room had general mechanical ventilation, and the mixer had an LEV system around the rim of the mixer (external LEV). Good housekeeping procedures were in place at this facility, with the workroom kept clean and the empty bags from which the nanopowders were poured disposed of in a special waste. One worker was present in the area during the task. The worker wore PPE for the duration of the task, specifically an apron, gloves and a filter face mask.

Details of the nanomaterials poured into the mixer during this task are given in Table 2.9.

In addition to these nanomaterials, other powder and liquid materials were also poured into the mixer during the task. In total, 3446.8 kg of powder was added to the mixer and blended over a period of approximately 6 hours. The duration of each addition ranged from approximately

TABLE 2.9 Details of Nanomaterials Used in CES 4.1

Nanomaterial	Shape/Form	Particle Size	Quantity	Method of Addition
Titanium dioxide (TiO$_2$)	Perfect crystals; idiomorphic	220 nm	2675 kg (added in batches of 500 kg and 25 kg)	Big bags (500 kg) handled with forklift; small bags (25 kg) added by hand
Talc	Flake agglomerates	Unknown	280 kg (added in batches of 25 kg)	By hand
Silica	Nanostructured aggregates	Unknown	175 kg (added in batches of 25 kg)	By hand

30 seconds to 1 minute for small bags to approximately 10 minutes for big bags.

EXPOSURE ESTIMATION

Exposure measurements were taken at the source point (near-field) and approximately 5 meters away (far-field) for the entire duration of the activity. In addition, background measurements were taken during the night prior to the activity.

The triplex cyclone PM1 was used to take both personal measurements and measurements at the source point. The total personal mass concentration (within the breathing zone of the worker) was determined to be 0.65 mg/m^3 across the duration of the task, excluding three breaks. These data were not corrected for the background. Total mass concentration at the source point was determined to be 0.09 mg/m^3 across the duration of the task, as compared to a background measurement before the task of 0.05 mg/m^3 and a far-field measurement during activity of 0.07 mg/m^3.

Particle number concentration was measured using a Fast Mobility Particle Sizer (FMPS, model 3091, TSI Inc., range 5.6–560 nm). The mean particle number concentration measured at the source point across the entire duration of the task was 79,000 particles/cm^3, compared to a background mean of 2500 particles/cm^3 prior to the task and a far-field mean of 4000 particles/cm^3. Particle size distribution was measured using the FPMS and an optical particle counter (OPC). The size distribution was observed to peak in the range of 200 nm, compared to a background range of 60–400 nm prior to the task. It is important to note that there was a leak in the mixing equipment at the time of measurement, which could have resulted in increased particle concentrations above normal operation.

CES.4.2. Pouring and Blending Powders in a Mixer with Internal LEV (in the Mixer)

OPERATIONAL CONDITIONS AND RISK MANAGEMENT MEASURES

In CES 4.2, batches of powders, including nanopowders, were manually poured sequentially into an automatic mixer and blended. This task was undertaken in a sectioned work area (approximate floor space 140 m^2) in a big hall of >5000 m^3 in volume. There were other mixing chambers located in the work area, but only one was in use during the monitoring period. The work room had general mechanical ventilation, and the mixer had an internal LEV system under the mixer (internal LEV). The task was undertaken in the open air, but the lid of the mixer was closed between additions when the powders were being blended. Good housekeeping procedures were in place at this facility, with the mixer cleaned with a hose in between each addition and the empty bags from which the nanopowders were poured disposed of in a special waste. The workers wore gloves during the task.

The primary nanomaterial used in this task was pure TiO$_2$ powder of particle size 220 nm (according to manufacturer specifications). TiO$_2$ was manually poured into the mixer in batches of 25 kg, such that 1640 kg in total was added during the task. In addition, the following substances were also added, including calcium carbonate (CaCO$_3$), nepheline syenite and diatomite. The total amount of all powders added was 3339 kg over a total task duration of 2.7 hours.

EXPOSURE ESTIMATION

Exposure measurements were taken at the source point (near-field) and approximately 10.5 meters away (far-field) for the entire duration of the activity. In addition, background measurements were taken during the night prior to the activity.

The triplex cyclone PM1 was used to take both personal measurements and measurements at the source point. The total personal mass concentration (within the breathing zone of the worker) was determined to be 0.16 mg/m^3 across the duration of the task. These results were not corrected for background levels. Total mass concentration at the source point was determined to be 0.10 mg/m^3 across the duration of the task, as compared to a background measurement before the task of 0.04 mg/m^3 and a far-field measurement during activity of 0.07 mg/m^3. Particle number concentration was measured using an FMPS (model 3091, TSI Inc., range 5.6—560 nm). The mean particle number concentration measured at the source point across the entire duration of the task was 24,000 particles/cm^3, compared to a background mean of 2300 particles/cm^3 prior to the task and a far-field mean of 7500 particles/cm^3. Particle size distribution was measured using the FPMS and an optical particle counter (OPC). The size distribution was in the range 80–400 nm, with a peak in the range of 200–300 nm, compared to a background range of 80–400 nm prior to the task.

2.7 SUMMARY

Exposure scenarios play an important role in exposure assessment and communication of exposure information to stakeholders. This chapter described the development of occupational exposure scenarios for engineered nanomaterials across the life-cycle of these materials. Harmonization of the information collected during the measurement surveys and description of scenario details will aid in the interpretation of the results. It will also allow others to apply the information to similar exposure scenarios. Within the FP7 project 'MARINA' (Managing Risks of Nanomaterials), an extended exposure scenario library is being developed, which will be used as part of a tiered exposure assessment approach. The exposure scenario library will allow stakeholders to search and review exposure scenarios to provide an initial, tier 1, exposure estimate for the scenario that they are reviewing. The success of such a library will depend on i) the quality of the contextual information as well as the exposure assessment and ii) sharing of exposure scenario descriptions between stakeholders. Currently, most exposure information for engineered nanomaterials and nano-objects is related to the primary synthesis of nanomaterials and production of nano-objects, rather than downstream use. There is an urgent need for generation and sharing of exposure data for all scenarios across the entire life-cycle of engineered nanomaterials.

Acknowledgements

The authors of this chapter thank the following members of the NANODEVICE Consortium for their contribution towards development of the case examples: Elżbieta Jankowska (Central Institute for Labour Protection, Warszawa, Poland), Christof Asbach (Institute of Energy and Environmental Technology e.V., Duisburg, Germany), Carsten Möhlmann (Institute for Occupational Safety and Health, Sankt Augustin, Germany), Keld Jensen and Ismo Koponen (National Research Centre for the Working Environment, Copenhagen, Denmark).

References

[1] Faraday M. Experimental relations of gold (and other metals) to light. Philos Trans R Soc London 1857;147:145.
[2] Zsigmondy R. Zur Erkenntnis der Kolloide. Germany: Salzwasser-Verlag Gmbh; 1905.
[3] Aitken RJ, Creely KS, Tran CL. Nanoparticles: An occupational hygiene review. Research Report 274; 2004.
[4] World Health Organisation International Agency for Research on Cancer. IARC Monographs on the Evaluation of Carcinogenic Risks to Humans Volume 93: Carbon Black, Titanium Dioxide, and Talc; 2010.

[5] Binnig G, Rohrer H. Scanning tunneling microscopy. IBM J Res Dev 1986;30(4):355.

[6] Kroto HW, Heath JR, Obrien SC, et al. C-60-Buckminsterfullerene. Nature 1985; 318(6042):162—3.

[7] Iijima S. Helical microtubules of graphitic carbon. Nature 1991;354(6348):56—8.

[8] Wijnhoven SWP, Dekkers S, Kooi M, et al. Nanomaterials in consumer products - Update of products on the European market in 2010. RIVM Report 340370003/2010; 2010.

[9] Roco MC. Nanoscale Science and Engineering: Unifying and Transforming Tools. Am Inst Chem Eng J 2004;50(5):890—7.

[10] Sengül H, Theis T, Ghosh S. Towards sustainable nanoproducts: An overview of nanomanufacturing methods. J Ind Ecol 2008;12(3):329—59.

[11] Bergeson L, Auerbach B. Reading the small print. Environ Forum 2004;21(2):30—40.

[12] International Organization for Standardization. Nanotechnologies - Occupational risk management applied to engineered nanomaterials - Part 1: Principles and approaches. ISO 2012. 12901—1.

[13] Nowack B, Brouwer C, Geertsma RE, et al. Analysis of the occupational, consumer and environmental exposure to engineered nanomaterials used in 10 technology sectors. Nanotoxicology 2012. Epub ahead of print.

[14] Keles Y, Morales A, Escolano C. Science and Technology Assessment: Automotive & Aerospace. Report of the ObservatoryNANO Project; 2009.

[15] Chaudhry Q, Scotter M, Blackburn J, et al. Applications and implications of nano-technologies for the food sector. Food Addit Contam 2008;25(3):241—58.

[16] Robinson DKR, Morrison MJ. Nanotechnology Developments for the Agrifood Sector. Report of the ObservatoryNANO Project; 2009.

[17] Heugens EHW, Wijnhoven SWP, Bleeker EAJ. Environment, Health and Safety (EHS) Impacts Technology Sector Evaluation: Agrifood. Report of the ObservatoryNANO Project; 2010.

[18] Gleiche M. Construction General Sector Report. Report of the ObservatoryNANO Project; 2008.

[19] ObservatoryNANO. ObservatoryNANO Factsheets. Report of the ObservatoryNANO Project; 2012.

[20] Ross BL, Aitken RJ. Environmental, Health and Safety Impact Evaluation of the Construction Technology Sector. Report of the ObservatoryNANO Project; 2010.

[21] Bertoldi O, Berger S. Report on Energy. Report of the ObservatoryNANO Project; 2009.

[22] Mueller NC, Nowack B. Report on Nanotechnology in the Technology Sector Enviroment. Report of the ObservatoryNANO Project; 2009.

[23] Rickerby D, Morrison M. Report from the Workshop on Nanotechnologies for Environmental Remediation. 16th-17th April 2007; JRC Ispra; 2007.

[24] Nowack B. Environment, Health and Safety (EHS) Impacts Technology Sector Evaluation: Environment. Report of the ObservatoryNANO Project; 2010.

[25] ObservatoryNANO. Health, Medicine and Nanobio Full Report. Report of the ObservatoryNANO Project; 2009.

[26] Geertsma RE, Brouwer C, Wijnhoven SWP, Sips JAM. Environment, Health and Safety (EHS) Impacts Technology Sector Evaluation: Health Medicine & Nanobio. Report of the ObservatoryNANO Project; 2011.

[27] Toufekstian M. Environment Health and Safety (EHS) Impacts Technology Sector Evaluation: Information & Communication. Report of the ObservatoryNANO Project; 2010.

[28] Singh KA. Security Technology Sector First Year Progress Report. Report of the ObservatoryNANO Project; 2009.

[29] Mantovani E, Zappelli P. Report on Textiles Technology Sector. Report of the ObservatoryNANO Project; 2009.

[30] Ross BL, Aitken RJ. Environmental, Health & Safety (EHS) Impacts Technology Sector Evaluation: Textiles. Report of the ObservatoryNANO Project; 2011.

[31] British Standards Institution. Guide to assessing airborne exposure in occupational settings relevant to nanomaterials. BSI PD 2010:6699–3.

[32] van Tongeren MJ, Kromhout AH, Gardiner K. Trends in levels of inhalable dust exposure, exceedance and overexposure in the European carbon black manufacturing industry. Ann Occup Hyg 2000;44(4):271–80.

[33] European Commission. REGULATION (EC) No 1907/2006 OF THE EUROPEAN PARLIAMENT AND OF THE COUNCIL of 18 December 2006 concerning the Registration, Evaluation, Authorisation and Restriction of Chemicals (REACH), establishing a European Chemicals Agency, amending Directive 1999/45/EC and repealing Council Regulation (EEC) No 793/93 and Commission Regulation (EC) No 1488/94 as well as Council Directive 76/769/EEC and Commission Directives 91/155/EEC, 93/67/EEC, 93/105/EC and 2000/21/EC. Official J Eur Union 2007. L 1 36/3.

[34] European Chemicals Agency. Guidance on information requirements and chemical safety assessment Exposure Scenario Format in Part D: Exposure scenario building; in Part F: CSR format; 2012. ECHA-10-G-11-EN.

[35] European Commission. (2008). Follow-up to the 6th Meeting of the REACH Competent Authorities for the implementation of Regulation (EC) 1907/2006 (REACH), 15–16 December 2008: Nanomaterials in REACH. CA/59/2008 rev. 1.

[36] National Institute for Occupational Safety and Health. Occupational Exposure to Carbon Nanotubes and Nanofibers. NIOSH Draft Curr Intell Bull DHHS (NIOSH) 2010. Publication No. 2010–XXX.

[37] National Institute for Occupational Safety and Health. Occupational Exposure to Titanium Dioxide. NIOSH Curr Intell Bull 2011;63. DHHS (NIOSH) Publication No. 2011–160.

[38] Roco MC, Mirkin CA, Hersam MC. Nanotechnology Research Directions for Societal Needs in 2020-Retrospective and Outlook. WTEC Panel Report; 2010.

[39] NANEX. NANEX Project Publishable Summary; 2011.

[40] Kuhlbusch TAJ, Asbach C, Fissan H, et al. Nanoparticle exposure at nanotechnology workplaces: A review. Part Fibre Toxicol 2011;8(22):2–18.

[41] Kuhlbusch TAJ, Fissan H. Particle characteristics in the reactor and pelletizing area of carbon black production. J Occup Environ Hyg 2006;3:558–67.

[42] Fujitani Y, Kobayashi T, Arashidani K, et al. Measurement of physical properties of aerosols in a fullerene factory for inhalation exposure assessment. J Occup Environ Hyg 2008;5:380–9.

[43] Tsai SJ, Hofmann M, Hallco M, et al. Characterization and evaluation of nanoparticle release during the synthesis of single-walled and multiwalled carbon nanotubes by chemical vapor deposition. Environ Sci Technol 2009;43(6017):6023.

[44] Brouwer D, Berges M, Virji MA, et al. Harmonization of measurement strategies for exposure to manufactured nano-objects; report of a workshop. Ann Occup Hyg 2012;56(1):1–9.

[45] European Chemicals Agency. Guidance on information requirements and chemical safety assessment Part D: Exposure Scenario Building; 2012. ECHA-12-G-17-EN.

[46] European Agency for Safety and Health at Work. Case studies: Non-functionalised and functionalised silica nanospheres; 2011.

[47] Jankowska E, Zielecka M. Emission of nanosized particles during weighing and mixing silica nanospheres with chemicals in the process of nanocomposites production. INRS

Occupational Health Research Conference; 5th-7th April 2011; Palais des Congres, Nancy, France; 2011.

[48] Jankowska E, Zielecka M. Potential exposure to silica nanoparticles during production of nanocomposites. INRS Occupational Health Research Conference; 5th-7th April 2011; Palais des Congres, Nancy, France; 2011.

[49] Wang J, Asbach C, Fissan H, et al. Emission measurements and safety assessment for the production process of silicon nanoparticles in a pilot-scale facility. J Nanopart Res. 2012;14:759.

[50] Möhlmann C, Pelzer J, Berges M, et al. Exposure to carbon nano-objects in research and industry. INRS Occupational Health and Research Conference; 5th-7th April 2011; Nancy, France; 2011.

[51] Kristensen HV, Hansen SB, Holm GR, et al. Nanopartikler i arbejdsmiljøet-Viden og inspiration om håndtering af nanomaterialer; 2010.

3

Nanomaterials and Human Health

Harri Alenius [1], *Julia Catalán* [1], *Hanna Lindberg* [1],
Hannu Norppa [1], *Jaana Palomäki* [1], *Kai Savolainen* [2]

[1] Systems Toxicology, Finnish Institute of Occupational Health,
Helsinki, Finland
[2] Nanosafety Research Centre, Finnish Institute of Occupational Health,
Helsinki, Finland

3.1 INTRODUCTION

Nanotechnologies are already a part of our everyday lives, with workers being in contact with engineered nanomaterials (ENM) in their workplaces and consumers coming into contact through several market products incorporating engineered nanomaterials. Hence, the nanotechnologies impact with us on a daily basis. The rapid development that has taken place in commercialization of engineered nanomaterials can be traced to the promising characteristics of these materials. For example, they allow the introduction of novel beneficial properties into existing and new consumer products. Nanotechnologies also provide possibilities for scientific breakthroughs in many areas including medicine and new generation of consumer products including mobile phones and tablets with novel properties. They are also being used in energy production as well as appearing in novel materials such as light and strong polymers and have found applications in innovative manufacturing technologies (see [1]).

Hence, it is not surprising that engineered nanomaterials and nanotechnologies have received a remarkable amount of attention. The unique technological and beneficial properties of engineered nanomaterials are due to their small size, large surface-to-volume ratio, high surface area, reactivity, often excellent electrical conductivity, persistency and high tensile strength and their potential to form highly resistant, durable, reactive and self-cleaning surfaces [1]. Several of the novel applications of engineered nanomaterials can be encountered in a variety of consumer products, e.g., sunscreens with better protection against ultraviolet (UV) radiation now contain nano-scale titanium dioxide, which is an effective UV filter. Engineered nanomaterials are also present in many consumer products such as cosmetics, clothes, cleaning materials such as bleach, sportswear and other sports products such as cross-country skis and tennis rackets [2]. The economically most important applications of engineered nanomaterials are found in industry; these include various coatings, opto- and printed electronics, applications of nanocellulose in packaging and as food additives. Engineered nanomaterials are also being used as additives to oil (nano-diamonds) and gasoline (cerium oxide), where they reduce friction (diamonds) or promote the release of carbon dioxide (cerium oxide improving the burning of gasoline) (see [1]). Nanotechnologies are increasingly being used in the production of solar energy as well as other energy applications, clean water and even purification of soils [1,3].

However, it has also been noted that these materials do not only possess beneficial properties but in some cases can cause harm to human health or the environment. In fact, many of the typical properties of engineered nanomaterials, e.g., their small size — at least one dimension between 1—100 nm — their large surface area per weight and their reactivity, i.e., the properties that make them technologically so valuable, are also the likely reasons for their potentially harmful effects [4—8]. Figure 3.1 depicts various mechanisms through which these materials may evoke unwanted effects inside cells [5].

There are several hundreds of thousands of different engineered nanomaterials that have been synthesized in laboratories. In fact, there are more than 50,000 different forms of one single group of these materials, carbon nanotubes (see [9,10]). It is noteworthy that these carbon-containing materials, in spite of many marked similarities, exhibit clearly distinct material properties and possess distinct potentials to harm human health or damage the environment. In humans, different materials target different human organs, and they can affect different environmental compartments should leaks into the environment take place. The most widely used engineered nanomaterials include carbon black, which is used as a filler in car tires, amorphous silica used in construction materials, nano-scale titanium dioxide in paints and coatings or consumer products, cerium oxide in gasoline and several metal particles such as silver, copper and gold, which have a variety of industrial applications such as electronics. Carbon-based materials such as fullerenes, graphene, single- (SWCNT) and multi-walled carbon nanotubes (MWCNT) or carbon nanowires are widely used in electronics and semiconductors (see [11]).

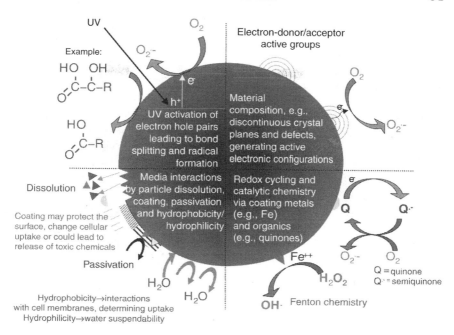

FIGURE 3.1 **The many mechanisms through which engineered nanomaterials can evoke potential harmful effects in cells and organs when coming into contact with living organisms.** *From Nel A et al. (2006) Toxic potential of materials at the nanolevel. Science 311, 622-627* [5]. *Reprinted with permission from AAAS.*

One important challenge in our attempts to reliably assess the potential risks and safety of engineered nanomaterials is the lack of systematic knowledge on exposure to these materials in different environments, or on their risks to human health or the environment. There has been very little systematic research conducted into hazards of engineered nanomaterials on human health or the environment, and most of it covers only a few nanomaterials (see [7,12−15]). Unfortunately, the amount of data on exposure in workplaces and the general environment to these materials is even more limited [4,13,14,16,17]. Without this information, the assessment of risks of exposure to different nanomaterials is not practically feasible.

Data on both hazards and exposure are needed if one wants to conduct a quantitative risk assessment as indicated by the well-known equation (see [18]):

$$\text{hazard} \times \text{exposure} = \text{risk}$$

The steps of the process of chemical risk assessment of the currently used risk assessment paradigm as initially formalized by the US National

Research Council [18] are depicted in Figure 3.2. Many of the details of the initial process have been upgraded to meet the current practices; the picture shown here is adapted from reference [19].

Even though quite a few of the engineered nanomaterials have been identified as hazardous [6,8,11,20—22], as with other chemicals, it is likely that many of engineered nanomaterials are harmless or only slightly harmful [23]. It may be a fact that some of the engineered nanomaterials are hazardous and some of them less hazardous or even harmless. However, because of their material characteristics, this does not mean that they are similar to or comparable with soluble chemicals or even with larger though chemically identical particles [23]. It is also important to remember that the currently used risk assessment paradigm was developed for soluble chemicals that readily dissolve or disperse in solvents such as water, alcohol or organic solvents; they were not initially meant to be used with particulate matters, and hence there are many unknowns that complicate the adoption of the same methods [24] and principles in the assessment of engineered nanomaterials in terms of risks and safety. Indeed, one can ask whether the currently used chemical hazard assessment methods are really suitable for the hazard assessment of engineered nanomaterials. Thus, one cannot conclude that because some of the engineered nanomaterials are hazardous and some harmless just like soluble chemicals that this means that these materials are similar to other chemicals; this judgment would be premature and lacking any scientific justification [23].

For the above reasons, our abilities to differentiate the harmful engineered nanomaterials from their harmless counterparts are very limited, and this greatly complicates the assessment of safety and risks of engineered nanomaterials. The available methods do not enable us to distinguish between harmful and harmless engineered nanomaterials in different exposure settings due to limitations of the current methods applied for the testing of hazards of and the exposure to engineered nanomaterials. Subsequently, in the absence of validated hazard and exposure assessment methods, which are prerequisites for a chemical risk assessment, the reliable assessment and management of risks of these materials remain fraught with difficulty. It is also noteworthy that the currently used risk assessment concepts have not undergone any thorough evaluation with regard to their suitability for the assessment of particulate matter such as engineered nanomaterials.

The current situation has caused concerns of unexpected events with unpredictable, potentially harmful consequences for humans or the environment due to exposure to engineered nanomaterials. The uncertainties about the safety of engineered nanomaterials may create a demand to formulate a new regulatory paradigm for these materials, and this might lead to unexpected regulatory actions. This is an issue

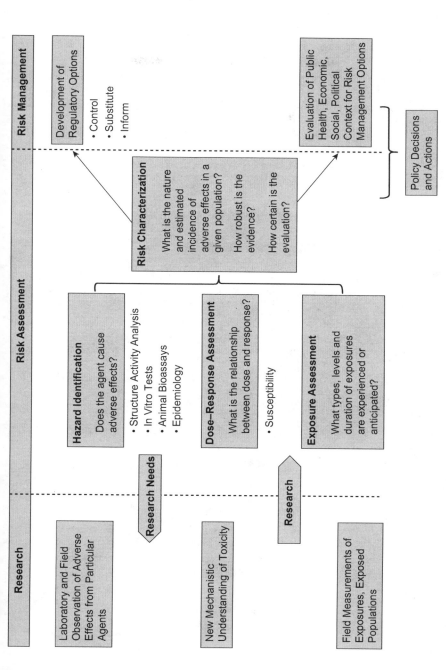

FIGURE 3.2 This framework shows the four key steps of risk assessment: hazard identification, dose–response assessment, exposure assessment and risk characterization. The depicted and updated framework also shows the role of research in the various steps of the risk assessment process and the relationship of the process with risk management important in the risk management process, which optimally is a knowledge-based policy decision-making process with a foundation in scientific risk assessment [19].

that alarms the producers of engineered nanomaterials and the downstream industries incorporating these materials into novel products. The uncertainty around the safety and health of engineered nanomaterials is a major source of ungovernable risk. In fact, this may be one reason for reduced investments into nano-based innovations and production of materials and products. These concerns have been identified by the European Commission in the 2nd regulatory review of REACH legislation, where they were recognized as a major obstacle for the effective use of these materials in conjunction with existing and emerging technologies [23]. This is especially concerning because nanotechnologies have been identified by the EC to be in the six key enabling technologies that should serve as major drivers of European well-being in the future [25].

The health concerns around engineered nanomaterials started to appear as the use of these materials became more widespread. First Oberdörster et al. [26] and then Elder et al. [27] showed in experimental animals that nasal exposure to engineered nanomaterials could lead to the uptake of the particles, notably manganese oxide [26] and iron oxide [27], by axons of the olfactory nerve in the olfactory epithelium of the nose and lead to transportation of the particles into the olfactory bulb under the frontal cortex (see Figure 3.3). Later studies have shown that these materials can be certainly distributed to other parts of the brain after their uptake through the nose, but the potential effects

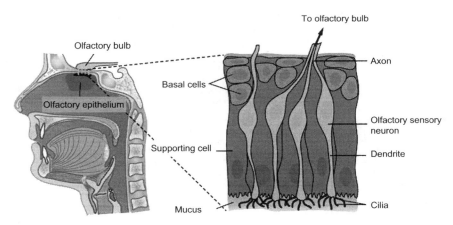

FIGURE 3.3 This figure depicts the anatomical features of the nasal cavity and of the olfactory epithelium in the nasal cavity. The figure also illustrates the intimate relationship between the olfactory epithelium and the axons of the olfactory nerve in the epithelium, and the olfactory bulb at the bottom of the frontal lobe of the brain. Only a few millimeters separate the nerve endings in the olfactory epithelium from the olfactory bulb. *Adapted from Duodecim, 2010 [36] and Oberdörster et al., 2004 [26].*

remain to be elucidated (see [27,28]). There are now several studies demonstrating that nanoparticles that have reached the olfactory bulb can be distributed to other parts of the brain [29,30]. Furthermore, a plethora of later studies have demonstrated that titanium dioxide nanoparticles may induce pulmonary inflammation, and exposure to single-walled carbon nanotubes (SWCNT) can evoke progressive pulmonary fibrosis [20]. Multi-walled carbon nanotubes (MWCNT) have been associated with pulmonary granulomas and inflammation [31] and even asbestos-like changes in the peritoneal cavity of experimental animals, e.g., mesothelial thickening and granulomas, and markedly elevating the incidence of mesotheliomas, even exceeding the potential of crocidolite to induce these tumors in a sensitive mouse model [32—35].

The predominant challenge associated with the use of engineered nanomaterials is the uncertainty that is associated with their potential health and environmental effects. Hence, one of the major challenges of nanotechnologies is to alleviate the uncertainties associated with the use of these materials in consumer products and industrial applications. Until this challenge has been resolved through more research into the safety of these materials, combined with more appropriate and affordable approaches for their risk assessment, this uncertainty hangs like the sword of Damocles into the foreseeable future not only in Europe but globally as indicated by the active discussion on the topic [1,7,37—40]. Hence, attempts to remove these uncertainties are being encouraged by the European Commission, interested other parties in Europe [23,41] and outside Europe, e.g., the United States [3,42].

In the following sections, the biological effects of different types of engineered nanomaterials will be dealt with in some detail. The focus of Section 3.2 is on the biokinetics, cardiovascular effects and neurotoxicity of engineered nanomaterials. Section 3.3 focuses on the inflammatory effects of engineered nanomaterials, Section 3.4 on their pulmonary effects, especially pulmonary inflammation and its mechanisms, and Section 3.5 on the dermal effects of engineered nanomaterials. Section 3.6 will provide an introduction to the genotoxicity mechanisms in general and has detailed information on the genotoxic and carcinogenic effects of several different engineered nanomaterials. The topic of Section 3.7 is risk assessment and management and associated topics predominantly from the health point of view, the importance of safety for the use of engineered nanomaterials and the development of nanotechnologies and nano-enabled consumer and industrial applications. In the concluding Section 3.8, the goal is to deal with the future outlook of engineered nanomaterials and nanotechnologies by considering the significance of the safety and health of these materials.

3.2 BIOKINETICS, CARDIOVASCULAR AND NEUROTOXICITY OF ENGINEERED NANOMATERIALS

The focus of this chapter is on the biokinetics (absorption, distribution, biotransformation, excretion) and system target organ toxicity of chosen key target organs of engineered nanomaterials, notably the effects of engineered nanomaterials on the central nervous system and the vasculature.

3.2.1 Biokinetics of Engineered Nanomaterials

When assessing hazards of engineered nanomaterials, it is crucial to have a thorough understanding of their biokinetics and translocation. This discussion will briefly deal with general issues of biokinetics and translocation and then focus on metal and metal oxide nanoparticles.

The rates and mechanisms of deposition, bio-accessibility, route-specific absorption, the blood and/or lymph flow-dependent distribution throughout the body and the passive or active uptake into cells and tissues and the breakdown (metabolism) and the excretion of engineered nanomaterials are known to some extent for a very limited number of specific engineered nanomaterials. Currently, there is very little information on how to integrate knowledge of the kinetic and toxicological effects, with data emerging from toxicity testing of engineered nanomaterials. The lack of integration leaves open many questions such as whether the data obtained during kinetics testing can be applied to the results obtained from toxicity testing studies. Most information is available for the inhalation exposure, which indicates that there is a complex deposition of ENM based on their size distribution. For the oral route, little is known about how the various physicochemical conditions (pH, protein, lipids) encountered during passage in the gastrointestinal tract affect absorption. For the dermal route, the release rates of cosmetic products and consumer products are largely unknown, as is the influence of sweat, temperature, shear force and even their interaction with water as can occur during showering and hand washing (bio-nano-interactions). Dermal toxicity and the biokinetics of engineered nanomaterials in the skin will be dealt with in Section 3.6.

Absorption is usually defined as the passive and/or active crossing of cellular outer membranes. Currently, information is scarce on rates of absorption of ENM after exposure via the three most relevant routes: oral, dermal and inhalation. There are likely to be major differences in rates of absorption between routes and between engineered nanomaterials, even between engineered nanomaterials and their chemically identical but non-nano counterparts. The information is still very limited due to the scarcity of routine analytical equipment with which to conduct particle analysis in these systems. Intravenous studies have indicated that engineered nanomaterials distribute rapidly from blood to tissues, mainly to those organs that contain phagocytic cells (see [43]). Their apparent very short blood/plasma half-life is in contrast to their long whole-body half-life. Distribution profiles may differ between engineered nanomaterials with different characteristics, and different distribution profiles of engineered nanomaterials complicate attempts to combine hazard and exposure information.

The kind of metabolism undergone by classical soluble chemicals is most likely to be irrelevant to most engineered nanomaterials. These materials are usually too large to fit into the active sites of the enzymes responsible for biotransformation. It is thought possible that some oxidation reactions may take place on the outer surface of engineered nanomaterials, but even these reactions may be either enzymatic or non-enzymatic in nature. This may indeed be related to the Fenton chemistry of metals (see [5]). Excretion of engineered nanomaterials via the usual routes, notably via biliary, urinary, mammary glands or salivary routes is generally unknown. The rare studies available, however, suggest that excretion is very slow, and this may be the reason why so little is known. The longer the half-life, the longer the study needed to acquire a reliable assessment of the rate of excretion. While it makes these studies relatively expensive, nevertheless they are very relevant if one wishes to assess the terminal half-life, i.e., the longer the half-life, the higher the risk for bioaccumulation.

In the following section, mainly the biokinetics of metal and metal oxide nanoparticles will be discussed. In fact, there is very little information on the biokinetics of the other types of engineered nanomaterials. More complex materials such as carbon nanotubes or graphene will remain outside this discussion. In particular, the longer (>10−15 μm) carbon nanotubes are likely to be trapped in the interstitial tissue of the lungs, whereas the shorter carbon nanotubes can be removed from the lungs via lymphatic drainage or phagocytosis by immunological cells that are moved from the lungs by the mucociliary escalator to the oral cavity and then swallowed or exhaled as a result of coughing. However, it is also worth stressing that much of the discussion on metal oxide nanoparticles is likely to be applicable also to metal nanoparticles (see [6,8,44]).

Several metal or metal oxide nanoparticles are believed to become dissolved upon exposure, releasing free ions [45]. For example, nano-silver readily dissolves, and hence it is difficult to assess whether its possible effects are due to the nano-sized particles or to the released silver ions [46]. Zinc oxide is another example of a metal oxide that readily dissolves, rendering the assessment of the role of the particles or ions problematic, and the same applies to nano-copper [47,48]. However, it is noteworthy that even though iron oxide is less soluble than zinc oxide, as much as 81% of iron oxide was dissolved in monocytes within 24 h. Likewise, cadmium containing quantum dots can disintegrate due to surface oxidation, leading to their dissolution. There are also poorly soluble metal oxides, e.g., ceria (cerium oxide, CeO_2), nano-titania (titanium dioxide, TiO_2) and nano-zirconia (zirconium dioxide, ZrO_2), which all are insoluble in water [45]. Nano-titania is insoluble through a pH range of 3–11, and nano-ceria is insoluble at pH 4.5 [49]. Gold nanoparticles have also been considered to be insoluble, as demonstrated in citrate-stabilized aqueous dispersion. The benefits of using insoluble or poorly soluble engineered nanoparticles in biokinetic studies is obvious because one does not have to consider their dissolution into biological media and, for example, analytical techniques such as ICP-MS can provide reliable detection of these particles without the confounding effects of the ions being released from soluble engineered nanomaterials. For example, ceria is a very good choice for biokinetic studies due to its poor solubility and widespread commercial use, e.g., as a gasoline additive ([50]; see [51]).

3.2.1.1 Absorption

The most relevant routes for metal and metal oxide nanoparticles are inhalation and oral exposure. It is noteworthy that due to their tendency to agglomerate, the half-life of the primary particles is usually very short, and agglomeration takes place within minutes. Therefore, it is not surprising that metal oxide nanoparticles often express a bimodal size distribution with peaks at 200–400 and 2000–3000 nm, respectively [14,52]. Some engineered nanomaterials have been shown to be absorbed from the lungs into the systemic circulation, this being followed by translocation to other organs in a fashion similar to that seen after intravenous administration of these materials. This has been found for carbon-, metal- and metal-oxide-based engineered nanomaterials. These materials mainly accumulate as agglomerates in liver and spleen. In the case of ceria, after intratracheal ceria administration, the concentration of the material was highest in the liver followed by spleen, kidney, heart, testis and brain. About 0.1–1% of the 7-nm ceria administered to the lungs via intratracheal instillation could still be found in these organs one month after administration. This is in agreement with the results obtained after intravenous administration of the same compound, and hence, at least in this case, exposures and subsequent translocations into different organs of this material markedly resembled each other (see [51]).

The absorption from the nasal cavity olfactory epithelium via the axons (see also Figure 3.3) of the olfactory nerve to the olfactory bundle under the frontal globe and other parts of the brain including hippocampus, cerebral cortex and cerebellum has been described for colloidal silver-coated gold (50 nm) particles (see [51]), ^{13}C-particles (35 nm) [26], Mn (30 nm) [27] and iron oxide (280 nm) [53]. In the same way, there is a possibility for absorption of nano-sized particles in the maxillary region of the trigeminal nerve by direct uptake via the axonal nerve endings.

The extent of gastrointestinal or oral absorption is usually inversely associated with the size of the engineered nanoparticle, meaning that a smaller particle size favors better absorption [54]. For example, when ceria was used as a model metal oxide nanoparticle, 0.00001% of the orally administered dose was found after one week in spleen, lung, heart, testis and brain and 0.01% in the liver. Based on this example alone, it seems that in some cases the gastrointestinal tract is likely to play a minor role as a route for engineered nanomaterials into the body. However, there may be marked differences in this regard between different engineered nanoparticles (see [51]).

When compared with the lungs and the gastrointestinal tract, intact skin seems to provide a much better barrier against absorption of engineered nanomaterials into the body. Engineered nanomaterials applied on the skin in the absence of organic solvents did not penetrate through the intact skin as far as the stratum granulosum or stratum spinosum (the outermost layers of the skin) [55], which are areas from which the engineered nanomaterials could enter the lymphatic or blood circulation. Quantum dots have shown deeper penetration but have not been shown to penetrate beyond the stratum corneum [56]. One exception seems to be 4-nm nano-TiO$_2$, which, when repeatedly administered onto pig ear skin, penetrated deep into the basal cell layer of the epidermis [57]. After 60 days of dermal exposure of nano-TiO$_2$ (10–60 nm) with hairless mice, Ti concentrations rose in the internal organs, indicative of systemic absorption [57].

There is one human study that used a highly purified Zn isotope to reveal penetration of this Zn isotope via the lung into the blood and urine. A small number of male and female volunteers were exposed to a sun-block cream containing 18% of this highly purified Zn in the form of ZnO. The cream was applied onto the skin of the back for 5 days, with the duration of each exposure being 30 min. After the cessation of the exposures, the Zn isotope was analyzed in the blood and urine; the isotope was found in all cases in urine and blood in both males and females. When compared with ZnO paste with coarse ZnO particles, there was no difference in penetration through the skin of males, but there was a clear difference in females, perhaps due to their thinner skin. The study could not demonstrate whether Zn penetrated the skin as a nanoparticle or as an ion, but it did reveal that when ZnO nanoparticle-containing sun-block cream was applied to human skin, Zn rapidly gained access to the systemic circulation and was rapidly excreted in the urine [58].

3.2.1.2 Distribution

There is very little information available on the distribution of engineered nanomaterials in the blood. It has been shown that 25-nm gold and 22-, 25- and 80-nm titania have entered red blood cells, i.e., evidence that they are capable of crossing the cell membrane [59]. The distribution of 5—55-nm ceria in blood was explored in rats after repeated intravenous dosing. Ten minutes after the ceria infusion, 2% of the 13—55-nm ceria remained in the circulating blood as compared with 30% of the 5-nm ceria. Surprisingly, the concentrations of ceria of various sizes increased 2—4 hours after the cessation of the infusion, suggesting a rapid distribution of a wide size range of ceria to body compartments from where they were subsequently released into the systemic circulation. The results also indicated that the size of the ceria had an impact on the binding of serum proteins on the surface of the ceria particles. This kind of protein corona formation around 15—30-nm ceria particles may explain why they were tightly attached to the clot fraction of blood over time as compared with the smaller and larger ceria particles [60].

3.2.1.3 Metabolism

If one excludes the dissolution of engineered nanomaterials, then the opsonization of the particles, or the formation of the protein corona around the nanoparticles [61,62], is the most important process that modulates engineered nanomaterials in the body. Of course, dissolution can be considered to be one form of biotransformation of engineered nanomaterials, but it is a more passive process and it largely depends on the pH surrounding the particles. It has been emphasized that the corona around the particles may be important for their recognition and processing by various cells and subsequent fate of the particles [61]. It has been proposed that the shape and size of the nanoparticle can influence which proteins can coat onto the nanomaterial [61]. Since the corona also influences the apparent size of the nano-sized particles, the corona has an impact not only on their metabolism or biotransformation but also most likely on their distribution, i.e., penetration through biological barriers [62].

3.2.1.4 Excretion

Renal glomerular filtration takes place for globular proteins and engineered nanomaterials with a hydrodynamic diameter ≤ 6 nm. For example, 40—80% of quantum dots with a diameter of 4.4—5.5 nm were excreted in urine within 4 hours after intravenous injection [63], and 8.6% of 1.4-nm gold nanoparticles were excreted in urine and 5% in feces within 24 hours. In contrast, less than 0.5% of 30-nm ceria were eliminated in urine and feces within 2 weeks. In the latter case the question arises about the fate of ceria — if it is not eliminated via excretion or biotrans-formation, what is the fate of the material after it has entered the systemic circulation? In studies in which rats were given ceria for 1, 7, 30 or 90 days, and 17 different organs were monitored for 90 days for their ceria content, no changes in this content were seen. It seemed that the material was present in the immunological cells, i.e., the macrophages, of the reticu-loendothelial system as agglomerates that were resistant to elimination from these cells. The presence of this material in the cells was associated with morphological alterations in the tissue such as granulomas, but the reason for this phenomenon is not known. The findings with ceria do agree with reports with other poorly soluble or insoluble nano-sized metal oxides such as titania [64], gold [63] and functionalized quantum dots [65]. It is clear that our understanding of the biokinetics and trans-location of engineered nanomaterials does not allow us to draw any clear conclusions on the possible association between biokinetics and the effects of engineered nanomaterials in living organisms.

3.2.2 Effects of Engineered Nanomaterials on the Central Nervous System

One important issue has been whether engineered nanoparticles can cross the blood—brain barrier (BBB) and hence reach perhaps the best protected cells in the human body, the neurons. Levels of some poorly soluble engineered nanoparticles have been found in the brain, e.g., ceria (30 nm), gold (1.3–100 nm), silica (50 and 100 nm), titania (24 nm) and quantum dots (13 nm). The reports have not demonstrated, though, whether these materials are in the endothelial cells of the cerebral vasculature, the cerebral phagocytic cells, e.g., glial cells, or within the actual neurons [66]. In fact, it has not been demonstrated that any of these materials can penetrate the BBB [67]. Most of these studies have been conducted in mature rodents, and the number of studies carried out with pregnant rodents is limited. However, it does seem that the mature BBB is a very effective barrier against the entry of engineered nanomaterials into the adult brain, and there is also no convincing evidence that these par-ticles could penetrate the developing BBB under normal conditions [68].

It has been shown, though, that many engineered nanomaterials can be transported from the nasal cavity olfactory epithelium through the axons of the olfactory nerve to the olfactory bulb at the bottom of the frontal cortex. This has been demonstrated for C^{13} nanoparticles [26], gold [66], manganese oxide [27] and iron oxide [53]. In fact, this is an old observa-tion, as it also applies to viruses. By utilizing this route, particles do not have to pass through the BBB, as they are directly taken up by the nerve endings of the axons of the olfactory nerve, which are not protected by the epithelium making up the BBB (see also [28]). For details, see Figure 3.3, which illustrates this issue.

For the above reasons, this slightly unusual exposure route has been used to explore whether different engineered nanomaterials can enter the brain when there is no BBB protection, whether they can be translocated to other parts of the brain via this route and whether they can evoke neurotoxicity when exposure to the brain takes places through this route (see [69]).

For example, nasal instillation of Cu nanoparticles for 14 days in mice at a relatively high dose (40 mg/kg) led to particle accumulation and caused several morphological lesions in the olfactory bulb. However, there were no other morphological lesions in other parts of the brain [70], but there were marked changes in the levels of monoamine neurotransmitter levels in the olfactory bulb of mice exposed to Cu nanoparticles. There were also changes in the neurotransmitter levels outside the olfactory bulb, most notably in hippocampus, striatum, cerebral cortex and cerebellum. Wang et al. [71] found that nasally instilled titania nanoparticles could be translocated into the murine brain and evoked pathological lesions in the hippocampus and caused alterations in neurotransmitter levels also in other parts of the brain. In another study, Wang et al. [72] found that low concentrations of nano-sized titania (500 µg/mouse) caused neurotoxicity in the murine brain and that there were minor changes in the neurotoxicity induced by anatase and rutile, respectively. Zhang et al. [73] have shown that this murine neurotoxicity has been associated with exposure to nano-sized particles via the nose, whereas a similar exposure scenario of mice to larger, micro-sized particles did not cause any effects. Altogether these observations provide evidence that intranasally introduced copper and titanium dioxide nanoparticles have been translocated to the olfactory bulb and to other parts of the brain. These particles have evoked both morphological lesions in those areas to which they have been translocated and have induced alterations in neurotransmitter levels. Surface chemistry has had an influence on the neurotoxic effects of these materials (see, e.g., [72]).

There have also been reports suggesting that some engineered nanomaterials may evoke phototoxicity when they come into contact with the retina, the only part of the central nervous system that is in contact with light. Several engineered nanomaterials have been designed to be photoactive, with TiO_2 being the main example. Photoactivity and photocatalytic reactions may lead to phototoxicity. The hydroxyl radicals formed can reduce oxygen molecules, i.e., producing superoxide radicals that can initiate a chain reaction of events leading to the formation of increased amounts of reactive oxygen species. Titanium dioxide nanoparticles have been shown to induce phototoxicity in both *in vitro* model systems. Once taken up by retinal cells or cells modeling the retina, titanium dioxide nanoparticles can induce oxidative damage after the cells are exposed to visible light or ultraviolet radiation [66].

It seems that some engineered nanomaterials are likely to be neurotoxic when they can enter neuronal cells. However, passage of the engineered poorly soluble nanomaterials through the BBB and to control neurons seems unlikely. Exposure through the nasal cavity and the olfactory epithelium olfactory nerve axonal endings is of course possible, though not likely to be a significant exposure route associated with any major hazards in the human setting. Nonetheless, the neurotoxicity of the especially poorly soluble engineered nanomaterials should merit attention in the future.

Neurotoxic effects are not likely after systemic exposure to these materials via inhalation or oral routes via the systemic circulation. Some of the reactive engineered nanomaterials may also induce phototoxicity that may be relevant for ocular toxicity in cases where reactive engineered nanomaterials come into contact with the eye, especially the retina.

3.2.3 Effect of Engineered Nanomaterials on the Vascular System

A number of studies with concentrated ambient particles and epidemiological studies on the same topic support the potential contribution of particles as being causal factors in cardiovascular diseases. Indeed, there are studies that demonstrate the potential of air pollution to evoke cardiovascular effects [74,75]. Hence, it is not surprising that there are concerns about the effects of engineered nanomaterials on blood circulation and the distribution of engineered nanomaterials to distant organs. A strong association has been shown between human cardiac mortality and elevated levels of micro-sized ambient particles in urban air [74]. Similar results have been obtained by using concentrates of ambient particles in experimental studies [76]. Cardiovascular diseases are of course a large entity, a family of diseases of which perhaps atherosclerosis is the most relevant in this content. The gold standard for assessing the likelihood of the potential of particles to induce atherosclerosis is the Apo $-/-$ mouse model. Kang et al. [76] showed that inhalation of nickel hydroxide nanoparticles for 5 months at a dose corresponding to 80 µg nickel/m^3 accelerated the atherosclerotic pathology in the aortic arch and induced aortic mitochondrial DNA damage. The authors claimed that the atherosclerosis was causally and mechanistically linked with the DNA damage. In addition, similar patterns of gene expression changes were found in the lungs, aorta and spleen tissue. Furthermore, inflammatory changes were observed in the lungs. Atherosclerosis of the vasculature was also found after exposure of mice to carbon black and single-walled carbon nanotubes [77].

An important sign of a harmful effect on vasculature is endothelial dysfunction. In a series of studies with nano-sized titanium dioxide, it was found that a dose level of 6 mg/m^3 in an *ex vivo* model caused flow-induced dilation in subepicardial arterioles and that endothelial relaxation of coronary arterioles by acetylcholine was impaired [78,79]. In these types of short-term or single-dose studies it was apparent that nano-sized titanium dioxide was more active than its micro-sized counterparts [80].

On the other hand, there has been speculation about whether engineered nanomaterials can be utilized as drug carriers, and therefore the effects of engineered nanomaterial on platelet activation and thrombus formation as well as engineered nanomaterial effects on microcirculation have been investigated *in vivo* [81]. The microcirculation studies are based on an epi-illumination technique with an intravital video-fluorescence microscope [82]. Engineered nanomaterials as well as possible fluorescent dyes were injected into the blood circulation via the left carotid artery, and mean arterial blood pressure was constantly recorded. With blood samples being taken via the same route and after a certain period of time, the epi-illumination was performed. When investigating platelet function, fluorescently labeled platelets were then injected into the blood circulation via the left carotid artery to evaluate the platelet—endothelial cell interactions, platelet rolling and their velocity [82]. A video analysis was also performed off-line to analyze quantitatively the microcirculatory parameters such as venular and arteriolar diameters and macromolecular leakage [83]. It was postulated that ultrafine particles originating from ambient pollution could cause pro-thrombotic but not pro-inflammatory effects in the liver microvessels [82]. Bihari and colleagues conducted studies into engineered nanomaterials and showed that SWCNT could interfere with platelet function and trigger pro-thrombic effects on microcirculation, whereas diesel exhaust and titanium dioxide did not have any effects on blood vessels [81]. The effect of different engineered nanomaterials in blood vessels and their distribution to distant organs will need to be studied more extensively in the future.

In line with the pro-thrombotic effects of engineered nanomaterials reported above, pro-thrombotic effects subsequent to short-term inhalation exposure of mice to nano-sized carbon black have also been reported. These effects included platelet accumulation in the hepatic microvasculature and fibrinogen deposition in hepatic and cardiac microvessels [84]. Nano-sized carbon black also affected the plasma thrombin—antithrombin complex and the levels of fibrinogen in spontaneously hypertensive rats after a short-term exposure, whereas no such effects were found in healthy rats [85,86].

In addition to these observations, changes in neuronal transmission due to nanoparticle exposure and local pulmonary inflammation have also been proposed as causes for such alterations in vascular dysfunctions and diseases [87]. One can conclude that our knowledge and understanding of the cardiovascular effects of engineered nanomaterials are very limited, and hence this area merits much more attention in the future.

3.3 INFLAMMATORY EFFECTS OF ENGINEERED NANOMATERIALS

3.3.1 Introduction to the Immune System

A human body is protected from invading pathogens by innate and adaptive immune responses [88]. The immune system is a complex network consisting of a wide range of specialized effector cells and molecules. A substance inducing an immune reaction is called an antigen, and proteins produced by specialized cells that recognize antigens are called antibodies. The immune system has to be able to mount an effective immune response in order to protect the body from invading pathogens or foreign materials [88].

Innate immunity is a nonspecific system consisting of both physical and biochemical barriers inside and outside of the body as well as the immune cells themselves. It does not create an immunological memory but is a rapid and effective system for eliminating pathogens. Innate immunity is the first line of defense against invading pathogens, and it is responsible for inducing the adaptive immune response. Skin and the surfaces of the respiratory tract and digestive tracts are the most important parts of the epithelial defense. These mechanical barriers formed by mucus, cilia and epithelial cells are not normally considered as being a part of the immune system. On the other hand, macrophages are important scavengers of the innate immunity, phagocytosing and killing of invading microbes and particles. They produce large amounts of toxic chemicals, degradative enzymes and pro-inflammatory factors after their activation. The phagocytic cells, like dendritic cells (DCs), are also able to prime and activate other immune cells, and thus they link innate and adaptive immunity [89,90]. Macrophages and DCs are classified as professional antigen presenting cells (APCs), and they respond to microbes by secreting cytokines [88].

The adaptive immunity is a dedicated system of tissues, cells and molecules acting together to provide specific immune defenses. If the primary defenses of innate immunity are breached, then an adaptive immune response is activated. The three main characteristics of adaptive immunity are memory, specificity and discrimination between self and non self. T cells are the major effector cells in cell-mediated immunity, and B cells are the major effector cells in humoral immunity. They both require cell—cell interactions with APC to become activated and functional [91]. The specific T cell receptor or the B cell receptor on the lymphocyte surface recognizes the particular antigen and focuses the immune response on that antigen. There is also a large number of rearranged receptors, which ensure maintenance of an immunological memory. The successful elimination of invaliding pathogens requires several levels of interactions between innate and adaptive immunity, e.g., innate immunity determines the nature of the adaptive T cell and B cell immune responses [92].

3.3.2 Protection and Defense Cells in the Immune Response

The first line of defense is formed by phagocytic cells, which clean body cavities, and epithelial cells, which form a tight barrier to protect tissues. Professional phagocytic cells can either cleanse the site by ingesting ambient materials or induce inflammatory effects by secreting inflammatory mediators to activate other inflammatory cells. The most important function of epithelial cells is their barrier formation. The more specific functions of these first-line defense cells will be described below, and a simplified view of protection barrier is represented in Figure 3.4.

Epithelial cells (EC) are the first cell types to come into contact with external stimuli, such as chemicals and particulate materials. EC form the mechanical barrier of body cavities and protect it from exposure to foreign materials. EC are also able to engulf apoptotic cells and secrete anti-inflammatory cytokines [93], recruit dendritic cells (DC) to the site of injury and produce a cytokine microenvironment triggering DC maturation [94]. It has also been speculated that EC might be able to promote allergic sensitization by enhancing the DC response [95]. However, if foreign material crosses the barrier formed by EC, it might penetrate into the bloodstream and cause effects on the cardiovascular system and reach distant organs [75].

Macrophages mature from monocytes circulating in blood, and they then migrate into the tissues. Monocytes and macrophages are phagocytic cells resident in almost all tissues; they are long-lived cells with roles in both innate and adaptive immunity. The most important functions of macrophages are to phagocytose invading microorganisms and particles, to orchestrate the immune response by secreting inflammatory proteins and to remove dead cells and debris via their scavenger activity [88]. Macrophages are mainly located in the lymphoid organs, in connective tissues and in body cavities. Alveolar macrophages in lungs sense and kill microbiological agents early in the course of the infection to prevent the pathogens from entering the circulation. They produce pro-inflammatory mediators, reactive oxygen species (ROS) and proteolytic enzymes, all of which can eliminate pathogens. Exposure to different agents transported to lungs may prime or activate alveolar macrophages to express a distinctive inflammatory profile [89,96].

Phagocytic dendritic cells (DC) are ubiquitous throughout the body, being found in lungs, skin, liver and body cavities, and have a role in the priming of adaptive immune responses, in the induction of self-tolerance and in elimination of viral infections [97]. The DCs in the epidermis and in the gut mucosa are called Langerhan's cells (LC), whereas alveolar DCs

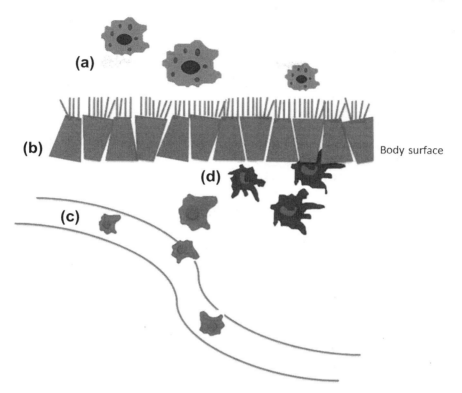

FIGURE 3.4   A simplified view of the protection barrier and defense cells in the immune response. (a) Macrophages located in the body cavities are the first line of defense, acting as scavengers and secreting cytokines and chemokines to activate inflammatory machinery. (b) Epithelial cells form a mechanical barrier to protect the body from foreign matter and secrete inflammatory mediators to attract other immune cells. (c) Immature dendritic cells are recruited from the bloodstream to the site of injury or antigen entry by secreted mediators of inflammation. (d) After antigen sensing and uptake, mature dendritic cells migrate to the local lymph nodes to activate efficiently an adaptive immune response. *Courtesy of Jaana Palomäki.*

are located in the lungs [97,98]. DCs are located in the lumen and beneath the epithelium in the lung, where they are highly likely to encounter pathogens and particles carried into the lungs [92]. DCs engulf pathogens in tissues by phagocytosis or macropinocytosis and degrade and process the pathogenic antigens, which are then presented on their cell surface in order to activate T cells. During this process DCs migrate to lymph nodes, allowing T cells to come into contact with peripheral antigens. This whole process is called maturation. When they become mature, DCs lose their capability to phagocytose pathogens and sample antigens but will efficiently present antigens, adhesion and co-stimulatory molecules and prime the adaptive immune response [99]. The signals evoking DC maturation may be triggered by microbial patterns, danger signals or inflammatory cytokines. DCs are known to have a unique and very potent capacity to provide antigen-specific activation as well as co-stimulatory signals to naïve T cells and possibly also to B cells [92]. Therefore, DCs have the potential to interact with lymphocytes and to regulate their function and produce distinctive lymphocyte responses [97].

3.3.2.1 Engineered Nanomaterials and the First-Line Defense Cells

In vitro experiments utilizing epithelial cells, macrophages or dendritic cells are the most common cell models used to assess the toxicological and immunological effects of ENM. There has been some debate about whether *in vitro* cell models can predict the effects that ENM evoke in living organisms, and the need for standardized protocols for testing the immunotoxicity of ENM is well recognized [100]. In addition, there is a clear need to utilize adequate and specific positive controls in every immunotoxicological assay. The amount of data collected from different kinds of experiments can be enormous, and more effective tools to collect and analyze the data, e.g., by bioinformatics tools, would be beneficial. A brief introduction to the objectives and results of *in vitro* studies performed within these three different cell types will be given below.

Human airway epithelial cell lines such as Calu-3 [101] and A549 [102] have been widely used to test how barrier cells respond to ENM exposure. It has been postulated that different CNT can affect the barrier function of epithelial cells, whereas carbon black NP do not cause alterations in the maintenance of tight junctions [101]. Exposure to aluminium and aluminium oxide NP altered the immunological response of A459 cells to a pathogen [102]. The toxicological information given by studies performed with epithelial cells is limited, and there has been a recommendation to use epidermal cells in the cocultures with APC [103].

In the immunological *in vitro* testing of ENM, human macrophages have been the most widely utilized cells. Cell lines such as THP-1 cells [104] and human primary monocyte-derived macrophages [22] have been used in ENM testing. Recently, the studies have concentrated on cell—cell communication with THP-1 monocultures and cocultures with epithelial cells. The hope has been that a comparison of these responses can be helpful in developing a reliable and standardized tool for ENM risk assessment [104]. At present, the most advanced method based on macrophage monoculture is associated with the assessment of activation of the pro-inflammatory machinery by NLRP3 inflammasome activation. It has been shown that the NLRP3 inflammasome is activated after exposure to silica crystals, crocidolite asbestos and long, rigid CNT, but it is not activated if exposed to carbon black NP, short CNT or long, tangled CNT [22,105]. These results resemble the data collected from *in vivo* studies, and it has been suggested to utilize NLRP3 inflammasome activation studies in the risk assessment of fibrous NM [22]. With spherical NP, more studies will be required to assess the predictability of the assay and its usability in the NP risk assessment. Figure 3.5 provides a conceptual representation of the NLRP3 inflammasome and its activation pathways induced by particulate materials, especially rigid carbon nanotubes, which have been demonstrated to cause a robust secretion of the IL-family cytokines [22,106].

Human primary monocyte-derived dendritic cells [103] and mouse bone-marrow-derived DC [107] have been used in ENM immunotoxicology testing. Dendritic cells are most commonly combined with co-cultures with macrophages and epithelial cells, but they have also been used as monocultures. Mouse bone-marrow-derived dendritic cells showed high sensitivity to TiO_2, ZnO and CNT ENM, i.e., there was extensive cell death and those cells were also activated by exposure with ZnO but not with different types of short CNT in the study by Palomäki et al. [107]. It has also been shown that human primary DC are able to phagocytose NP, although not as efficiently as macrophages [103], and there has been speculation whether this property could be utilized in therapeutic drug delivery [108].

As mentioned above, cocultures that combine different cell types and that therefore resemble real tissue barriers have also been used in the immunotoxicological assessment. It is believed that these three-dimensional cocultures permit cell—cell interactions, enable communication between different cell types and mimic tissue responses more efficiently than a one-cell culture [109]. It was observed that cytotoxicity and the pro-inflammatory potential of SiO_2 and TiO_2 NP and SWCNT differed when tested in monoculture as compared to two- or three-cell cocultures [104]. A human three-cell coculture, said to resemble the epithelial barrier (macrophages/epithelial cells/dendritic cells), has been developed and used in the investigation of particle effects on body surfaces such as lung cavities [103,110]. The possibilities to utilize coculture models in ENM risk assessment should be studied in greater detail in order to standardize the techniques so that they can be incorporated into ENM high-throughput toxicity testing *in vitro*.

A summary of the different cell types used in the ENM immunotoxicity assessment is given in Table 3.1.

3.4 NANOMATERIAL-INDUCED PULMONARY INFLAMMATION

3.4.1 Defense Mechanisms in the Lung

The average person inhales about 10,000 liters of gas every day. The upper and lower airways together represent the largest epithelial surface

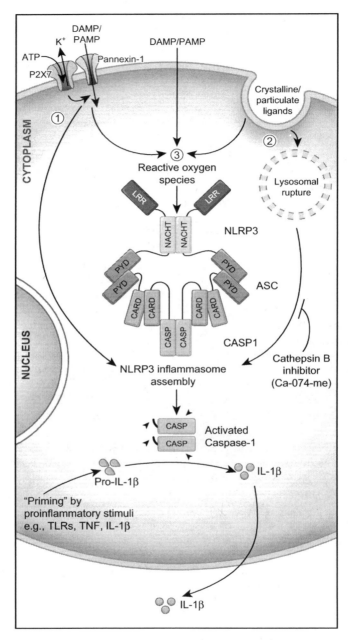

FIGURE 3.5   Three major models for NLRP3 inflammasome activation are shown in the field, but these may not be the only possible routes: (1) The NLRP3 agonist, ATP, triggers P2X$_7$-dependent pore formation by the pannexin-1 hemichannel, allowing

TABLE 3.1   Examples of Different Cell Types Used in the NM Toxicity Assessment

	Reference
EPITHELIAL CELLS	
Calu-3 cell line	Rotoli et al. 2008 [101]
A549 cell line	Braydich-Stolle et al. 2010 [102]
MACROPHAGES	
THP-1 cell line	Dekali et al. 2012 [104]
Human monocyte-derived primary macrophages	Palomäki et al. 2011 [22]
DENDRITIC CELLS	
Human monocyte-derived primary DC	Rothen-Rutishauser et al. 2005 [103]
Mouse bone-marrow derived DC	Palomäki et al. 2010 [107]
COCULTURES	
Three-cell coculture (human primary macrophages/A549/ human primary DC)	Rothen-Rutishauser et al. 2005 [103] Clift et al. 2011 [110]
Two-cell coculture (A549/THP-1)	Dekali et al. 2012 [104]

exposed to the outside environment. The inspired air is the source of oxygen for the body, but it also contains numerous particles and microorganisms. In order to allow for gas exchange, foreign substances and microorganisms need to be stopped and removed. Fortunately, several immune mechanisms function in the lung to help maintain its sterility (an overview of innate and adaptive immunity is given in Chapter 4). The innate immune system acts as the first line of defense of the lung. Its component parts are mucus, which provides a physical barrier that coats the airway epithelium; opsonins, such as immunoglobulins and

◄————

extracellular NLRP3 agonists to enter the cytosol and directly engage NLRP3. (2) Crystalline or particulate NLRP3 agonists are engulfed, and their physical characteristics lead to lysosomal rupture. The NLRP3 inflammasome senses the lysosomal content in the cytoplasm, for example, via cathepsin-B-dependent processing of a direct NLRP3 ligand. (3) All danger-associated molecular patterns (DAMPs) and pathogen-associated molecular patterns (PAMPs), including ATP and particulate/crystalline activators, trigger the generation of reactive oxygen species (ROS). An ROS-dependent pathway triggers NLRP3 inflammasome complex formation. Caspase-1 clustering induces autoactivation and caspase-1-dependent maturation and secretion of proinflammatory cytokines, such as interleukin-1β (IL-1β). *Reprinted from Schroder and Tschopp, Cell 2010;140(6):821−32 with permission* [106].

complement; and phagocytic cells, such as alveolar macrophages and neutrophils (Figure 3.6).

3.4.1.1 Mechanical and Physical Defense

The nasopharyngeal anatomic barriers are important since they efficiently prevent the penetration of particles or organisms into the lower airways. Coughing is another nonspecific defense mechanism of the airways. A forced sudden expiration triggers turbulence in the airways, and this extrudes particles, infected mucus or products of epithelial damage. The mucus gel on the airway epithelium acts as a barrier against microbial attack. Approximately 90% of inhaled particles with a diameter greater than 2–3 µm are deposited on the mucus overlying the ciliated epithelium. The mucociliary transport allows impacted particles to be removed

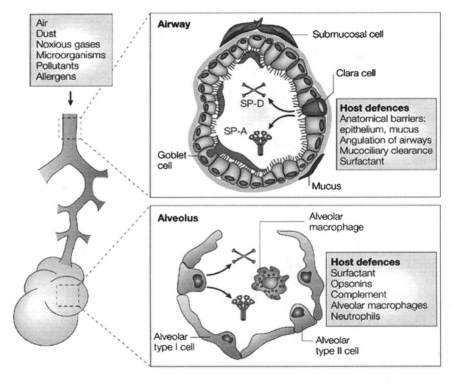

FIGURE 3.6   Overview of the defense machinery of the lungs. Airways are exposed to various inflammatory stimuli including environmental particles, allergens, noxious gases and microbes. The defense system of the upper airway includes anatomical barriers, mucus production and mucociliary clearance. The aleveolar macrophages, surfactant, opsonins and neutrophils act together to create the major defense system of the alveoli region in the lower airways. *Reprinted by permission from Macmillan Publishers Ltd: Nat Rev Immunol. 5(1):58–68, copyright (2005) [111].*

from the bronchioles to the trachea by the ciliary beats of epithelial cells of the bronchi. The mucociliary system can be considered as a 'mucociliary escalator' that moves particles up to the mouth, where they can be swallowed. The physical properties of mucus are provided mainly by mucins, which are glycoproteins and proteoglycans secreted from the surface of epithelial cells.

3.4.1.2 Defense System of the Lower Airways

Despite the efficient clearance system described above, small particles may still reach the alveolar gas-exchange regions of the lung. The host-defense functions in the peripheral air spaces include surfactants, other opsonins (such as immunoglobulins) and innate immune cells (including alveolar macrophages and neutrophils). The presence of immunoglobulin A (IgA) can block the entry of bacteria across the epithelium by preventing microbial adherence to the epithelial surface. IgA also neutralizes toxins and viruses and binds to pathogens, evoking phagocytosis and antibody-dependent cell-mediated cytotoxicity. The pulmonary surfactants reduce surface tension at the air–liquid interface of the lung. In the absence of a surfactant, surface tension would be extremely high and this would make breathing difficult. Surfactants also contain proteins (e.g., SP-A and SP-D), which play an important role in the immune functions.

3.4.1.3 Important Cells in Pulmonary Defense

Dendritic cells and macrophages are the first cellular line of defense in recognizing various pathogens, and both play an important role in defending the airways (for introduction see Section 3.3). Macrophages reside in the alveoli and in lung interstitium. Alveolar macrophages are normally the only phagocytic cells present within the lower respiratory tract. The macrophages engulf and digest inhaled particles, and they are the main source of cytokines, chemokines and other inflammatory mediators that trigger or in some situations may suppress the immune response. Following an insult, macrophages and epithelial cells secrete chemokines and cytokines, promoting neutrophil accumulation and local inflammation. During the acute phase of inflammation, neutrophils leave the circulation and migrate toward the site of the inflammation. After phagocytosis, neutrophils kill ingested microbes with reactive oxygen species, antimicrobial proteins and degradative enzymes.

3.4.2 Pulmonary Inflammation Induced by Nanomaterials

Exposure to engineered nanomaterials during the production or use of nanomaterials is most likely to occur via the lungs. There is good evidence that ultrafine ambient air particles (e.g., diesel exhaust particles) pose

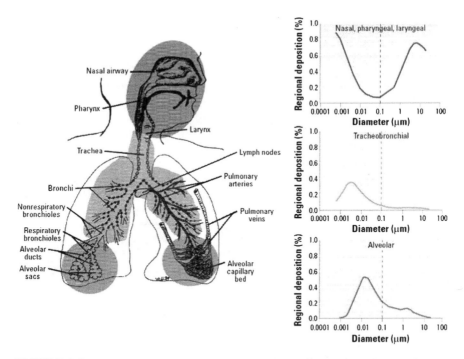

FIGURE 3.7 **Predicted fractional deposition of inhaled particles in the nasopharyngeal, tracheobronchial and alveolar region of the human respiratory tract during nasal breathing.** *Reprinted with permission from Environ Health Perspect 2005 113(7): 823—839. Published online 2005 March 22. doi: 10.1289/ehp.7339 [112].*

considerable health risks if they gain access to the airways. The symptoms can include rhinitis, bronchitis and pulmonary inflammation. However, there is considerably less information on the potential health effects of engineered nanomaterials on airways. This information has originated essentially from rodent *in vivo* studies; in fact nothing about human health effects is known at present. However, evidence exists that nano-sized particles may be more potent than larger particles in evoking inflammatory and toxic responses in the lungs since these nanoparticles will be more efficiently retained in the alveolar region (see Figure 3.7).

3.4.2.1 *Effects of Metal/Metal Oxide Nanoparticles*

Titanium dioxide (TiO_2) nanoparticles are widely used in the fields of life sciences, biotechnology, pharmaceuticals, cosmetics and the textile industries. Oberdörster et al. [113] reported that ultrafine anatase TiO_2 particles (primary particle size of 20 nm), when instilled intratracheally into rats or mice, induced a much greater pulmonary-inflammatory neutrophil response than fine anatase TiO_2 (primary particle size of 250 nm) when both types of particles were instilled at the same mass dose. However, when the instilled dose was expressed as particle surface area, then the neutrophil response in the lung for both ultrafine and fine TiO_2 fitted onto the same dose—response curve, suggesting that the particle surface area for particles of different sizes but of the same chemical composition, such as TiO_2, is a better dose metric than simply particle mass or particle number [113].

In contrast, Rossi et al. [21] investigated the inflammatory potential of several types of TiO_2 in mice using inhalation exposure. Only SiO_2-coated rutile TiO_2 particles were able to elicit significant pulmonary neutrophilia. Pristine rutile and anatase TiO_2 as well as the nano-size SiO_2 used for coating did not evoke inflammation. Moreover, pristine TiO_2 particles were found to be located inside membrane-bound lysosomes within lung macrophages, whereas exposure to SiO_2-coated TiO_2 revealed that the great majority of the nanoparticles were present as agglomerates, and those were found to be located freely in the macrophage cytosol, suggesting that they had escaped from the lysosomal compartments. The authors speculated that this release of TiO_2 nanoparticles from membrane-bound organelles may interfere with cell signalling, leading ultimately to cell activation. No association was found between lung inflammation and the surface area of the particles, their primary or agglomerate particle size, or their capabilities for free radical formation. These results indicated that alterations of nanoparticles, e.g., by surface coating, may drastically change their toxicological properties.

Zinc oxide nanoparticles (ZnONP) are utilized in many commercial products including cosmetics, paints, textiles, food additives and personal hygiene products and sunscreens. It has been previously reported that airway exposure to ZnO nanoparticles could induce eosinophil infiltration and pulmonary fibrosis in mouse models. Cho et al. [114] investigated the molecular and cellular mechanisms involved in ZnO-induced inflammation. Airway exposure to ZnONP induced eosinophilia, proliferation of airway epithelial cells, goblet cell hyperplasia and pulmonary fibrosis and serum IgE levels were upregulated in conjunction with the eosinophilia. The instillation of dissolved Zn^{2+} into rat lungs caused similar pathologies as were elicited by ZnONP. Lysosomal stability was decreased and cell death resulted following treatment of macrophages with ZnONP *in vitro*. These findings suggest that the rapid, pH-dependent dissolution of ZnONP inside the phagosomes is likely to be the main cause of the many ZnONP-induced progressive, severe lung injuries (see Figure 3.8).

3.4.2.2 Effects of Carbon Nanomaterials

Carbon nanotubes, fullerenes and mesoporous carbon structures constitute a new class of carbon nanomaterials that have many

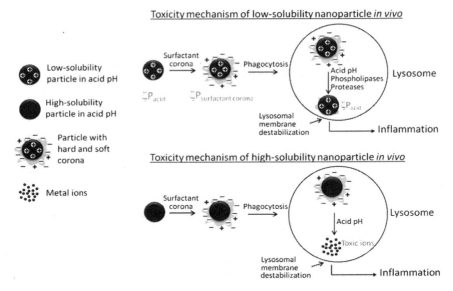

FIGURE 3.8   Diagram of the hypothetical mechanism of lung inflammation caused by metal/metal oxide NP. Low-solubility NP in the acidic lysosomal environment lose their surfactant corona, revealing the ζPacid of the NP, which can destabilize the lysosomes, resulting in inflammation. High-solubility NP in the acidic lysosomal environment cause lysosomal destabilization by dependence on the ionic toxicity. *Reprinted from Toxicological Sciences 126(2), 469—477 (2012) with permission [115].*

applications, for example in semiconductors, controlled drug delivery/release, batteries, data storage, waste recycling and thermal protection. Although health effects of carbon nanomaterials are not well characterized at the moment, there is some preliminary evidence that exposure to needle like fibrous multi-walled carbon nanotubes (MWCNT) can induce severe adverse effects in rodents. The needle like shape of certain types of CNTs has been compared to asbestos, raising concerns that their widespread use may lead to pleural fibrosis and/or mesothelioma (cancer of the lining of the lung), which today are diseases mostly associated with exposure to asbestos.

It has been reported that exposing the mesothelial lining of the body cavity of mice, as a surrogate for the mesothelial lining of the chest cavity, to long MWCNT can result in asbestos-like, length-dependent, inflammatory behavior, for example the formation of inflammatory lesions known as granulomas [33]. A more recent study by Ryman-Rasmussen [116] revealed that MWCNT can reach the subpleura in mice after inhalation exposure. Wang et al. [30] reported that the dispersal state of MWCNT affected the profibrogenic cellular responses, and these correlated with the extent of pulmonary fibrosis and could be of potential value in predicting pulmonary toxicity. Murray et al. [117] reported that specific surface area and particle number did not necessarily predict the biological/toxicological responses to carbonaceous fibrous nanomaterials in living organisms and therefore they could not be considered as the most appropriate dose metric for agglomerating nanomaterials. Instead, they proposed that metrics assessing the effective accessible surface area of agglomerated ENM could be useful for prediction of biologic responses, for example pulmonary inflammation and fibrosis.

Graphene is a new nanomaterial with several unusual and useful physical and chemical properties. Schinwald et al. [118] investigated commercially available graphene-based nanoplatelets, which consisted of several sheets of graphene (few-layer graphene). Aerodynamic modeling showed that the nanoplatelets, which were up to 25 μm in diameter, were respirable and thus could deposit beyond the ciliated airways following their inhalation. These large but respirable graphene plates (GP) were inflammogenic in both the lung and the pleural space in mice when delivered by either pharyngeal aspiration or direct intrapleural installation. Based on the results, the authors concluded that nanoplatelets may represent a novel nanohazard with a distinctive structure—toxicity relationship, highlighting the complexity of nanoparticle toxicology.

3.4.3 Effects of Nanomaterial Exposure on Allergic Asthma

The great majority of studies exploring potential pathogenic effects of ENMs have been conducted in study settings mimicking the effects of these agents in healthy individuals, but only part of the world's population can be categorized into this group (Figure 3.9). Many individuals suffer from pulmonary diseases (such as asthma or COPD), and these may make them more susceptible to experience health problems from particulate exposure. There is recent evidence indicating that exposure to environmental particulate matter can exacerbate the symptoms of asthma. It is therefore important to investigate the effects of nanomaterial exposure in animal models that mimic the conditions of asthmatic individuals.

Rossi et al. [120] reported recently that ovalbumin (OVA) protein-induced allergic pulmonary inflammation was dramatically suppressed in asthmatic mice that were exposed to nano-sized or fine TiO_2 particles — i.e., the levels of leukocytes, cytokines, chemokines and antibodies that are typically found in allergic asthma were substantially decreased. These results indicated that repeated airway exposure to TiO_2 particles could modulate the extent of airway inflammation depending on the immunological status of the exposed mice. In line with this, Park [121] found that

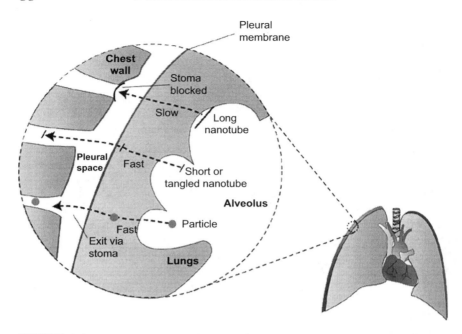

FIGURE 3.9   Movement of particles through the lung. The pleural space is located between the two pleural membranes (red lines), which contains fluid that lubricates the movement of lungs. Small particles (blue circles) and short or tangled nanotubes (black lines) can move rapidly into the pleural space and then drain into the lymphatic system through pores in the chest wall (3–8 μm diameter), whereas this excretion route is not possible for long, fibrous nanotubes. Long nanotubes become lodged in the drainage points and block them, and this can cause inflammation, triggering pathological changes over time. *Reprinted by permission from Nature Nanotechnology 4(11), 708–710, copyright 2009 [119].*

OVA-induced levels of inflammatory cells, airway hyperresponsiveness, and levels of Th2 cytokines in lungs were all significantly reduced after the administration of silver nanoparticles. In addition, they also found that the normally elevated intracellular reactive oxygen species (ROS) levels in bronchoalveolar lavage fluid after OVA inhalation could be decreased by the exposure to silver nanoparticles. These results indicate that the antioxidant effect of silver nanoparticles may provide a potential molecular mechanism for preventing or treating asthma.

In contrast to these potential beneficial responses, adverse effects on asthma have also been reported after nanomaterial exposure. Exposure to certain chemicals in occupational settings has induced asthmatic symptoms that are not mediated by Th2 type immunity, which is the dominating type in classical allergic asthma. Isocyanates are widely used in many industrial and consumer products, and they are a major cause of chemical-induced occupational asthma throughout the world. Hussain et al. [122] reported that exposure to Au and TiO_2 nanoparticles could aggravate airway hyperreactivity and the inflammatory response in toluene diisocyanate (TDI)-sensitized animals, whereas TiO_2 nanoparticles aggravated only the inflammatory response. These results point to the possibility of aggravation of chemically induced occupational asthma by simultaneous nanoparticle exposure. Ryman-Rasmussen et al. [123] examined the effects of airway exposure to MWCNT in an OVA-induced asthma model. Their results demonstrated that MWCNT could potentiate airway fibrosis in a murine asthma model, suggesting that individuals with preexisting allergic inflammation may be susceptible to airway fibrosis from inhaled MWCNT. One caveat of that study was that the high dose of MWCNT was not likely to mimic occupational exposure scenarios in real life and therefore the conclusions drawn need to be considered with caution.

3.5 NANOMATERIAL-INDUCED SKIN INFLAMMATION

The skin is crucial in acting as a physical, immune and sensing barrier for chemicals in our environment to which we may become exposed (Figure 3.10). Some chemicals may actually be absorbed if they make contact with the skin accidently after environmental, occupational or recreational exposure, etc. or, in the case of a cosmetic or dermatological product, applied to the skin deliberately. It is well known that small (<600 Da) lipophilic molecules can readily penetrate through the skin in a passive manner [124]. A variety of factors can influence the extent of the dermal uptake: the skin barrier integrity, the anatomical site and the presence of skin diseases such as contact dermatitis, atopic eczema and psoriasis. Moreover, mechanical insult, irritant detergents and chemicals can all increase absorption through the skin.

3.5.1 Effects of Nanomaterial Exposure on Skin Penetration and Inflammation

The skin is the largest organ of the body and can serve as one of the principal portals by which nanomaterials can enter the body. It also has a relatively large surface area for exposure. At present very little is known about nanomaterial-induced skin inflammation. The skin is often considered as being less permeable, and the risk for exposure via this route is considered to be very low (reviewed in [125]). However, in the literature there are studies that suggest that the skin can be an important

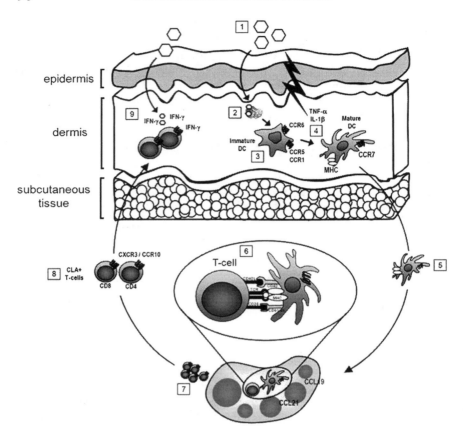

FIGURE 3.10 Schematic model of chemical-induced T-cell-mediated skin inflammation. Small and lipophilic chemicals penetrate epidermis (1) and interact with skin proteins, forming chemical–protein complexes (2). Complexes are internalized by immature DCs (3). The chemical-induced skin insult results in secretion of proinflammatory cytokines, which in turn induces DC maturation (4) and enhances expression of homing molecules in the draining lymph nodes (5). Adaptive immune responses will take a place in the draining lymph node (6) in which proliferation of chemical specific T-cells takes place (7). A repeated encounter with the same chemical can result in recruitment of chemical-specific T-cells (8) to the site of exposure and initiation of skin inflammation (9). (Courtesy of Harri Alenius).

route of entry for nanoparticles in both occupational and consumer settings. In this review, only studies derived from *in vivo* models have been included.

Alvarez-Roman et al. [126] examined the penetration of polystyrene nanoparticles containing covalently bound FITC dye in a porcine ear skin model. Confocal laser scanning microscopy images clearly revealed that the nanoparticle systems preferentially accumulated in the follicular openings of the skin. This localization appeared to be dependent on particle size, with smaller nanoparticles demonstrating higher accumulation in the follicular regions. No evidence for uptake of the particles away from the follicles was observed, and overall penetration was clearly impeded by the stratum corneum, the outmost layer of the skin.

Kim et al. [127] found that near-infrared fluorescent type II quantum dot (QD) nanoparticles administered in the dermis migrated to regional lymph nodes, potentially via skin macrophages and Langerhans cells, raising potential concern for immunomodulation. It should be noted that these QD were intradermally injected into the skin, and therefore they were not subject to the natural penetration barrier of the skin, i.e., stratum corneum. Ryman-Rasmussen et al. [128] employed a porcine skin flow-through diffusion cells model in conjunction with confocal microscopy and reported that QD of different sizes, shapes and surface coatings could penetrate through normal skin, indicating that exposure of intact skin to QD in an occupationally relevant dose could result in dermal penetration. However, discrepant findings have also been reported. Zhang et al. [129] reported that porcine skin penetration of different-type QD was minimal and limited primarily to the outer stratum corneum layers and near hair follicles. These results suggest that differences in the QD studied as well as methodology used can greatly influence the results.

There is no published literature on dermal absorption of CNTs. However, Rouse et al. [130] investigated the influence of mechanical flexion on dermal absorption of fullerene amino-acid-derivatized peptide nanoparticles using dermatomed porcine skin. The authors reported that the fullerenes in the unflexed skin were mostly localized in the epidermal layer, while those in the flexed samples had achieved greater epidermal and dermal penetration. They also reported, from the ability of the fullerenes to enter the skin's dermis, that nanomaterials could potentially be absorbed through capillaries.

Metal oxide nanoparticles such as ZnO and TiO_2 are commonly used in personal-care formulations as protective agents against exposure to ultraviolet radiation. In the normal pigment size ranges (150–300 nm for TiO_2 and 200–400 nm for ZnO) these particles reflect and scatter light, making the sunscreens appear white. However, as particles decrease in size (between 20–150 nm for TiO_2 and 40–100 nm for ZnO) they absorb and scatter UV radiation and largely absorb visible wavelengths, making the sunscreens appear transparent on skin and thus more aesthetically pleasing. Several reviews have concluded that metal oxide nanoparticles do not penetrate the stratum corneum [128]. In contrast, Gulson et al. [58] utilized highly sensitive stable Zn isotopes as tracers and reported that small amounts of Zn from ZnO particles in sunscreens could pass through the protective layers of skin in a real-life environment and can be detected in blood and urine. The levels of Zn isotope in blood and urine from females receiving the nano sunscreen appeared to be higher than in males, which may be due to differences in skin thickness — on average females have thinner skins than males. However, the total amounts Zn ions absorbed and detected in blood and urine were small when compared with the amounts of natural Zn normally present in the human body.

Amorphous silica (SiO_2) is commonly used in semiconductors and has also become a popular material in microelectronics and chromatography. Nabeshi et al. [131] demonstrated that daily applications of nano-sized amorphous silica (particle size of 70 nm) to the ears of BALB/c mice for 28 days resulted in accumulation of silica NP not only in the skin, the regional lymph nodes and the liver but also in the cerebral cortex and hippocampus. Exposure to amorphous nanosilica also caused systemic exposure in mouse and induced mutagenic activity *in vitro*.

3.5.2 Effects of Nanomaterial Exposure on Allergic Skin Diseases

Atopic dermatitis (AD) is a common inflammatory skin disease with the lifetime prevalence estimated at 15–30% in children and 2–10% in adults. The incidence of AD has increased by two- to threefold during the past three decades in industrialized countries. Hirai et al. [132] reported that intradermal injections of silica nanoparticles (SPs) together with Der p1 allergen in a mouse model aggravated the AD-like skin lesions in a size-dependent manner. These effects were correlated with the excessive induction of total IgE and a more intense systemic Th2 response. Further studies will be needed to investigate whether SPs are the risk factor of AD associated with the use of topical applications of SPs, which is a more likely mode of exposure to NMs.

Approximately 10% of the population in Europe and also in the USA suffer from nickel allergy and are therefore unable to wear jewelery and other nickel-containing products. Vemula et al. [133] demonstrated that applying a layer of emollient containing $CaCO_3$ or $CaPO_4$ nanoparticles on the skin of mice could prevent the penetration of nickel ions into the skin. Both their *in vitro* and *in vivo* results indicate that nanoparticles could effectively prevent the penetration of nickel into the skin and may therefore abrogate nickel-induced contact dermatitis. Thus, the use of $CaCO_3$ or $CaPO_4$ nanoparticles within topical compositions may represent an effective approach to limit skin exposure to metal ions, which should be beneficial both occupationally and socially to individuals who suffer from metal-induced contact dermatitis.

3.6 GENOTOXICITY OF ENGINEERED NANOMATERIALS

3.6.1 Brief Description of Genotoxicity

Genotoxicity describes any alterations that affect the integrity of the genetic material. The primary lesion leading to a genotoxic effect can occur in DNA or in proteins (damage to the mitotic or meiotic apparatus). If DNA damage is not properly repaired, it can lead to alterations in gene sequence ($\rightarrow$ gene mutations) and to structural chromosomal aberrations ($\rightarrow$ chromosomal mutations). Mutations in cancer-associated genes can contribute to carcinogenesis. Defects of the mitotic or meiotic apparatus can lead to errors in chromosomal segregation between daughter nuclei, giving rise to abnormalities in the chromosome number ($\rightarrow$ genomic mutations). Genotoxicity is of particular concern, because mutations play an important role in the initiation and progression of carcinogenesis and in reproductive and developmental abnormalities.

The assessment of genotoxicity is generally based on various genotoxicity assays that detect major types of genetic damage, such as DNA damage, gene mutations and structural and numerical chromosomal aberrations. Genotoxicity tests, also called short-term assays (for the detection of carcinogens), have been shown to be good indicators of the potential carcinogenicity of chemicals, since most known human carcinogens are genotoxic (about 80%). Although genotoxicity assays reveal genotoxic carcinogens, it should be kept in mind that carcinogens acting via other (non-genotoxic) mechanisms are not detected.

Table 3.2 gives examples of the genotoxicity assays currently used in toxicological testing for evaluating different types of genetic damage.

One of the most common DNA damage assays is the single-cell gel electrophoresis assay, popularly known as the comet assay. As shown in Figure 3.11, negatively charged fragmented (or looped) DNA runs quicker than intact DNA during the electrophoresis of lysed agarose-embedded cells, resembling a comet tail. Depending on the pH being used, the comet assay is able to detect a wide variety of DNA alterations such as single- and double-strand breaks, incomplete excision repair sites, cross links (by decreased comet tail) and alkali-labile sites (e.g., abasic sites). The DNA lesions detected can still be repaired by cellular DNA repair systems. If the lesions are misrepaired or if unrepaired lesions cause replication errors, then gene mutations and chromosomal mutations can be formed. A more detailed description of the comet assay and its applications in studies of nanomaterial genotoxicity can be found in a recent review [134].

Gene mutation assays detect changes in the base sequence of the DNA that have resulted from misrepair of DNA or DNA replication errors. As

TABLE 3.2 Examples of Genotoxicity Assays Currently Used in Toxicological Testing

Endpoint, Assay	Alterations Detected	Repair	Relevance
DNA DAMAGE			
Single-cell gel electrophoresis (comet) assay	Single- and double-strand breaks, incomplete excision repair sites, alkali-labile sites (e.g., abasic sites) and (reduction of comet tail) DNA cross-links	DNA bases or single-strand breaks can efficiently be repaired by excision repair; double-strand breaks can be repaired by error-prone repair systems	Misrepair of lesions and replication errors arising from unrepaired lesions can create gene mutations or chromosomal aberrations; the cell may die or cease to proliferate
GENE MUTATIONS			
Bacterial mutation (Ames) test	Specific alteration in base sequence reverting the gene function	Cannot be repaired	Mutations in critical genes can impede a cell's ability to carry out its function and appreciably increase the likelihood of tumor formation
Mammalian Hprt (hypoxanthine guanine phosphoribosyltransferase) gene mutation assay	Any forward mutation in Hprt gene impairing enzyme function	Cannot be repaired	
CHROMOSOME DAMAGE			
Chromosomal aberration assay	Chromosome breakage or rearrangement (clastogenic effect)	Cannot be repaired	As a part of a chromosome or an entire chromosome is lost, gained or misplaced, the effect is similar to multiple mutations and may contribute to tumor formation; the cell may die or cease to proliferate
Micronucleus assay	Chromosome breakage (clastogenic effect) or loss/ gain by micronucleation (aneugenic effect)	Cannot be repaired	

FIGURE 3.11   Single-cell gel electrophoresis (comet) assay. A black and white fluorescence microphotograph of nucleoids of human bronchial epithelial BEAS 2B cells after electrophoresis. DNA is stained by ethidium bromide. The white arrow indicates a nucleoid from which damaged DNA has migrated out during the electrophoresis, forming a 'comet tail'. The amount of migrated DNA (e.g., percentage of DNA in tail) is a measure of DNA damage. *Courtesy of Satu Suhonen.*

mentioned before, mutations in critical genes may increase the likelihood of tumor formation. The classical bacterial mutation assay, the Ames test, is performed in *Salmonella typhimurium* or *Escherichia coli* bacteria. The original test is based on bacterial strains that have a mutation in the histidine gene and are thus unable to synthesize this amino acid, which is essential for bacterial replication [135]. The presence of a mutagenic agent enables the bacteria to reverse the mutation, resulting in histidine production and colony formation [136]. An example of mammalian gene mutation tests is the hypoxanthine-guanine phosphoribosyltransferase (*Hprt*) mutation assay, based on clonal selection for Hprt-deficient mutant cells resistant to thioguanine. Unlike normal cells, the mutants are unable to incorporate this toxic analogue of guanine to their newly synthesized DNA and hence survive at a thioguanine concentration lethal to normal cells [137]. In addition, transgenic mice, rats and cell lines, facilitating easier mutant detection, have been developed for use in mutagenicity testing (e.g. [138]).

Assays for chromosomal damage detect alterations in either chromosome structure — giving rise to chromatid-type and chromosome-type breaks and rearrangements — or chromosome number — producing aneuploid or polyploid cells. In the former case, we talk about a clastogenic effect and in the latter case about an aneugenic effect. Clastogenic events are conventionally detected by the chromosomal aberration assay, where cells are arrested in metaphase (Figure 3.12, left panel) and aberrations are scored and classified [139,140]. The micronucleus assay (Figure 3.12, right panel) detects both clastogenic and aneugenic effects. *In vitro*, micronuclei are usually detected by the cytokinesis-block technique where cytokinesis is prevented by cytochalasin B (Cyt-B), which allows the investigator to identify the cells that have divided once since they are binucleated. Whole chromosomes or chromosomal fragments

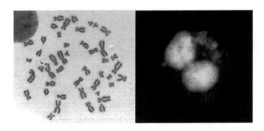

FIGURE 3.12   Left: Chromosomal aberration assay. A microphotograph of Giemsa-stained human lymphocyte metaphase showing a chromatid break (arrow). Right: Micronucleus assay. A fluorescence microphotograph of an acridine-orange-stained binucleate human bronchial epithelial BEAS 2B cell with a micronucleus. Nuclei (DNA) are green, cytoplasm (RNA) red. The cytoplasm contains nano-sized titanium dioxide, which also appears red *(Courtesy of Hilkka Järventaus).*

that have lagged behind in the previous mitosis can be identified as small nuclei, called 'micronuclei' [141].

Several of the genotoxicity assays have been standardized and internationally validated (see [142]), and many of them are compulsory tests that companies have to perform and report when registering chemicals under REACH [143], the European Union regulation that addresses the production and use of chemical substances and their potential impacts on both human health and the environment (http://www.echa.europa.eu/).

3.6.2 Genotoxic Mechanisms of Engineered Nanomaterials

According to the paradigm for particle genotoxicity [144], there are several mechanisms by which particulate materials could be genotoxic. Figure 3.13 illustrates, in a simplified manner, the most commonly postulated mechanisms. After inhalation, oral or dermal exposure, the particles may be able to gain access to the cell through a number of mechanisms (e.g., passive diffusion or endocytosis) and subsequently to the nucleus, either through diffusion across the nuclear membrane (if they are small enough) or via transport through the nuclear pore complexes. For instance, cerium oxide NP were found in the nuclei of BEAS 2B cells after a 12-h treatment [145]. Furthermore, during mitosis, when the nuclear envelope disappears, NP that are free in cytoplasm might be able to interact with the chromosomes, e.g., affect their segregation and become engulfed inside the nucleus when the envelope again develops. Once in the nucleus, NP (especially those that have primary genotoxic activity) could interact with the DNA molecule or DNA-related proteins, damaging the genetic material.

Another indirect, genotoxic mechanism is assumed to operate through oxidative stress, which is believed to be a major deleterious consequence

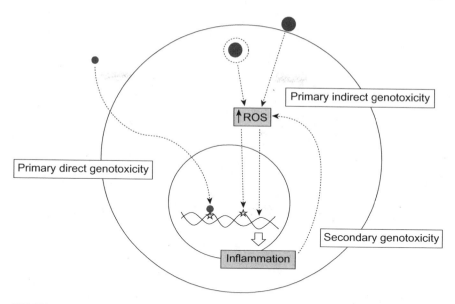

FIGURE 3.13   Possible mechanisms of genotoxicity for small particles in a mammalian cell. Red dots show particles, stars DNA damage. *Modified from Singh et al. [147].*

of exposure to nanomaterials [146]. Reactive oxygen species (ROS) or reactive nitrogen species (RNS) are usually generated inside the cells as a consequence of the metabolic processes, but particles could enhance their production or deplete the natural antioxidant pathways, giving rise to elevated ROS or RNS production. NP may increase ROS production, not only by interfering directly with antioxidant defense within mitochondria but also without entering the cells by interacting with the cell surface or (in the case of partly soluble particles) by releasing free ions, which could cross through the cell membrane. In addition, partly soluble particles may also be dissolved inside the cell (see below). The ROS can react with DNA and produce oxidative damage. This mechanism has been called primary indirect genotoxicity [147].

Furthermore, oxidative stress can trigger an inflammatory response, which may lead to further ROS release from inflammatory cells such as neutrophils, again inducing DNA damage. This kind of inflammation-mediated genotoxicity is called secondary genotoxicity. An example of the putative relationship between particle exposure, inflammation and mutagenicity was reported by Driscoll et al. [148], who observed an increased frequency of *Hprt* mutations in alveolar type II epithelial cells isolated from rat lungs 15 months after intratracheal instillation exposure to several particulate materials (quartz particles, nano-sized carbon black and fine TiO_2; Figure 3.14) in a dose- and material-dependent manner. In

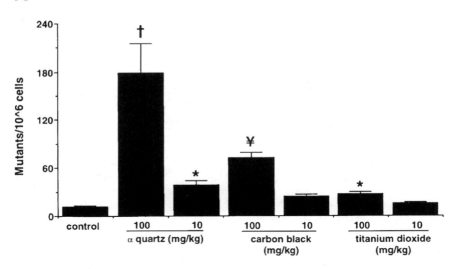

FIGURE 3.14   Effect of *in vivo* particle exposure on *Hprt* mutation frequency in rat alveolar epithelial cells. *Hprt* mutation frequency was determined in alveolar type II epithelial cells isolated from rat lungs 15 months after intratracheal instillation of saline or saline suspensions of 10 and 100 mg/kg fine α-quartz, nano-sized carbon black and fine titanium dioxide. Columns represent means, error bars SDs. *The symbols (*, †, ¥) show statistically significant differences (P < 0.05) from the control. *Reproduced with permission from Oxford University Press [148].*

addition, an inflammatory response was observed, suggesting that the observed mutagenic effect occurred through an inflammation-mediated process. To test this hypothesis, rat lung epithelial RLE-6TN cells were cocultured *in vitro* with bronchoalveolar lavage (BAL) cell populations extracted from quartz-exposed rats. In addition, the cells were treated with similar BAL cell populations that had been enriched for macrophages and neutrophils (Figure 3.15). The frequency of *Hprt* mutations was significantly increased by all the treatments, especially by the neutrophil-enriched populations. Since RLE-6TN cells treated *in vitro* with the particles themselves did not show any increase in the frequency of *Hprt* mutations, it was concluded that the observed genotoxicity depended on a neutrophilic inflammatory response and the production of oxidants by macrophages and neutrophils [148].

It has also been suggested that for some nanomaterials the secondary genotoxic mechanism is driven by the elicitation of an inflammatory response at the site of particle deposition. This phenomenon has still been relatively poorly studied, especially with respect to comparisons conducted between nano-sized and larger particles. For more detailed explanations of the different mechanisms, see Singh et al. [147] and Magdolenova et al. [149].

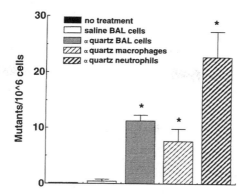

FIGURE 3.15   Effect of *in vitro* exposure to BAL cell populations from α-quartz-exposed rats on *Hprt* mutant frequency in rat lung epithelial cell line RLE-6TN. The initial α-quartz-elicited BAL cell population averaged 47% macrophages, 50% neutrophils and 3% lymphocytes, the macrophage-enriched population was 85% macrophages, 14% neutrophils and 1% lymphocytes and the neutrophil-enriched population was 82% neutrophils, 13% macrophages and 5% lymphocytes. BAL cells from rats exposed to saline were ≥97% macrophages. Asterisks show statistical significance in comparison with saline. *Reproduced with permission from Oxford University Press [148].*

The genotoxic mechanisms mentioned above do not only concern nanomaterials but may be common to any particulate material of a suitably small size. Nanomaterials have at least two characteristic properties that may be important when true nanospecific mechanisms of genotoxicity are considered: a) their extremely small size, which could allow them (i) access to compartments where larger particles cannot gain access and (ii) to interact with macromolecules in nano-scale, and b) their enhanced surface reactivity. In the following section, we will provide some examples of possible nano-specific genotoxic effects.

3.6.3 Nano-Specific Mechanisms of Genotoxicity

The specific mechanisms by which different types of nanomaterials can induce genotoxic effects in biological systems are still poorly understood. However, several nano-specific mechanisms of genotoxic action have been postulated. As nanomaterials often have many unique characteristics, it is not always easy to judge if an effect is due to one or several of such characteristics.

The small size of NP may allow them to gain access to the nucleus, where they could have direct or indirect effects depending on the material. The role of particle size was highlighted in a study with amorphous silica NP, where it was shown that 40-nm and 80-nm silica particles reached the nucleus of human epithelial HEp-2 cells, induced protein aggregation in the nucleoplasm and inhibited gene expression, whereas larger silica particles (>200 nm) did not alter nuclear structure and function [150]. In experiments *in vivo*, amorphous silica NP seemed to exert no significant genotoxic or inflammatory effects on mouse lungs in short-term exposure [151].

One example of an interaction with macromolecules on the nano-scale has been described for long, thin, tubular carbon nanotubes (CNTs), which are of a similar shape as cellular microtubules. This type of single-wall CNTs (SWCNTs) was suggested to interact with cellular biomolecules, such as the mitotic spindle and motor proteins, which separate chromosomes during cell division [152]. SWCNTs (diameter 1–4 nm, length 0.5–1 μm) appeared to be able to mechanically disrupt the mitotic spindle in primary human respiratory epithelial SAEC cells and bronchial epithelial BEAS 2B cells, inducing fragmented centrosomes, multipolar mitotic spindles, anaphase bridges and aneuploidy (Figure 3.16) [153]. Mitotic disruption occurred even when SAEC and BEAS 2B cells were exposed to levels SWCNT anticipated to be present in exposed workers; 95% of the disrupted mitoses showed multipolar mitotic spindles [154].

The large surface area of nanomaterials is associated with enhanced surface reactivity, which is also a possible nano-specific mechanism of genotoxicity. In A549 cells, ultrafine (20–80 nm) TiO_2 particles with a specific surface area of 50 m^2/g were able to elicit oxidative stress and inflammation, while fine (40–300 nm) TiO_2 particles with a surface area of 10 m^2/g were not active in this respect [155]. As stated above, oxidative stress may lead to genotoxic effects. In biological systems, nanomaterials are covered with proteins, and it is presently unclear how this protein corona could affect the genotoxicity of nanomaterials.

It has been proposed that some of the NP used in sunscreens, such as TiO_2 and ZnO, have photo-genotoxic properties. This could occur by the NP efficiently absorbing UV light and thus catalyzing the formation of superoxide and hydroxyl radicals, which could then initiate oxidation reactions. Sunlight-illuminated nano-sized TiO_2 was reported to possess a photo-oxidation potential and a DNA-damaging effect in human MRC-5 fibroblasts [156], and a combined treatment with UV light and nano-sized TiO_2 anatase (21 nm) induced a TiO_2-dose-dependent increase in chromosomal aberrations in Chinese hamster CHL-IU cells [157]. However, another study [158] could not detect any effect of UV on the induction of chromosomal aberrations in Chinese hamster ovary (CHO) cells by eight different forms of ultrafine TiO_2 (coated and uncoated rutile and anatase particles of 14–60 nm) used in sunscreens. One study reported that the clastogenic properties of 'microfine' ZnO (<200 nm) used in sunscreens could be enhanced by UV light irradiation before and during ZnO treatment [159]; as an increasing effect on ZnO-induced chromosomal aberrations was also seen in the pre-irradiated cultures (without irradiating the

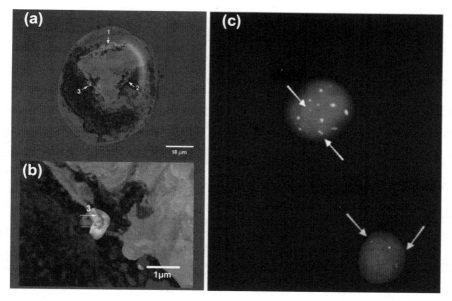

FIGURE 3.16 (a) A 3D reconstruction of a multipolar mitotic spindle with three poles (indicated by white arrows). DNA is blue, tubulin is red, centrosomes are green and SWCNTs black. The reconstructed image shows nanotubes in association with each centrosome fragment at the three spindle poles. Nanotubes were also integrated with microtubules and DNA. (b) A higher-resolution image of the centrosomes and the portion of the mitotic spindle region 3 in (a) shows details of the SWCNT association with centrosome and tubulin. Nanotubes can be seen both within the centrosome structure (white arrow) and in association with the microtubule (yellow arrow). (c) Aneuploidy in SAEC cells. A fluorescence micrograph of two cells treated with 0.24 μg/cm² of SWCNTs after centromeric fluorescence *in situ* hybridization of chromosomes 4 (green) and 1 (red). The cell indicated by white arrows has eight signals of chromosome 4 and five signals of chromosome 1. Yellow arrows show a normal cell with two green signals and two red signals. *Modified from Sargent et al. (2012) Mutat Res, 745, 28–37, reproduced with permission from Elsevier [154].*

ZnO), the finding could have represented sensitization of the cells by UV instead of actual photo-genotoxicity.

Another proposed nano-specific mechanism of genotoxicity concerns partly soluble particles that are taken up by the cell as particles but become dissolved in the acidic conditions inside lysosomes. This kind of Trojan horse mechanism could be more effective for NP than larger particles due to the size-specific uptake and more effective intracellular dissolution of smaller than larger particles. In human alveolar type-II-like epithelial A549 cells, a higher level of cytotoxic effects and DNA damage was observed for nano-sized than micro-sized CuO particles of the same chemical composition (Figure 3.17) [160,161]. In addition to differential

(a)   **(b)**

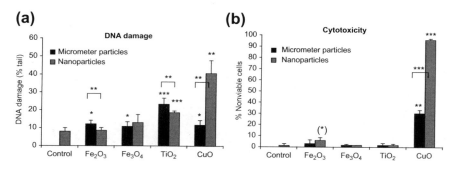

FIGURE 3.17 DNA damage as measured by the comet assay (a) and cell viability (b) in A549 cells after a 4-h (A) or 18-h (b) exposure to 40 µg/cm^2 nano- or micrometer particles of different composition. CuO nanoparticles were significantly more potent than the micrometer particles of CuO. Asterisks represent statistical significance in comparison with control or between the two sizes of the same material. *Modified from Karlsson et al. (2009) Toxicol Lett, 188, 112—118, reproduced with permission from Elsevier [161].*

intracellular dissolution, the size-dependent toxicity of the CuO NP could also have been due to surface reactivity, physical interaction with the mitochondrial membrane, generation of ROS in the cytoplasm or penetration of the CuO NP to the nucleus [160,161]. However, results with Fe$_2$O$_3$, Fe$_3$O$_4$ and TiO$_2$ particles (Figure 3.17) demonstrated that NP cannot generally be assumed to be more genotoxic than micrometer-sized particles; instead the question needs to be assessed on a case-by-case basis.

3.6.4 Genotoxicity of Different Types of Nanomaterials

During the last 5 years, a great number of studies have been published on the genotoxicity of various nanomaterials, and about ten reviews have addressed this question (see, for instance, [149]). Nevertheless, for several reasons it is difficult to draw conclusions from the literature, e.g., due to incomplete description and characterization of the materials, lack of standardized test methods and the multitude of different biological systems being used. Many nanomaterials of low toxicity exert slight genotoxic activity in some studies *in vitro* but are negative in others, probably reflecting the fact that weak positive effects are difficult to reproduce. In addition, numerous factors may affect the outcome of the tests, including particle composition, size, agglomeration state, shape, crystal structure, surface area, coating material and surface charge, as well as cell type, dispersion technique and exposure route [162]. Below, we will present some general considerations on the genotoxicity of two types of nanomaterials: metal-based NP and carbon-based nanomaterials.

3.6.4.1 Metal-Based Nanoparticles

Several metal compounds having particulate forms are well-established genotoxic and carcinogenic agents for humans (e.g., chromium and nickel), and the possibility that their nano-sized forms could have a higher genotoxic and carcinogenic potency than their non-nano-sized forms is an interesting question that has not been examined systematically. On the basis of cytotoxicity, metal-based nanoparticles can be divided roughly into the more toxic partly soluble NP and less toxic poorly soluble NP, and this division often holds with respect to their genotoxic potential as well, provided the ionic form is genotoxic and available to the cell, e.g., by the Trojan horse effect. Partly soluble metal NP could then act by a combination of ionic and particle effects.

A number of recent studies have shown that ZnO NP are able to produce DNA and chromosomal damage *in vitro* [163] (in fact the information on ZnO in Table report 3.3., based on a published in 2012, is already partly outdated). It is not yet clear how much of these findings are due to dissolution of the particles inside the cell (Trojan horse effect) or in the culture medium and if ROS play any role. Results from studies with experimental animals are still too sketchy to draw conclusions on the genotoxicity of ZnO NP *in vivo*. Similarly, there is insufficient information on the *in vivo* genotoxicity of Ag NP, despite the widespread commercial use of Ag NP and the potential human exposure. The existing *in vitro* studies do indicate that Ag NP are genotoxic, potentially inducing DNA damage, chromosomal aberrations and micronuclei, although substantial variation exists between the studies, probably partly reflecting the fact that various types of Ag NP, with or without different coatings, have been examined [164]. Some findings suggest that both Ag NP and the Ag ions have genotoxic effects, but the data are inconsistent. Oxidative stress may also contribute to Ag NP genotoxicity (see [162]).

Cobalt–chromium NP represent a rare case among partly soluble metal NP, since there are some human genotoxicity data available. Several studies have indicated that patients with Co–Cr hip replacements have increased levels of structural and numerical chromosomal aberrations in peripheral blood lymphocytes. However, it has not been possible to draw any firm conclusions with regard to the extent to which metal ions, NP or larger particles are responsible for the genotoxic effects observed (see [147]).

Poorly soluble metal-based NP (such as TiO_2) generally appear to be less genotoxic than partly soluble NP. When genotoxic effects are found, they usually occur at relatively high doses. The genotoxicity of TiO_2 NP has been extensively studied *in vitro*, and especially induction of DNA damage has been observed for most types of TiO_2 NP (see [162]). The crystalline form (phase) and coating (and possibly other surface characteristics) may influence the genotoxicity of TiO_2 NP, so that anatase and uncoated forms of TiO_2 appear to be more genotoxic and cytotoxic than rutile and coated NP ([165]; see [166]). There are only a few *in vivo* studies on TiO_2 NP, but positive results have been seen at high doses, possibly in connection with ROS formation and inflammation, pointing to an indirect or secondary mechanism that might have a threshold dose. However, pulmonary inflammation as such was not adequate to result in signs of genotoxicity in mice inhaling TiO_2 anatase-brookite [167]. Another example of the influence of coating concerns iron oxide NP: positively charged iron oxide NP increased DNA strand breaks in fibroblasts, while negatively charged NP were not active in this respect [168]. One possible explanation was that only the positively charged particles penetrated through the nuclear envelope and were able to interact with DNA. In addition, coatings may affect NP genotoxicity by affecting the particles' agglomeration status.

Some poorly soluble NP have been described to show 'positive' effects, such as suppression of ROS production described for cerium oxide (CeO_2) [166]. On the other hand, CeO_2 has been reported to induce DNA damage and micronuclei *in vitro*. The surface oxidation state of CeO_2 allows it either to scavenge ROS or act in a catalytic manner. Short-term inhalation studies have actually hinted that CeO_2 can lead to pulmonary inflammation with only partial recovery, which in principle might cause secondary genotoxic effects.

The great majority of genotoxicity studies conducted with metal-based NP have been performed *in vitro*, and it is unclear how well the *in vitro* results reflect the outcome of *in vivo* assays. Although there are indications of genotoxic effects also *in vivo* with some metal-based NP, the few studies available have been performed with different techniques, and are mutually comparable only to a limited extent [169].

3.6.4.2 Carbon-Based Nanomaterials

A number of studies have illustrated that SWCNTs and multi-wall CNTs (MWCNTs) possess the potential to be genotoxic, although the findings are still somewhat contradictory, possibly reflecting differences among the materials tested and the biological systems in use. Several *in vitro* studies with mammalian cells have shown that many types of CNTs can induce DNA strand breakage, DNA base modification and chromosomal damage (see [170,171]). *In vivo*, SWCNTs or MWCNTs have been reported to cause aortic mitochondrial DNA damage, mutations,

DNA damage, micronuclei and 8-oxo-2'-deoxyguanosine DNA adducts in several cell types obtained from rats and mice. However, the outcomes of some of the *in vitro* and *in vivo* studies with CNTs have been negative (see [170,171]).

There are indications that certain CNTs can exert both direct and indirect genotoxic effects. Two main modes of action have been proposed: generation of ROS — by the particles themselves, upon particle—cell contact or due to particle-elicited inflammation — or by mechanical interference with cellular components [144,172]. Oxidative stress and inflammation can ultimately lead to carcinogenesis [173], although the mechanisms involved and the actual role of genotoxic alterations are largely unknown. There is an extensive variability in the physical—chemical characteristics of different CNTs with respect to fiber length, diameter, rigidity, surface area, density, shape, contaminant metals and crystallinity, and their significance in carcinogenesis is poorly understood [174]. In particular, rigid CNTs with a high aspect ratio (length:thickness >3:1), consisting of long fibers (5—>20 μm), have been suggested to be of particular concern, since they may be able to induce lung cancer and mesothelioma in a manner similar to asbestos fibers [175].

Apart from CNTs, very few other types of fibrous carbon-based nanomaterials have been studied with respect to genotoxicity. For instance, carbon nanofibers (CNFs) in which the graphene plane and fiber axis do not align are widely used, but their (geno)toxicity has not extensively been evaluated. Based on the available data, there are indications that these materials may evoke pulmonary inflammation, and in one study evidence was found for a possible genotoxic effect of CNFs [176]. Long and rigid graphite nanofibers appeared to induce DNA damage and micronuclei *in vitro* in a similar manner as long and thin CNTs [170]. It is clearly important to carry out further studies of various fibrous carbon materials to clarify their toxicological properties.

Several *in vivo* and *in vitro* genotoxicity studies have been conducted using various types of fullerenes and their derivatives, especially C_{60} fullerene. Positive results *in vitro* have been obtained using the comet assay, gene mutation assays, transgenic mouse embryonic stem cells and micronucleus assay. *In vivo*, positive results have been shown with the comet assay, detection of oxidative DNA damage and chromosomal aberration assay. However, negative results have also been reported. There are inconsistencies in the reports investigating the genotoxicity of fullerenes and their derivatives, which makes it difficult to draw any firm conclusions [177].

Carbon black (CB) has been used widely as a particle control for *in vivo* toxicological evaluation of diesel exhaust particles, a model of urban air pollution particulate matter and as a reference material (i.e., Printex 90). CB can generate ROS in cellular and acellular systems. The genotoxic potential of CB has been demonstrated *in vitro* in several ways, e.g., by increases in DNA base oxidation, mutation frequency, strand breaks and micronucleus frequency in lung epithelial cells as well as increases in strand breaks in fibroblasts. CB exposures by inhalation or intratracheal instillation have been shown to result in dramatic pulmonary inflammatory responses in rodents, which can greatly exacerbate ROS generation via activation of polymorphonuclear granulocytes. Therefore, it has been speculated that CB can mediate secondary genotoxicity by means of inflammation and oxidative stress. A few studies in rats have demonstrated CB-induced DNA base oxidation and increased gene mutation frequency (see [178]). In mice, DNA damage (alkaline comet assay) induced by CB inhalation was independent of neutrophil infiltration, indicating that the inflammatory response was not a prerequisite for DNA damage [179]. However, instillation of CB to mouse lungs induced a persistent inflammation and DNA damage (comet assay) in the lungs, BAL cells and the liver at least for 28 days after the exposure; polymorphonuclear cell counts in BAL correlated strongly with oxidative DNA damage but not with DNA strand breaks in the lungs [178].

3.6.5 Suitability of Current Genotoxicity Test Methods for Engineered Nanomaterials

A central issue in genotoxicity testing of nanomaterials is whether the current test methods are suitable for nanomaterials. A few reviews have dealt with the question of the suitability of the assays that have been used for assessing nanomaterial genotoxicity in general [149,180], *in vitro* [181,182] and *in vivo* (metal oxides) [169,182]. All reviewers have come to similar conclusions: there are too few *in vivo* studies, very few assays on gene mutations have been performed (most of them in bacteria and with negative results) and the most prevalent test in the published literature is the *in vitro* comet assay, with the second most popular being the *in vitro* micronucleus assay. An illustrative summary of the genotoxicity tests that have commonly been used in testing nanomaterials is provided in Table 3.3, taken from Doak et al. [182].

The Ames test has given primarily negative results with all the nanomaterials included in Table 3.3. At the time of that review, 19 studies had been published where the Ames test had been utilized for testing the genotoxicity of engineered nanomaterials. Seventeen of these studies observed no mutagenicity, with weak mutagenic effects being reported in the remaining two reports. On the other hand, the same nanomaterials that were negative in the Ames test were mostly genotoxic in mammalian

TABLE 3.3 Summary of General Test Outcome for Some Nanomaterials Examined in Commonly Used *in vitro* and *in vivo* Genotoxicity Assays. Black Dots Represent Positive Results and White Dots Negative Results

Nanoparticles	Ames Test	Chromosomal Aberrations	Micronucleus Assay	Comet Assay	*In vivo* Genotoxicity
SWCNT	○	●	●	●	●
MWCNT	○	○	●	●	●
TiO$_2$	○	○	●	●	○
ZnO	○	●	○	●	○
Silver	○	●	●	●	●
Silica	○	●	●	●	●
Aluminium oxide	○	●	●	●	●
Iron oxide NPs	○	●	●	●	●

Reprinted from Doak, S.H., et al., In vitro Genotoxicity Testing Strategy for Nanomaterials and the Adaptation of Current OECD Guidelines, Mutat Res, 745, 104—111, Copyright (2012), with Permission from Elsevier [182].

in vitro assays, and many of them also gave positive results *in vivo*. Clearly, the Ames test does not appear to be suitable for assessing the genotoxicity of engineered nanomaterials, probably because of a lower uptake of nanomaterials by the bacterial cells than their human counterparts. Prokaryotes do not employ endocytosis, and their cell wall prevents simple diffusion of particles into the bacterial cell. This lack of uptake could potentially lead to false-negative results. Moreover, the antimicrobial activity of some nanomaterials such as nanosilver could inhibit bacterial growth, further complicating the use of the Ames test [182]. There is an intensive debate on whether the Ames test is a relevant short-term assay for identifying the potential carcinogenicity of nanomaterials [136]. This question is especially critical for the REACH evaluation process, where a mutagenicity test in bacteria is the only genotoxicity test requested if the production or commercialization of the substance is 1—10 tons per year (true for many nanomaterials in the EU). Consequently, the European Chemicals Agency [183] has recently advised that the bacterial mutation assay should be used in conjunction with mammalian gene mutation tests.

Table 3.3 also shows that the comet assay largely gave a positive result with nanomaterials. The same conclusion was stated already in an earlier review [180]. The comet assay has been criticized because of the high number of positive outcomes. The question is, whether the positive results are due to the sensitivity of the assay in detecting DNA lesions or whether some of the findings could represent false-positive results? In principle, residual nanomaterials could react with DNA after lysis of agarose-embedded cells (which could not occur in intact cells) [184]. However, when this possibility was examined by exposing lysed cells to high doses of several nanomaterials, no increase was observed in DNA strand breaks except with oleic acid-coated Fe_3O_4 NP, which, however, also induced DNA damage in whole cells [184]. Thus, the comet assay is probably not providing false-negative results with nanomaterials, but it is quite sensitive at detecting various types of DNA lesions, most of which are subject to DNA repair (H. Karlsson, personal communication; see [134]). The comet assay is not included among the OECD genotoxicity tests guidelines and was also not recommended by the OECD Working Party on Manufactured Nanomaterials [185] for assessing the genotoxicity of nanomaterials. Although the comet assay is probably not the test of choice for regulatory purposes, it remains a good research tool, for instance, in evaluating the timing and duration of a genotoxic insult. It is interesting that DNA damage associated with bio-durable nanoparticles such as CNTs or TiO_2 *in vitro* can be detected for a long time after the treatment, i.e., DNA damage seems to be occurring continuously (or initial damage is not repaired) in cells treated with the particles [165,170]. There is some evidence for such persistent DNA damage also *in vivo* [178]. In the ECHA guidance for testing of nanomaterials [183], the *in vivo* comet assay is presently listed among the applicable *in vivo* genotoxicity assays, as an alternative for an *in vitro* gene mutation assay and as a follow-up assay in cases showing positive *in vitro* evidence of clastogenicity or gene mutation induction.

In addition to the comet assay, the *in vitro* micronucleus test has been the most popular genotoxicity assay in studies on nanomaterials. Table 3.3 also shows primarily positive findings with nanomaterials in this assay, but, in general, the *in vitro* micronucleus assay yields a negative outcome more often than the *in vitro* comet assay. Some adjustments in applying the micronucleus assay to nanomaterials have been proposed. Cyt-B is known to inhibit endocytosis and might thereby prevent the cellular uptake of nanomaterials, and this could give rise to false-negative results, for example, if Cyt-B is added at the same time as the nanomaterial [184]. Thus, it is important to allow the cells to take up the nanomaterials before adding Cyt-B [172,184]. In some cases (bio-persistent nanomaterials of low toxicity), the presence of the nanomaterial on the microscopic slides used for the micronucleus test may disturb and (at high doses) even prevent the analysis (e.g., [165,170]), but there are modifications available for the technique to minimize these problems [164,186]. The *in vitro* micronucleus test is included among the genotoxicity assays of choice described in the ECHA [183] guidance for testing of nanomaterials, as is the *in vitro* chromosomal aberrations test, the classical assay for clastogenicity that has been applied to nanomaterials in some studies (e.g., [157−159,164,171]).

One clear conclusion from the literature is the fact that few *in vivo* studies are available — according to the latest review [149], only 22 of 112 genotoxicity studies included *in vivo* data — although the number of *in vivo* studies has increased during the last few years. *In vivo* assays are probably the only appropriate way to detect secondary genotoxicity, which cannot be assessed in the traditional *in vitro* assays. Clift et al. [136] proposed that assays based on multicellular systems could be used for investigating the effects of nanomaterials on biological systems *in vitro*. This could offer an alternative approach in genotoxicity testing of the nanomaterials used in cosmetics, where the use of animal experiments is banned in the EU. However, *in vivo* studies are still needed to be able to evaluate how well *in vitro* and *in vivo* data actually agree with each other, how the exposure route and animal species influence the results and to elucidate the *in vivo* genotoxic mechanisms. Of the *in vivo* genotoxicity assays applicable to nanomaterials, the comet assay, which can be used for various tissues, has often been used, and several studies have examined oxidative DNA damage, either using the enzyme comet assay or analysis of oxidative adducts in DNA.

The erythrocyte micronucleus assay, currently used as the primary *in vivo* genotoxicity test, may be suitable for examining soluble or partly soluble nanomaterials, if the bone marrow is adequately exposed to the soluble form of the material. However, with poorly soluble nano-materials, sufficient amounts of the material may not reach the target organ during this short-term test. Similar considerations apply to the *in vivo* bone marrow chromosomal aberration test and the liver unsched-uled DNA synthesis test. However, systemic genotoxic effects due to secondary mechanisms such as intense inflammation may, in principle, be detected, as has been indicated for TiO_2 [187]. Nevertheless, most nanomaterials are expected to exert their genotoxic effects mainly locally and therefore, assays that focus on target tissues are needed. In occu-pational exposure, inhalation is usually the most important route of exposure and the lungs are the primary target organ. Methods for assessing the genotoxic effects in the lungs, preferably in the target cells of lung carcinogenesis, should, therefore, be further developed and validated (e.g., [167,188,189]).

Due to the increasing number of different engineered nanomaterials, it is clear that it will be very difficult to assess all nanomaterials for geno-toxicity on a case-by-case basis. Therefore, there is an urgent need to identify the mechanisms of nanomaterial genotoxicity and carcinogenic-ity and to define those characteristics of nanomaterials that are relevant in this context. Furthermore, information on longer-term exposure, both *in vitro* and *in vivo*, is needed to understand better the consequences of the nanomaterial-induced genotoxic effects [162]. This is particularly impor-tant, as it is becoming increasingly obvious that many nanomaterials possess some genotoxic potential. Although the genotoxic effects observed have often been relatively weak, they may have significance in terms of carcinogenesis if the exposure continues for a prolonged time.

3.7 CARCINOGENICITY OF ENGINEERED NANOMATERIALS

Few studies have been published on the carcinogenicity of nano-materials. The examples given below concern bio-durable nanoparticles of low toxicity, particularly carbon black and TiO_2, which have been shown to induce lung cancer after pulmonary exposure in rats, and MWCNTs, which have been observed to produce mesothelioma after intraperitoneal (i.p.) or intrascrotal injection in mice and rats.

3.7.1 Bio-Durable Nanoparticles Without Significant Specific Toxicity

Pott and Roller [190] used repeated intratracheal instillation in female Wistar rats to study the carcinogenicity of 19 dusts, differing from each other in chemical composition, density, specific surface area and mean particle size. Sixteen of the dusts were defined as respirable granular bio-durable particles (GBP) without any known significant specific toxicity. Four of the GBP were 'ultrafine' i.e., nano-sized — carbon black (mean primary particle size 14 nm), hydrophilic TiO_2 anatase-rutile (21 nm), aluminium oxide (13 nm) and aluminium silicate (15 nm) — four were 'fine small' — carbon black (95 nm), hydrophilic TiO_2 anatase (200 nm), diesel soot (200 nm) and lung dust from a coal mine (200 nm) — and eight were 'fine large' (1.8—4 μm) — six coal mine dusts of variable compositions, test toner for a copier and kaolin. All 16 GBP, including the ultrafine carbon black, TiO_2, aluminium oxide and aluminium silicate particles, produced lung tumors. The study also included three known toxic dusts: crystalline silica, a human lung carcinogen with a powerful inflammatory and fibrotic potency (due to the presence of non-soluble toxic silanol groups on the surface; positive control), ultrafine (30 nm) TiO_2 with hydrophobic organic silicon coating (showing high acute toxicity reportedly due to dissolution of the coating) and non-bio-durable ultrafine pyrogenic amorphous silica, which can induce acute inflammation and fibrosis. The positive control crystalline silica was strongly carcinogenic, but the findings were not significant for the hydrophobic TiO_2 or amorphous silica. When non linear regression was used to study the dependence of lung tumor incidence on instilled GBP mass, volume, dust surface area and particle size, it was found that volume in connection with particle size was the most appropriate dose metric for assessing the carcinogenicity of the GBP. The ultrafine GBP were twice as effective as the small fine GBP and 5.5 times more effective than the large fine GBP. It was concluded that the differential distribution pattern of the GBP in epithelial and other cells, the alveoli, interstitium and lymph nodes was an important factor determining the carcinogenic potency of the dusts. This study indicated that all respirable granular bio-durable dust types possess a carcinogenic potential in the lungs of the rat, such that the carcinogenic potency increases with decreasing particle size.

When rat lung tumor responses of carbon black, TiO_2 and diesel particles were compared in inhalation and instillation experiments, it was noted that about 5—6-fold higher lung dust burden after one year of inhalation exposure corresponded to the lung burden produced by a few weekly instillations. The clearly higher tumor incidences after instillation were considered to be due to high exposure early in life. After instillation, the dust level in the lungs is high for a much longer time than can be achieved with inhalation, where the level of retained dust increases slowly [190,191].

The International Agency for Research on Cancer (IARC) has classified carbon black as possibly carcinogenic to humans (2B) [192]. In three studies where the carcinogenicity of carbon black was tested by inhalation exposure in rats, significant increases in the incidence of malignant lung tumors, benign tumors and lung tumors combined were observed in female rats (see [192]). Increased incidences were also observed for lesions described as benign cystic keratinizing squamous-cell tumors or squamous-cell cysts. In one study in female mice, inhalation of carbon black did not elevate the incidence of respiratory tract tumors. Three studies using intratracheal administration of carbon black or toluene extracts of carbon black to female rats reported increased incidences of benign and malignant lung tumors. Studies of dermal application of carbon black in mice showed no carcinogenic effect on the skin, although benzene extracts of carbon black induced skin tumors. Studies in mice with a carbon black that contained demonstrable quantities of carcinogenic polycyclic aromatic hydrocarbons produced local sarcomas, whereas a carbon black without any polycyclic aromatic hydrocarbons did not evoke any sarcomas. Following subcutaneous injection, solvent extracts of carbon black produced sarcomas in several studies in mice. Evidence obtained in experimental animals for the carcinogenicity of carbon black (and separately for carbon black extracts) was considered sufficient. However, IARC regarded the evidence for the carcinogenicity of carbon black in humans as inadequate [192].

Studies on the deposition and retention kinetics of inhaled carbon black particles after their intratracheal instillation or inhalation in rodents detected evidence of rapid clearance of inhaled carbon particles except when the exposure resulted in impaired clearance or accumulation of particles in the lungs (see [192]). This kind of overloading has been described in rats, mice and hamsters exposed to carbon black. Exposure to carbon black resulted in a dose-dependent impairment of alveolar macrophage-mediated clearance at lower mass doses of ultrafine particles than larger particles. Hamsters appeared to have more efficient clearance of carbon black particles than rats and mice. Pulmonary inflammation and epithelial injury increased with elevations in the retained lung dose of carbon black particles; these effects were more severe and prolonged in rats than in mice and hamsters. Fine and ultrafine carbon black particles could translocate from the lungs to other organs (see [192]).

The International Agency for Research on Cancer [192] has classified TiO_2 as possibly carcinogenic to humans (class 2B), with sufficient evidence in experimental animals. Furthermore, US National Institute for Occupational Safety and Health [193] has determined ultrafine TiO_2 as a potential occupational carcinogen. These evaluations primarily relied on several studies showing that inhaled or intratracheally administered nano- and fine-sized TiO_2 could induce lung cancer in rats. Nano-sized TiO_2 was a more potent carcinogen than fine TiO_2 and induced lung cancer in rats at lower exposure levels than larger particles (see [192,193]). The higher tumor potency of nano-scale than fine TiO_2 may be due to its greater translocation to the lung interstitium [190,191,193].

The few published studies on the carcinogenicity of TiO_2 in mice have been negative. The only inhalation study available on the carcinogenicity of ultrafine (14−40 nm) TiO_2 anatase/rutile (80%/20%; on the average 10 mg/m^3 for 13.5 months) in mice was likewise negative [194], but the NMRI mice used also had a 30% control incidence of lung tumors, which may have compromised the sensitivity of the bioassay [193]. Topical administration of non-coated or coated (Al(OH)$_3$ and stearic acid) TiO_2 NP of unknown phase did not produce skin cancer in mice when conducted in a medium-term two-stage skin carcinogenesis model for post-initiation carcinogenic potential [195], and intramuscularly implanted nano-sized and bulk TiO_2 was not found to be carcinogenic in rats [196]. However, poorly tumorigenic and non-metastatic fibrosarcoma QR-32 cells formed tumors and acquired a metastatic phenotype when subcutaneously transplanted into C57BL/6J mice at sites previously implanted with nano-sized $ZrO_2Al(OH)_3$-doped hydrophilic TiO_2 rutile; similar effects were not observed at sites implanted with hydrophobic nano-sized rutile with an additional stearic acid coating or if TiO_2 was co-transplanted with QR-32 cells [197]. The authors considered that their findings were due to the generation of ROS in the cells. Similarly, TiO_2 nanopowder (mixture of rutile and anatase) injected i.p. to C57BL/6J mice once a day for 28 days before subcutaneous implantation of B16F10 melanoma cells caused an enhancement of tumor growth, which was proposed to be due to an immunomodulatory effect of TiO_2 [198].

In rat lung, the carcinogenicity of TiO_2 and other poorly soluble, non-fibrous particles of low acute toxicity, has been suggested to be attributable to a high-dose overloading effect due to particle deposition and retention on the respiratory epithelium, resulting in impairment of the clearance mechanism of the lung, inflammatory response, production of reactive oxygen and nitrogen species, epithelial cell injury and proliferation and secondary genotoxic effects [199]. However, the mechanisms are not well understood, and the actual role of genotoxic events in this process is unclear. The fact that mutations were detected in rat lung epithelial cells 15 months after an intratracheal administration of fine TiO_2 (100 mg/kg body weight) and were associated with prolonged inflammation and oxidants produced by inflammatory cells appeared to point to the involvement of a secondary genotoxic mechanism [148]. Inflammatory effects were also assumed to be responsible for various types of genotoxic effects seen in mice after exposure to high doses of nano-sized anatase-rutile TiO_2 in their drinking water [187]. On the other hand, not all studies have detected a correlation between the DNA damaging and inflammatory effects of TiO_2. DNA damage (measured by the comet assay) was observed in BAL cells of female mice 24 h after a single intratracheal instillation of 54 µg (2.8 mg/kg) of both fine rutile (288 nm; coated with Al and polyalcohol) and nano-sized rutile (20.6 nm; coated with Al, Zr and polyalcohol), while only the latter produced pulmonary neutrophilia [200]. Conversely, similar treatment with uncoated anatase (9 nm; with 7.8% w/w rutile, 12 nm) did not induce DNA damage, although a clear increase was seen in the number of neutrophils and total cells in BAL fluid. Similarly, a 5-day inhalation exposure of mice to nano-sized anatase-brookite TiO_2 did not produce DNA damage in lung epithelial cells or micronuclei in bone marrow, although there was a clear increase in the numbers of inflammatory cells present in BAL [167].

3.7.2 Carbon Nanotubes

Long rigid MWCNTs, with median thickness of 90 nm (70–170 nm), median length of 2 µm (1–20 µm), have been demonstrated to induce mesothelioma in $p53^{+/-}$ mice after a single i.p. injection of 3 mg/animal (1×10^9 particles), with a mesothelioma incidence (14/16) comparable to that (14/18) induced by crocidolite asbestos (3 mg/animal; 1×10^{10} particles) but with an earlier onset [34]. In contrast, fullerene (3 mg/animal) did not induce mesothelioma. In Fischer 344 rats [201], an intrascrotal injection (1 mg/animal) of the same lot of MWCNTs as studied in mice evoked mesothelioma in nine of ten rats, while crocidolite asbestos (2 mg/animal) did not induce mesothelioma. These studies and the length-dependence and sustained inflammatory effect of CNTs [6,33,202,203] have raised concerns about the possible asbestos-like pathogenicity of CNTs.

Nagai et al. [204] compared mesothelioma induction by MWCNTs of different thickness in Fischer 344/Brown Norway F1 hybrid rats after i.p. injections. MWCNTs with high crystallinity ($\sim$50 nm thick; 1 mg/animal) induced malignant mesothelioma at a higher frequency and with an earlier progression than thicker ($\sim$150 nm; 1 mg/animal) or thinner tangled MWCNTs ($\sim$2–20 nm; 10 mg/animal). Histological analysis indicated that the mesotheliomas induced by MWCNTs were more aggressive than those induced by asbestos. The differential carcinogenicity of the different types of MWCNTs was claimed to be due to a diameter-dependent piercing effect on the mesothelial cell membrane. The rigid 50-nm MWCNTs pierced mesothelial cell membrane, was cytotoxic to mesothelial cells *in vitro* and induced inflammation and mesothelioma *in vivo*; the thick (150 nm) or thin tangled MWCNTs were not able to pierce the mesothelial cells, were not internalized and were less toxic, inflammogenic and carcinogenic. The interplay between direct mesothelial cell injury and macrophage activation by the MWCNTs was suggested to lead to persistent inflammation and the subsequent generation of mesothelioma. Mesotheliomas were not produced in Wistar rats treated i.p. with short (about 0.7 μm) or thin (geometric mean 11 nm) MWCNTs with (2 or 20 mg/animal) or without (20 mg) structural defects [205].

In a dose–response study [35], p53 heterozygous mice were given 300, 30 or 3 μg of MWCNTs (the same lot as used by Takagi et al. [34]) per animal i.p., which resulted in a cumulative mesothelioma incidence of 19/20, 17/20 and 5/20, respectively. A dose-dependent effect was also seen for the severity of peritoneal adhesion and granuloma formation. The mesotheliomas of the low-dose group were not accompanied by foreign body granulomas or fibrous scars. The time of tumor onset was independent of the dose. All mice in the low-dose group that survived until the end displayed atypical mesothelial hyperplasia regarded as a precursor lesion of mesothelioma, accompanied by an accumulation of mononuclear cells consisting of lymphocytes and macrophages. Some of the macrophages contained a MWCNT fiber in their cytoplasm. These findings were considered to be mediated by a process called frustrated phagocytosis, where the mesotheliomagenic microenvironment on the peritoneal surface is neither qualitatively altered by the density of the fibers per area nor by the formation of granulomas against agglomerates.

Long MWCNTs deposited in the distal alveoli could migrate to the intrapleural space, causing mesothelial proliferation, and MWCNTs injected into the intrapleural space were able to produce lesions in the parietal pleura ([206]; see [8]); long MWCNTs but not short and tangled MWCNTs induced an inflammatory response in the pleural cavity and lesions along the chest wall at 6 weeks after pharyngeal aspiration [207]. Studies with silver nanofibers of defined length, nickel nanofibers, and MWCNTs injected to the pleural space showed a threshold effect, demonstrating that only fibers longer than 4 μm were able to induce an inflammatory response in the pleura [208]. On the basis of *in vitro* studies, Murphy et al. [207] postulated a mechanism for the formation of meso-thelioma by MWCNTs, according to which long MWCNTs were being retained in the pleural space and incompletely phagocytized by macrophages (frustrated phagocytosis). This would elicit a modest pro-inflammatory cytokine response, which would then produce an amplified pro-inflammatory response in mesothelial cells. The authors also emphasized that the genotoxic effects of long fibers on mesothelial cells were important in themselves and possibly acted in concert with inflammation in the development of mesothelioma. It may be added that the pro-inflammatory response of the mesothelial cells might also result in a secondary genotoxic effect.

3.7.3 Carcinogenicity of Engineered Nanomaterials — Conclusions

Information of bio-durable nanoparticles without significant specific toxicity, such as carbon black and TiO_2, suggests that pulmonary exposure to nanoparticles of this type can induce lung cancer. The effect appears to depend on particle size, such that nanoparticles are more potent than larger particles. However, the studies are few in number, and, for instance, nanoparticles of different sizes have not systemati-cally been compared, although, in principle, their effects would be expected to depend on the ability of the particles to reach the target regions in the lungs. For nano-sized TiO_2, carcinogenicity has been shown only for uncoated mixtures of anatase and rutile with a primary particle size of 15—40 nm. Data are lacking on other crystalline forms (phases) and sizes of TiO_2 NPs. Nanoparticle surface characteristics and coating may exert a considerable influence on the particle distri-bution in the lungs and thus on the toxicity, as shown by Pott and Roller [190] for ultrafine TiO_2 anatase coated with hydrophobic organic silicon. Because of the higher toxicity, the doses of the coated TiO_2 that could be installed into rat lung were lower than in the case of the un-coated TiO_2, and furthermore, lung carcinogenicity could not be shown for the coated form. It is believed that the lung carcinogenicity associated with bio-durable particles without significant specific toxicity is due to overloading — impaired clearance of particles resulting in particle accumulation in the lungs, a prolonged inflammatory response, production of reactive oxygen and nitrogen species, epithelial cell injury and proliferation and secondary genotoxic effects. Lung cancer induction by bio-durable particles without significant specific toxicity has exclusively been observed in rats. However, overloading and the species specificity have been poorly documented for nanoparticles — which are expected to be carcinogenic at lower doses than the corresponding larger particles. A better understanding of the possible role of genotoxic events in the carcinogenicity of this type of nanoparticles would be an important part of risk assessment. If overloading is required for carcinogenicity, and genotoxic effects are only secondary, the carcinogenicity of bio-durable nanoparticles without significant specific toxicity is expected to have a threshold dose. If significant genotoxic effects are induced by bio-durable nanoparticles without overloading or inflammation, for instance, through continuous production of reactive species, a threshold dose would be less probable. Clearly, only the completion of well-conducted mechanistic studies can clarify these issues.

Our second case of nanomaterial carcinogenicity concerned long, rigid MWCNTs, which have been shown to induce mesothelioma when injected i.p. or intrascrotally in mice or rats. In contrast, shorter, thinner and tangled or thicker CNTs have shown lower or no activity. Although these findings are still based on few studies, they suggest that, in analogy with asbestos, long and thin but rigid CNTs may be carcinogenic. The mechanisms involved need to be investigated further. It is not yet known if mesothelioma could be induced by CNTs or other high-aspect-ratio nanomaterials after inhalation exposure. Unfortunately, there are no studies available on the lung carcinogenicity of CNTs.

In general, only a small number of carcinogenicity studies have been published on nanomaterials. The current lack of data makes it impossible to systematically evaluate the ability of short-term assays to correctly differentiate between carcinogenic and noncarcinogenic nanomaterials. Even for established carcinogenic metals, carcinogenicity studies of nano-sized forms are almost nonexistent, although it could be anticipated that nano-sized particles would be more efficiently distributed and dissolved in tissues than their larger counterparts. In rats, subcutaneously implanted ultrafine cobalt particles (50–200 nm) were able to induce malignant mesenchymal tumors, while bulk-sized cobalt particles were not active- in contrast to nickel particles which produced malignant rhabdomyosarcomas both in nano-sized and bulk form [196].

3.8 IMPLICATIONS OF HEALTH EFFECTS AND SAFETY OF ENGINEERED NANOMATERIALS FOR NANOTECHNOLOGIES

3.8.1 Take-Home Message on What We Know of the Health Effects of Engineered Nanomaterials

The previous sections (Sections 3.2–3.6) have dealt with key organ systems as potential targets of toxicity of different types of engineered nanomaterial. The take-home message of this analysis emphasizes that several organ systems can be targets of engineered nanomaterial toxicity. Recent research has clearly shown that exposure to some of the engineered nanomaterials can gain ready access to an organism and become widely distributed in various organs, e.g., mammalian body. The organ systems include the vasculature, especially the microcirculation, and central nervous system. Both can be affected harmfully by a range of engineered nanomaterials. One of the key target systems potentially affected by these materials is the immunologic system; its responses may be multifocal and have an impact on the whole organism. The pulmonary effects of engineered nanomaterials merit special attention because the lung is without doubt the most important target organ of these materials, which can induce inflammation, granulomas and subpleural thickenings, for example, after exposure to certain types of carbon nanotubes. The skin, the largest organ in the body, may also be affected by exposure to nanomaterials with a potential to penetrate the superficial layers of the skin. Cells and tissues can also show a genotoxic effect subsequent to exposure to engineered nanomaterials, and there is increasing evidence that some engineered nanomaterial may express carcinogenic potency. Overall, even though many or even most engineered nanomaterials may be slightly harmful or even harmless, several materials have been identified as hazardous or even highly hazardous, requiring immediate attention to protect humans from being exposed to these materials; the challenge remains how to best separate the innocent engineered nanomaterials from their harmful counterparts. At present, no fully reliable methods or working hazard, exposure or risk assessment concepts have become available, which means that development of novel hazard, exposure and risk assessment approaches requires an increasing amount of attention. This is especially important since these materials are now being increasingly used all around the world and their use expands dramatically each year [1, 3]. The strategic importance of nanotechnologies and materials is reflected by the European Union assessment that nanotechnologies belong to the key enabling technologies in Europe. Nanotechnological applications will become commonplace in the European and global markets, and consumer products incorporating nanomaterials or utilizing nano-enabled technologies will become routine [23].

Current trends suggest that the number of workers in companies using engineered nanomaterials and producing nano-enabled products worldwide will double every 3 years, with about 6 million jobs in 2020. Nanomaterials will represent a 3 trillion US dollar market [1]. For this reason, the importance of identification of those workplaces in which exposure to engineered nanomaterials will take place becomes increasingly important (see [10]). The different types of workplaces where exposure to engineered nanomaterials can take place have been identified in Figure 3.18.

Systematic control of matter at the nano-scale by using direct measurement and simulations will lead to the creation of advanced materials, devices and nano-systems by design. This development also challenges our abilities to assure the safety of engineered nanomaterials and requires the development of exposure-monitoring devices and technologies that allow the assessment of exposure to these materials with easy-to-use, affordable, even portable instruments that have on-line measurement capabilities, allowing prompt actions to be taken to reduce levels of exposures when deemed necessary (see [17]).

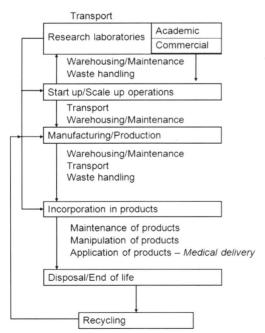

FIGURE 3.18 **Range of workplaces that could involve exposure to engineered nanoparticles.** *Adapted from Schulte et al., with permission from Scand J Work Environ Health [10].*

By 2020, it is predicted that it will be possible to incorporate engineered nanomaterials into a variety of nanotechnology-based products and services and to almost all consumer applications in most industrial sectors and medical fields. This development may lead to mass production of engineered nanomaterials and nano-enabled products. It is noteworthy that this development carries the potential of leaks of engineered nanomaterials from industrial processes, leading to increased exposure of humans and the environment. Hence, utilization of the benefits of nanotechnologies requires the parallel preparation for unexpected and unwanted effects of nanomaterial technologies, including the development of accurate exposure monitoring devices (see [1,12,52,209,210]). From a health and safety perspective, the challenges of the new materials and technologies are remarkable, and nonetheless, one needs to assure the regulators and the general public that they are safe.

The new technology applications not only should be safe themselves but should also offer substantial improvements to human health and environment protection. Due to the rapidly increasing production and use of engineered nanomaterials, the safety aspects must be fully understood and addressed. Successful promotion and growth of research on safety of engineered nanomaterials need to be able to predict promptly with accuracy and reliability the safety of novel nanomaterials and nanotechnologies for the next 10–20 years.

3.8.2 Future Scope and Outlook of Nanotechnologies

The assessment of risks and their management needs to become routine among regulators. Industry will need to adopt a way of working that assures the incorporation of safety perspective into the emerging nanomaterials and nano-based products and processes. This kind of development can exert a positive contribution to nanosafety throughout the world. In effect, this means that there has to be broad acceptance of the safe-by-design principle in the design of novel ENM and in the shaping of novel industrial processes utilizing these novel materials. One must hope that the knowledge and awareness related to engineered nanomaterials and nanotechnologies among different societies will have markedly improved, i.e., citizens will be able to understand the fundamental issues surrounding nano-based products and ENM. These processes will all depend on the neutral and reliable dissemination of information on ENM by regulators and academics, as well as the representatives of the manufacturing industries (see [1,3]).

3.8.3 Health and Safety of Nanotechnologies: Knowledge Needs and Gaps

The increased funding available for nanosafety research has greatly increased our understanding of many aspects of potential hazards of engineered nanomaterials (Figure 3.19), but it has resulted in little useful knowledge that can be used for their risk assessment and management (see [3,7,13,14,23]). The increased funding is clearly reflected in the numbers of scientific publications that have appeared during the last decade; these have increased in an exponential fashion (see [3]). However, only a limited number of these publications have provided results that actually support risk assessment and management of engineered nanomaterials. This has prompted funding agencies and stakeholders to seek new ways to improve the cost—benefit ratio of the research, and nowadays research serving risk assessment, management and governance is being strongly encouraged.

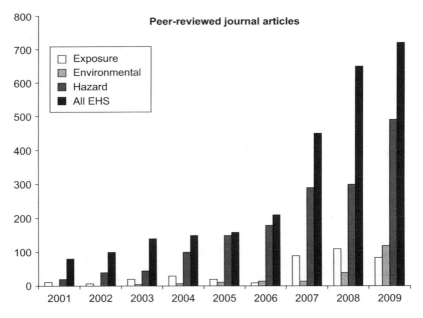

FIGURE 3.19 The dramatic increase in nanotoxicology, nanoecotoxicology and occupational-health-relevant publication over the years is shown. Even though the volume of research has dramatically increased during last 10 years, research that would support development of regulation has been very modest. Note also the so far very modest amount of research related to ecotoxicological effects of engineered nanomaterials. *Reproduced with permission from the National Academy of Sciences [3].*

Engineered nanomaterials are characterized by complexity in terms of their physicochemical characteristics, their distinct behavior in different media and the complex and unexpected interactions of these materials with living systems (see [5,176,211]). It is well justified to consider these materials conceptually as a paradigm for the 21st century, as recently discussed in depth by the nanosafety community [37–40,212]. Indeed, nanotechnologies call for creativity because at the nano-scale of matter, new laws of chemistry and physics apply, and at the nano-scale, matter has the potential to interact with living organisms at the cellular and even molecular level. Cells are in essence a biological microcosmos, the elements of which are expressed on a nano-scale, thereby bringing a qualitatively novel dimension to interplay between the materials and biology interplay [211,213].

To be able to respond to the risk assessment needs for safe nanomaterials and nanotechnologies at present, it will be necessary to complement valuable mechanistic studies on engineered nanomaterials with systematic short-term and long-term animal studies that would allow a reliable estimation of the possible hazards and risks of the engineered nanomaterials. Currently, the vast majority of the available studies that concentrate on a relatively small number of nanomaterials cannot be used for assessment of risks and safety. The European Union NanoSafety Cluster (NSC) has assessed the future research needs on engineered nanomaterials and nanotechnologies (http://www.nanosafetycluster.eu/) and identified the following areas of nanosafety research of strategic importance required to assure the safety and success of engineered nanomaterials and nanotechnologies. These research areas include 1) material characterization; 2) assessment of exposure, transformation and release of engineered nanomaterials throughout their entire life-cycle; 3) hazard mechanisms of the materials; and 4) risk assessment and control of engineered nanomaterials including risk management, risk communication, promotion of safety-by-design culture throughout the industries and stakeholder involvement. In addition, the NSC also identified a number of cross-cutting issues that are of high importance for the successful execution of strategic research into the safety of engineered nanomaterials. Such cross-cutting issues include a regulatory framework favoring this kind of research, existing infrastructures, capabilities for dissemination of results to obtain impact among stakeholders, and international collaboration because research on nanosafety and nanotechnologies is a global activity [37].

In line with the proposals put forward by the NSC (www.nanosafetycluster.eu), Hartung [214] called for a departure from the existing experimental animal-testing paradigm and to move to the use of high-throughput testing approaches. This proposal, which focused on testing and assessing chemical safety in general, was supported by the recent contribution by Nel et al. [213], also emphasizing the importance of using novel methods including high-throughput technologies and omics methodologies in the safety and toxicity testing of novel nanomaterials. From the regulatory point of view, both proposals would require a remarkable amount of research in order to reveal the association between alterations at gene, molecular and cellular levels in such a way that they would have a predictive potential regarding human health and clarify how disease and toxicity attributable to the engineered nanomaterials could be made relevant to their risk assessment. As long as there is no clear evidence of the predictive power of these approaches for *in vivo* systems, their acceptance in becoming a part of the regulatory framework of engineered nanomaterials is unlikely — today they are associated with too many uncertainties.

3.9 CONCLUSIONS

The challenges associated with the safety of engineered nanomaterials and nanotechnologies are global. The obvious benefits of engineered nanomaterials and nanotechnologies are threatened by the potentially harmful characteristics of these materials; often these characteristics are intimately associated with the technological benefits. Hence, the success of these materials hangs in the balance until there are assurances about their safety. Key challenges today include identifying and eliminating inappropriate hazard, exposure and risk assessment tools. In some cases, these tools are inappropriate since they are unable to provide reliable information, and in other cases they may be so laborious and expensive that their widespread use would overwhelm available resources, thus impeding the assessment of safety and risks of these materials and even drawing resources away from developing more advanced and appropriate technologies. Current resources or testing tools are not likely to be able to undertake the safety assessment of the novel nanomaterials that are flooding onto the markets. This means that new risk/safety assessment paradigms will need to be developed during the coming years in order to solve this problem. At the same time it is important to maintain the level of current resources, utilizing the current means of risk and safety assessment of existing engineered nanomaterials, and to support regulators and other stakeholders.

References

[1] Roco MC, Mirkin CA, Hersam MC. Nanotechnology research directions for societal needs in 2020: retrospective and outlook summary. NSF/WTEC (National Science Foundation/World Technology Evaluation Center) report (summary of the full report published by Springer); 2010.
[2] PEN. The Project of Emerging Nanotechnologies. http://www.nanotechproject.org/.

[3] National Academy of Sciences (NAS). A Research Strategy for Environmental, Health, and Safety Aspects of Engineered Nanomaterials. Natl Res Counc Natl Academies 2012:1–211.

[4] Borm PJA, Robbins D, Haubold S, et al. The potential risks of nanomaterials: a review carried out for ECETOC. Part Fibre Toxicol 2006;3:11.

[5] Nel A, Xia T, Mädler L, Li N. Toxic potential of materials at the nanolevel. Science 2006;311:622–7.

[6] Donaldson K, Murphy FA, Duffin R, Poland CA. Asbestos, carbon nanotubes and the pleural mesothelium: a review of the hypothesis regarding the role of long fibre retention in the parietal pleura, inflammation and mesothelioma. Part Fibre Toxicol 2010a;7:5.

[7] Savolainen K, Alenius H, Norppa H, et al. Risk assessment of engineered nano-materials and nanotechnologies-A review. Toxicology 2010;269:92–104.

[8] Castranova V, Schulte PA, Zumwalde RD. Occupational Nanosafety Considerations for Carbon Nanotubes and Carbon Nanofibers. Acc Chem Res 2012 [Epub ahead of print].

[9] NIOSH. Occupational Exposure to Carbon Nanotubes and Nanofibers. Curr Intell Bull 2010;161-A:1–149.

[10] Schulte P, Geraci C, Zumwalde R, et al. Sharpening the focus on occupational safety and health in nanotechnology. Scand J Work Environ Health 2008;34:471–8.

[11] Gulumian M, Kuempel ED, Savolainen K. Global challenges in the risk assessment of nanomaterials: Relevance to South Africa. S Afr J Sci 2012;108. Art. #922, 9 pages. http://dx.doi.org/10.4102/sajs. v108i9/10.922

[12] Auffan M, Rose J, Bottero JY, et al. Towards a definition of inorganic nanoparticles from an environmental, health and safety perspective. Nat Nanotechnol 2009;4: 634–41.

[13] Kahru A, Savolainen K. Potential hazard of nanoparticles: from properties to bio-logical and environmental effects. Toxicology 2010;269:89–91.

[14] Brouwer D. Exposure to manufactured nanoparticles in different workplaces. Toxi-cology 2010;269:120–7.

[15] Bouwmeester H, Lynch I, Marvin HJ, et al. Minimal analytical characterization of engineered nanomaterials needed for hazard assessment in biological matrices. Nanotoxicology 2011;5:1–11.

[16] Maynard AD, Aitken RJ, Butz T, et al. Safe handling of nanotechnology. Nature 2006;444:267–9.

[17] Schneider T, Brouwer DH, Koponen IK, et al. Conceptual model for assessment of inhalation exposure to manufactured nanoparticles. J Expo Sci Environ Epidemiol 2011;21(5):450–63. http://dx.doi.org/10.1038/jes.2011.4.

[18] National Research Council (NRC). Risk Assessment in the Federal Government: Managing the Process. Committee on the Institutional Means for Assessment of Risks to Public Health. Natl Res Counc. Available at: http://www.nap.edu/openbook.php?isbn=0309033497; 1983.

[19] Klaassen CD. Casarett and Doull's Toxicology: The Basic Science of Poisons. 6th ed. New York: McGraw-Hill; 2001. p. 1236.

[20] Shvedova AA, Sager T, Murray AR, et al. Critical Issues in the Evaluation of Possible Adverse Pulmonary Effects Resulting from Airborne Nanoparticles. In: Monteiro-Riviere NA, Tran CL, editors. Nanotoxicology - Characterization, Dosing and Health Effects. New York: Informa Healthcare; 2007. p. 225–36.

[21] Rossi EM, Pylkkänen L, Koivisto AJ, et al. Airway exposure to silica-coated TiO_2 nanoparticles induces pulmonary neutrophilia in mice. Toxicol Sci 2010;113:422–33.

[22] Palomäki J, Välimäki E, Sund J, et al. Long, needle-like carbon nanotubes and asbestos activate the NLRP3 inflammasome through a similar mechanism. ACS Nano 2011;5(9):6861−70.

[23] European Commission COM (2012) 572 final. (2012). Second regulatory review on nanomaterials. Available at http://eur-lex.europa.eu/LexUriServ/LexUriServ.do?uri=COM:2012:0572:FIN:EN:PDF

[24] OECD. Guidelines for the testing of chemicals. Available at, http://www.oecd-ilibrary.org/content/package/chem_guide_pkg-en; 1981.

[25] Horizon 2020. http://ec.europa.eu/research/horizon2020/index_en.cfm; 2013

[26] Oberdörster G, Sharp Z, Atudorei V, et al. Translocation of inhaled ultrafine particles to the brain. Inhal Toxicol 2004;16:437−45.

[27] Elder A, Gelein R, Silva V, et al. Translocation of inhaled ultrafine manganese oxide particles to the central nervous system. Environ Health Perspect 2006;114:1172−8.

[28] Oberdörster G, Elder A, Rinderknecht A. Nanoparticles and the brain: cause for concern? J Nanosci Nanotech 2009;9:4996−5007.

[29] Chen J, Dong X, Zhao J, Tang G. *In vivo* acute toxicity of titanium dioxide nano-particles to mice after intraperitoneal injection. J Appl Toxicol 2009;29:330−7.

[30] Wang X, Xia T, Ntim SA, et al. Dispersal state of multiwalled carbon nanotubes elicits profibrogenic cellular responses that correlate with fibrogenesis biomarkers and fibrosis in the murine lung. ACS Nano 2011;5(12):9772−87.

[31] Shvedova AA, Kisin ER, Porter D, et al. Mechanisms of pulmonary toxicity and medical applications of carbon nanotubes: Two faces of Janus? Pharmacol Ther 2009;121:192−204.

[32] Kane AB, Hurt RH. Nanotoxicology: the asbestos analogy revisited. Nat Nanotechnol 2008;3:378−9.

[33] Poland CA, Duffin R, Kinloch I, et al. Carbon nanotubes introduced into the abdominal cavity of mice show asbestos-like pathogenicity in a pilot study. Nat Nanotechnol 2008;3:423−8.

[34] Takagi A, Hirose A, Nishimura T, et al. Induction of mesothelioma in p53+/− mouse by intraperitoneal application of multi-wall carbon nanotube. J Toxicol Sci 2008;33:105−16.

[35] Takagi A, Hirose A, Futakuchi M, et al. Dose-dependent mesothelioma induction by intraperitoneal administration of multi-wall carbon nanotubes in p53 heterozygous mice. Cancer Sci 2012;103:1001−4.

[36] Savolainen K, Vainio H. Health risks of engineered nanomaterials and nanotechnol-ogies (in Finnish). Duodecim 2011;127(11):1097−104.

[37] Nature Nanotechnology Editorial. Join the dialogue. Nat Nanotechnol 2012;7:545.

[38] Nature Nanotechnology Editorial. The dialogue continues. Nat Nanotechnol 2013;8:69.

[39] Fadeel B, Savolainen K. Broaden the discussion. Nat Nanotechnol 2013;8:71.

[40] Savolainen K, Alenius H. Disseminating widely. Nat Nanotechnol 2013;8:72.

[41] European Academies Science Advisory Council (EASAC). Impact of Engineered Nanomaterials on Health: Considerations for Benefit-Risk Assessment. EASAC Policy Report No. 15/JRC Reference Report, EUR 24847 EN. European Commission, Luxembourg; 2011.

[42] National Nanotechnology Initiative (NNI). Environmental, Health, and Safety (EHS) Research Strategy. Available at: http://www.nano.gov/sites/default/files/pub_resource/nni_2011_ehs_research_strategy.pdf; 2011.

[43] Kreyling WG, Semmler-Behnke M, Möller W. Ultrafine particle-lung interactions: does size matter? J Aerosol Med 2006;19(1):74−83.

[44] Donaldson K, Schinwald A, Murphy F, et al. The Biologically effective dose in inhalation nanotoxicology. Acc Chem Res 2012. 2012 Sep 24. [Epub ahead of print].

[45] Brunner TJ, Wick P, Manser P, et al. *In vitro* Cytotoxicity of oxide nanoparticles: comparison to asbestos, silica, and the effect of particle solubility. Environ Sci Technol 2006;40:4374–81.

[46] Johnston HJ, Hutchison G, Christensen FM, et al. A review of the in vivo and in vitro toxicity of silver and gold particulates: particle attributes and biological mechanisms responsible for the observed toxicity. Crit Rev Toxicol 2010;40(4):328–46.

[47] Kasemets K, Ivask A, Dubourguier HC, Kahru A. Toxicity of nanoparticles of ZnO, CuO and TiO$_2$ to yeast Saccharomyces cerevisiae. Toxicol In Vitro 2009;23(6): 1116–22.

[48] Bai YH, Li JY, Xu JJ, Chen HY. Ultrasensitive electrochemical detection of DNA hybridization using Au/Fe3O4 magnetic composites combined with silver enhancement. Analyst 2010;135(7):1672–9.

[49] Schmidt J, Vogelsberger W. Dissolution kinetics of titanium dioxide nanoparticles: the observation of an unusual kinetic size effect. J Phys Chem B 2006;110(9):3955–63.

[50] Yokel RA, Florence RL, Unrine JM, et al. Biodistribution and oxidative stress effects of a systemically-introduced commercial ceria engineered nanomaterial. Nanotoxicology 2009;3:234–48.

[51] Jepson MA. Gastrointestinal tract. In: Fadeel B, Pietroiusti A, Shvedova AA, editors. Adverse effects of engineered nanomaterials: Exposure, Toxicology, and Impact on Human Health. San Diego: Academic Press; 2012. p. 209–24.

[52] Seipenbusch M, Binder A, Kasper G. Temporal evolution of nanoparticle aerosols in workplace exposure. Ann Occup Hyg 2008;52:707–16.

[53] Wang J, Zhou G, Chen C, et al. Acute toxicity and biodistribution of different sized titanium dioxide particles in mice after oral administration. Toxicol Lett 2007;168:176–85.

[54] Hillyer JF, Albrecht RM. Gastrointestinal persorption and tissue distribution of differently sized colloidal gold nanoparticles. J Pharm Sci 2001;90:1927–36.

[55] Yokel RA, Macphail RC. Engineered nanomaterials: exposures, hazards, and risk prevention. J Occup Med Toxicol 2011;6:7.

[56] Zhang LW, Monteiro-Riviere NA. Assessment of quantum dot penetration into intact, tape-stripped, abraded and flexed rat skin. Skin Pharmacol Physiol 2008;21:166–80.

[57] Wu J, Liu W, Xue C, et al. Toxicity and penetration of TiO$_2$ nanoparticles in hairless mice and porcine skin after subchronic dermal exposure. Toxicol Lett 2009;191(1):1–8.

[58] Gulson B, McCall M, Korsch M, et al. Small amounts of zinc from zinc oxide particles in sunscreens applied outdoors are absorbed through human skin. Toxicol Sci 2010;118(1):140–9.

[59] Rothen-Rutishauser BM, Schurch S, Haenni B, et al. Interaction of fine particles and nanoparticles with red blood cells visualized with advanced microscopic techniques. Environ Sci Technol 2006;40:4353–9.

[60] Dan M, Wu P, Grulke EA, et al. Ceria-engineered nanomaterial distribution in, and clearance from, blood: size matters. Nanomedicine (Lond) 2012;7(1):95–110.

[61] Monopoli MP, Bombelli FB, Dawson KA. Nanobiotechnology: nanoparticle coronas take shape. Nat Nanotechnol 2011;6:11–2.

[62] Monopoli MP, Åberg C, Salvati A, Dawson KA. Biomolecular coronas provide the biological identity of nanosized materials. Nat Nanotechnol 2012;7:779–86.

[63] Choi HS, Liu W, Misra P, et al. Renal clearance of quantum dots. Nat Biotechnol 2007;25(10):1165–70.

[64] Fabian E, Landsiedel R, Ma-Hock L, et al. Tissue distribution and toxicity of intravenously administered titanium dioxide nanoparticles in rats. Arch Toxicol 2008;82:151–7.

[65] Yang RSH, Chang LW, Wu JP, et al. Persistent tissue kinetics and redistribution of nanoparticles, quantum dot 705, in mice. Environ Health Perspect 2007;115: 1339–43.

[66] Ceccatelli S, Bardi G. Neurological system. In: Fadeel B, Pietroiusti A, Shvedova AA, editors. Adverse effects of engineered nanomaterials: Exposure, Toxicology, and Impact on Human Health. San Diego: Academic Press; 2012. p. 157–68.

[67] Hardas SS, Butterfield DA, Sultana R, et al. Brain distribution and toxicological evaluation of a systemically delivered engineered nanoscale ceria. Toxicol Sci 2010; 116:562–76.

[68] Yamashita K, Yoshioka Y, Higashisaka K, et al. Silica and titanium dioxide nanoparticles cause pregnancy complications in mice. Nat Nanotechnol 2011;6:321–8.

[69] Boyes WK, Chen R, Chen C, Yokel RA. The neurotoxic potential of engineered nanomaterials. Neurotoxicology 2012;33(4):902–10.

[70] Zhang L, Bai R, Liu Y, et al. The dose-dependent toxicological effects and potential perturbation on the neurotransmitter secretion in brain following intranasal instillation of copper nanoparticles. Nanotoxicology 2012;6(5):562–75.

[71] Wang J, Liu Y, Jiao F, et al. Time-dependent translocation and potential impairment on central nervous system by intranasally instilled TiO(2) nanoparticles. Toxicology 2008;254(1-2):82–90.

[72] Wang J, Chen C, Liu Y, et al. Potential neurological lesion after nasal instillation of TiO(2) nanoparticles in the anatase and rutile crystal phases. Toxicol Lett 2008;183(1-3): 72–80.

[73] Zhang L, Bai R, Li B, et al. Rutile TiO_2 particles exert size and surface coating dependent retention and lesions on the murine brain. Toxicol Lett 2011;207(1):73–81.

[74] Dockery DW, Pope CA, Xu X, et al. An association between air pollution and mortality in six U.S. Cities. N Engl J Med 1993;329:1753–9.

[75] Brook RD. Cardiovascular effects of air pollution. Clin Sci (Lond) 2008;115:175–87.

[76] Kang GS, Gillespie PA, Gunnison A, et al. Comparative pulmonary toxicity of inhaled nickel nanoparticles; role of deposited dose and solubility. Inhal Toxicol 2011;23(2): 95–103.

[77] Niwa Y, Hiura Y, Murayama T, et al. Nano-sized carbon black exposure exacerbates atherosclerosis in LDL-receptor knockout mice. Circ J 2007;71(7):1157–61.

[78] LeBlanc AJ, Cumpston JL, Chen BT, et al. Nanoparticle inhalation impairs endothelium-dependent vasodilation in subepicardial arterioles. J Toxicol Environ Health A 2009;72(24):1576–84.

[79] LeBlanc AJ, Moseley AM, Chen BT, et al. Nanoparticle inhalation impairs coronary microvascular reactivity via a local reactive oxygen species-dependent mechanism. Cardiovasc Toxicol 2010;10(1):27–36.

[80] Nurkiewicz TR, Porter DW, Hubbs AF, et al. Nanoparticle inhalation augments particle-dependent systemic microvascular dysfunction. Part Fibre Toxicol 2008;5:1.

[81] Bihari P, Holzer M, Praetner M, et al. Single-walled carbon nanotubes activate platelets and accelerate thrombus formation in the microcirculation. Toxicology 2010;269:148–54.

[82] Khandoga A, Stampfl A, Takenaka S, et al. Ultrafine particles exert prothrombotic but not inflammatory effects on the hepatic microcirculation in healthy mice *in vivo*. Circulation 2004;109:1320–5.

[83] Bajory Z, Hutter Jr, Krombach F, Messmer K. Microcirculation of the urinary bladder in a rat model of ischemia-reperfusion-induced cystitis. Urology 2002;60(6):1136–40.

[84] Khandoga A, Stoeger T, Khandoga AG, et al. Platelet adhesion and fibrinogen deposition in murine microvessels upon inhalation of nanosized carbon particles. J Thromb Haemost 2010;8:1632–40.

[85] Kodavanti UP, Schladweiler MC, Ledbetter AD, et al. Pulmonary and systemic effects of zinc-containing emission particles in three rat strains: multiple exposure scenarios. Toxicol Sci 2002;70(1):73–85.

[86] Elder A, Gelein R, Finkelstein JN, et al. Effects of subchronically inhaled carbon black in three species. I. Retention kinetics, lung inflammation, and histopathology. Toxicol Sci 2005;88(2):614–29.

[87] Hubbs AF, Mercer RR, Benkovic SA, et al. Nanotoxicology—a pathologist's perspective. Toxicol Pathol 2011;39(2):301–24.

[88] Murphy K, Travers P, Walport M, editors. Janeway's Immonobiology. 7th ed. New York: Garland Science, Taylor & Francis group, LLC; 2008.

[89] Stampfli MR, Anderson GP. How cigarette smoke skews immune responses to promote infection, lung disease and cancer. Nat Rev Immunol 2009;9(5):377–84.

[90] Barton GM. A calculated response: control of inflammation by the innate immune system. J Clin Invest 2008;118(2):413–20.

[91] Nathan C. Points of control in inflammation. Nature 2002;420(6917):846–52.

[92] Kang SM, Compans RW. Host responses from innate to adaptive immunity after vaccination: molecular and cellular events. Mol Cells 2009;27(1):5–14.

[93] Juncadella IJ, Kadl A, Sharma AK, et al. Apoptotic cell clearance by bronchial epithelial cells critically influences airway inflammation. Nature 2013;493(7433):547–51.

[94] Reibman J, Hsu Y, Chen LC, et al. Airway epithelial cells release MIP-3alpha/CCL20 in response to cytokines and ambient particulate matter. Am J Resp Cell Mol Biol 2003;28(6):648–54.

[95] Lambrecht BN, Hammad H. Biology of lung dendritic cells at the origin of asthma. Immunity 2009;31(3):412–24.

[96] Joshi GN, Knecht DA. Silica phagocytosis causes apoptosis and necrosis by different temporal and molecular pathways in alveolar macrophages. Apoptosis 2013;18(3): 271–85.

[97] Banchereau J, Steinman RM. Dendritic cells and the control of immunity. Nature 1998;392(6673):245–52.

[98] Novak N, Bieber T. Dendritic cells as regulators of immunity and tolerance. J Allergy Clin Immunol 2008;121(Suppl 2.):S370–4; quiz S413.

[99] Vermaelen K, Pauwels R. Pulmonary dendritic cells. Am J Resp Crit Care Med 2005;172(5):530–51.

[100] Clift MJ, Gehr P. Nanotoxicology: a perspective and discussion of whether or not in vitro testing is a valid alternative. Arch Toxicol 2011;85:723–31.

[101] Rotoli BM, Bussolati O, Bianchi MG, et al. Non-functionalized multi-walled carbon nanotubes alter the paracellular permeability of human airway epithelial cells. Toxicol Lett 2008;178(2):95–102.

[102] Braydich-Stolle LK, Speshock JL, Castle A, et al. Nanosized aluminum altered immune function. ACS Nano 2010;4(7):3661–70.

[103] Rothen-Rutishauser BM, Kiama SG, Gehr P. A three-dimensional cellular model of the human respiratory tract to study the interaction with particles. Am J Resp Cell Mol Biol 2005;32(4):281–9.

[104] Dekali S, Divetain A, Kortulewski T, et al. Cell cooperation and role of the P2X(7) receptor in pulmonary inflammation induced by nanoparticles. Nanotoxicology 2012 [Epub ahead of print] doi:10.3109/17435390.2012.735269.

[105] Hornung V, Bauernfeind F, Halle A, et al. Silica crystals and aluminum salts activate the NALP3 inflammasome through phagosomal destabilization. Nat Immunol 2008;9(8):847–56.

[106] Schroder K, Tschopp J. The inflammasomes. Cell 2010;140(6):821–32.

[107] Palomaki J, Karisola P, Pylkkänen L, et al. Engineered nanomaterials cause cytotoxicity and activation on mouse antigen presenting cells. Toxicology 2010;267(1-3): 125–31.

[108] McCullough KC, Bassi I, Démoulins T, et al. Functional RNA delivery targeted to dendritic cells by synthetic nanoparticles. Ther Deliv 2012;3(9):1077–99.

[109] Lee J, Lilly GD, Doty RC, et al. *In vitro* toxicity testing of nanoparticles in 3D cell culture. Small 2009;5(10):1213—21.

[110] Clift MJ, Foster EJ, Vanhecke D, et al. Investigating the interaction of cellulose nanofibers derived from cotton with a sophisticated 3D human lung cell coculture. Biomacromolecules 2011;12(10):3666—73.

[111] Wright JR. Immunoregulatory functions of surfactant proteins. Nat Rev Immunol 2005;5(1):58—68.

[112] Oberdorster G, Oberdorster E, Oberdorster J. Nanotoxicology: an emerging discipline evolving from studies of ultrafine particles. Environ Health Perspect 2005;113(7): 823—39. http://dx.doi.org/10.1289/ehp.7339.

[113] Oberdörster G, Finkelstein JN, Johnston C, et al. Acute pulmonary effects of ultrafine particles in rats and mice. Res Rep Health Eff Inst 2000;(96):5—74 disc. 75—86.

[114] Cho WS, Duffin R, Howie SE, et al. Progressive severe lung injury by zinc oxide nanoparticles; the role of Zn^{2+} dissolution inside lysosomes. Part Fibre Toxicol 2011;8:27.

[115] Cho WS, Duffin R, Thielbeer F, et al. Zeta potential and solubility to toxic ions as mechanisms of lung inflammation caused by metal/metal oxide nanoparticles. Toxicol Sci 2012;126(2):469—77.

[116] Ryman-Rasmussen JP, Cesta MF, Brody AR, et al. Inhaled carbon nanotubes reach the subpleural tissue in mice. Nat Nanotechnol 2009;4(11):747—51.

[117] Murray AR, Kisin ER, Tkach AV, et al. Factoring-in agglomeration of carbon nanotubes and nanofibers for better prediction of their toxicity versus asbestos. Part Fibre Toxicol 2012;9:10.

[118] Schinwald A, Murphy FA, Jones A, et al. Graphene-based nanoplatelets: a new risk to the respiratory system as a consequence of their unusual aerodynamic properties. ACS nano 2012;6(1):736—46.

[119] Donaldson K, Poland CA. Nanotoxicology: new insights into nanotubes. Nature Nanotech 2009;4(11):708—10.

[120] Rossi EM, Pylkkänen L, Koivisto AJ, et al. Inhalation exposure to nanosized and fine TiO_2 particles inhibits features of allergic asthma in a murine model. Part Fibre Toxicol 2010;7:35.

[121] Park HS, Kim KH, Jang S, et al. Attenuation of allergic airway inflammation and hyperresponsiveness in a murine model of asthma by silver nanoparticles. Int J Nanomedicine 2010;5:505—15.

[122] Hussain S, Vanoirbeek JA, Luyts K, et al. Lung exposure to nanoparticles modulates an asthmatic response in a mouse model. Eur Respir J 2011;37(2):299—309.

[123] Ryman-Rasmussen JP, Tewksbury EW, Moss OR, et al. Inhaled multiwalled carbon nanotubes potentiate airway fibrosis in murine allergic asthma. Am J Respir Cell Mol Biol 2009;40(3):349—58.

[124] Barry BW. Is transdermal drug delivery research still important today? Drug Discov Today 2001;6(19):967—71.

[125] Crosera M, Bovenzi M, Maina G, et al. Nanoparticle dermal absorption and toxicity: a review of the literature. Int Arch Occup Environ Health 2009;82(9):1043—55.

[126] Alvarez-Roman R, Naik A, Kalia YN, et al. Skin penetration and distribution of polymeric nanoparticles. J Control Release 2004;99(1):53—62.

[127] Kim S, Lim YT, Soltesz EG, et al. Near-infrared fluorescent type II quantum dots for sentinel lymph node mapping. Nat Biotechnol 2004;22(1):93—7.

[128] Ryman-Rasmussen JP, Riviere JE, Monteiro-Riviere NA. Penetration of intact skin by quantum dots with diverse physicochemical properties. Toxicol Sci 2006;91(1):159—65.

[129] Zhang LW, Yu WW, Colvin VL, Monteiro-Riviere NA. Biological interactions of quantum dot nanoparticles in skin and in human epidermal keratinocytes. Toxicol Appl Pharmacol 2008;228(2):200—11.

[130] Rouse JG, Yang J, Ryman-Rasmussen JP, et al. Effects of mechanical flexion on the penetration of fullerene amino acid-derivatized peptide nanoparticles through skin. Nano Lett 2007;7(1):155−60.

[131] Nabeshi H, Yoshikawa T, Matsuyama K, et al. Systemic distribution, nuclear entry and cytotoxicity of amorphous nanosilica following topical application. Biomaterials 2011;32(11):2713−24.

[132] Hirai T, Yoshikawa T, Nabeshi H, et al. Amorphous silica nanoparticles size-dependently aggravate atopic dermatitis-like skin lesions following an intradermal injection. Part Fibre Toxicol 2012;9:3.

[133] Vemula PK, Anderson RR, Karp JM. Nanoparticles reduce nickel allergy by capturing metal ions. Nat Nanotechnol 2011;6(5):291−5.

[134] Karlsson HL. The comet assay in nanotoxicology research. Anal Bioanal Chem 2010; 398:651−66.

[135] Ames BN, Lee FD, Durston WE. An improved bacterial test system for the detection and classification of mutagens and carcinogens. Proc Natl Acad Sci USA 1973;70:782−6.

[136] Clift MJD, Raemy DO, Endes C, et al. Can the Ames test provide an insight into nano-object mutagenicity? Investigating the interaction between nano-objects and bacteria. Nanotoxicology 2012 [Epub ahead of print] doi:10.3109/17435390.2012.741725.

[137] Compton PJE, Hooper K, Smith MT. Human somatic mutation assays as biomarkers of carcinogenesis. Environ Health Perspect 1991;94:135−41.

[138] Jacobsen NR, White PA, Gingerich J, et al. Mutation spectrum in FE1-MUTA(TM) Mouse lung epithelial cells exposed to nanoparticulate carbon black. Environ Mol Mutag 2011;52:331−7.

[139] Savage JRK. Classification and relationship of induced chromosomal structural changes. J Med Genet 1975;12:103−22.

[140] Savage JRK. Reflections and meditations upon complex chromosomal exchanges. Mutat Res 2002;512:93−109.

[141] Fenech M. The cytokinesis-block micronucleus technique and its application to genotoxicity studies in human populations. Environ Health Perspect 1993;101:101−7.

[142] OECD Guidelines for the Testing of Chemicals. Available at http://www.oecd.org/env/ehs/testing/oecdguidelinesforthetestingofchemicals.htm.

[143] European Commission Regulation no 1907/2006 of the European Parliament and of the Council of 18 December 2006 concerning the Registration, Evaluation, Authorisation and Restriction of Chemicals (REACH). Available at http://eur-lex.europa.eu/LexUriServ/LexUriServ.do?uri=oj:l:2006:396:0001:0849:en:pdf

[144] Donaldson K, Poland CA, Schins RP. Possible genotoxic mechanisms of nanoparticles: criteria for improved test strategies. Nanotoxicology 2010b;4:414−20.

[145] Park EJ, Choi J, Park YK, Park K. Oxidative stress induced by cerium oxide nanoparticles in cultured BEAS-2B cells. Toxicology 2008;245:90−100.

[146] Shvedova AA, Kagan VE, Fadeel B. Close encounters of the small kind: adverse effects of man-made materials interfacing with the nano-cosmos of biological systems. Annu Rev Pharmacol Toxicol 2010;50:63−88.

[147] Singh N, Manshian B, Jenkins GJS, et al. NanoGenotoxicology: The DNA damaging potential of engineered nanomaterials. Biomaterials 2009;30:891−914.

[148] Driscoll KE, Deyo LC, Carter JM, et al. Effects of particle exposure and particle-elicited inflammatory cells on mutation in rat alveolar epithelial cells. Carcinogenesis 1997; 18:423−30.

[149] Magdolenova Z, Collins AR, Kumar A, et al. Mechanisms of genotoxicity. Review of recent in vitro and in vivo studies with engineered nanoparticles. Nanotoxicology 2013 [Epub ahead of print] doi:10.3109/17435390.2013.773464.

[150] Chen M, von Mikecz A. Formation of nucleoplasmic protein aggregates impairs nuclear function in response to SiO_2 nanoparticles. Exp Cell Res 2005;305:51−62.

[151] Sayes CM, Reed KL, Glover KP, et al. Changing the dose metric for inhalation toxicity studies: short-term study in rats with engineered aerosolized amorphous silica nanoparticles. Inhal Toxicol 2010;22:348−54.
[152] Sargent LM, Shvedova AA, Hubbs AF, et al. Potential pulmonary effects of engineered carbon nanotubes: in vitro genotoxic effects. Nanotoxicology 2010;4:396−408.
[153] Sargent LM, Shvedova AA, Hubbs AF, et al. Induction of aneuploidy by single-walled carbon nanotubes. Environ Molec Mutag 2009;50:708−17.
[154] Sargent LM, Hubbs AF, Young SH, et al. Single-walled carbon nanotube-induced mitotic disruption. Mutat Res 2012;745:28−37.
[155] Singh S, Shi T, Duffin R, et al. Endocytosis, oxidative stress and IL-8 expression in human lung epithelial cells upon treatment with fine and ultrafine TiO$_2$: role of the specific surface area and of surface methylation of the particles. Toxicol Appl Pharmacol 2007;222:141−51.
[156] Dunford R, Salinaro A, Cai L, et al. Chemical oxidation and DNA damage catalysed by inorganic sunscreen ingredients. FEBS Lett 1997;418:87−90.
[157] Nakagawa Y, Wakuri S, Sakamoto K, Tanaka N. The photogenotoxicity of titanium dioxide particles. Mutat Res 1997;394:125−32.
[158] Theogaraj E, Riley S, Hughes L, et al. An investigation of the photo-clastogenic potential of ultrafine titanium dioxide particles. Mutat Res 2007;634:205−19.
[159] Dufour EK, Kumaravel T, Nohynek GJ, et al. Clastogenicity, photo-clastogenicity or pseudo-photo-clastogenicity: genotoxic effects of zinc oxide in the dark, in pre-irradiated or simultaneously irradiated Chinese hamster ovary cells. Mutat Res 2006;607:215−24.
[160] Karlsson HL, Cronholm P, Gustafsson J, Moller L. Copper oxide nanoparticles are highly toxic: a comparison between metal oxide nanoparticles and carbon nanotubes. Chem Res Toxicol 2008;21:1726−32.
[161] Karlsson HL, Gustafsson J, Cronholm P, Möller L. Size-dependent toxicity of metal oxide particles—a comparison between nano- and micrometer size. Toxicol Lett 2009;188:112−8.
[162] Xie H, Mason MM, Wise JP. Genotoxicity of metal nanoparticles. Rev Environ Health 2011;26:251−68.
[163] SCCS (Scientific Committee on Consumer Safety). Opinion on ZnO (nano form); 2012. 18 September 2012.
[164] Nymark P, Catalán J, Suhonen S, et al. Genotoxicity of 2-vinylpyrrolidone-coated silver nanoparticles in BEAS 2B cells. Toxicology 2013. doi:10.1016/j.tox.2012.09.014.
[165] Falck GC-M, Lindberg HK, Suhonen S, et al. Genotoxic effects of nanosized and fine TiO$_2$. Hum Exp Toxicol 2009;28:339−52.
[166] Schrand AM, Rahman MF, Hussain SM, et al. Metal-based nanoparticles and their toxicity assessment. WIREs Nanomed Nanobiotechnol 2010;2:544−68.
[167] Lindberg HK, Falck GC-M, Catalán J, et al. Genotoxicity of inhaled nanosized TiO$_2$ in mice. Mutat Res 2012;745:58−64.
[168] Hong SC, Lee JH, Lee J, et al. Subtle cytotoxicity and genotoxicity differences in superparamagnetic iron oxide nanoparticles coated with various functional groups. Int J Nanomed 2011;6:3219−31.
[169] Klien K, Godnic-Cvar J. Genotoxicity of metal nanoparticles: focus on in vivo studies. Arh Hig Rada Toksikol 2012;63:133−45.
[170] Lindberg HK, Falck GC, Suhonen S, et al. Genotoxicity of nanomaterials: DNA damage and micronuclei induced by carbon nanotubes and graphite nanofibres in human bronchial epithelial cells *in vitro*. Toxicol Lett 2009;186:166−73.
[171] Catalán J, Järventaus H, Vippola M, et al. Induction of chromosomal aberrations by carbon nanotubes and titanium dioxide nanoparticles in human lymphocytes *in vitro*. Nanotoxicology 2012;6:825−36.

[172] Gonzalez L, Lison D, Kirsch-Volders M. Genotoxicity of engineered nanomaterials: A critical review. Nanotoxicology 2008;2:252–73.

[173] Aschberger K, Johnston HJ, Stone V, et al. Review of carbon nanotubes toxicity and exposure—appraisal of human health risk assessment based on open literature. Crit Rev Toxicol 2010;40:759–90.

[174] Nagai H, Toyokuni S. Biopersistent fiber-induced inflammation and carcinogenesis: lessons learned from asbestos toward safety of fibrous nanomaterials. Arch Biochem Biophys 2010;502:1–7.

[175] Linton A, Vardy J, Clarke S, van Zandwijk N. The ticking time-bomb of asbestos: Its insidious role in the development of malignant mesothelioma. Crit Rev Oncol Hematol 2012;84:200–12.

[176] Kisin ER, Murray AR, Sargent L, et al. Genotoxicity of carbon nanofibers: are they potentially more or less dangerous than carbon nanotubes or asbestos? Toxicol Appl Pharmacol 2011;252:1–10.

[177] Ema M, Tanaka J, Kobayashi N, et al. Genotoxicity evaluation of fullerene C60 nanoparticles in a comet assay using lung cells of intratracheally instilled rats. Regul Toxicol Pharm 2012;62:419–24.

[178] Bourdon JA, Saber AT, Jacobsen NR, et al. Carbon black nanoparticle instillation induces sustained inflammation and genotoxicity in mouse lung and liver. Part Fibre Toxicol 2012;9:5.

[179] Saber AT, Bornholdt J, Dybdahl M, et al. Tumor necrosis factor is not required for particle-induced genotoxicity and pulmonary inflammation. Arch Toxicol 2005;79: 177–82.

[180] Landsiedel R, Kapp MD, Schulz M, et al. Genotoxicity investigations on nanomaterials: Methods, preparation and characterization of test material, potential artifacts and limitations-Many questions, some answers. Mutat Res 2009;681:241–58.

[181] Roller M. *In vitro* genotoxicity data of nanomaterials compared to carcinogenic potency of inorganic substances after inhalational exposure. Mutat Res 2011;727: 72–85.

[182] Doak SH, Manshian B, Jenkins GJS, Singh N. *In vitro* genotoxicity testing strategy for nanomaterials and the adaptation of current OECD guidelines. Mutat Res 2012; 745:104–11.

[183] ECHA. Guidance on information requirements and chemical safety assessment. Appendix R7–1 Recommendations for nanomaterials applicable to Chapter R7a Endpoint specific guidance. Eur Chem Agency. Available at: http://echa.europa.eu/documents/10162/13632/appendix_r7a_nanomaterials_en.pdf; 2012.

[184] Magdolenova Z, Lorenzo Y, Collins A, Dusinska M. Can standard genotoxicity tests be applied to nanoparticles? J Toxicol Environ Health 2012;A 75(13-15):800–6.

[185] OECD (2009). Preliminary Review of OECD Test Guidelines for Their Applicability to Manufactured Nanomaterials. Environment Directorate, Organisation for Economic Co-operation and Development, Paris. OECD Environment, Health and Safety Publications Series on the Safety of Manufactured Nanomaterials, No. 15.

[186] Lindberg HK, Falck GC-M, Singh G, et al. Genotoxicity of single-wall and multi-wall carbon nanotubes in human epithelial and mesothelial cells *in vitro*. Toxicology 2012 [Epub ahead of print] doi: 10.1016/j.tox.2012.12.008.

[187] Trouiller B, Reliene R, Westbrook A, et al. Titanium dioxide nanoparticles induce DNA damage and genetic instability *in vivo* in mice. Cancer Res 2009;69:8784–9.

[188] Muller J, Decordier I, Hoet P, et al. Clastogenic and aneugenic effects of multi-wall carbon nanotubes in epithelial cells. Carcinogenesis 2008;29:427–33.

[189] Lindberg HK, Falck GC-M, Catalán J, et al. Micronucleus assay for mouse alveolar type II and Clara cells. Environ Molec Mutag 2010;51:164–72.

[190] Pott F, Roller M. Carcinogenicity study with nineteen granular dusts in rats. Eur J Oncol 2005;10:249—81.

[191] Roller M, Pott F. Lung tumor risk estimates from rat studies with not specifically toxic granular dusts. Ann N Y Acad Sci 2006;1076:266—80.

[192] International Agency for Research on Cancer (2010). IARC monographs on the evaluation of carcinogenic risks to humans, vol. 93, Carbon black, titanium dioxide, and talc, Lyon, France.

[193] National Institute for Occupational Safety and Health. Occupational exposure to titanium dioxide. Department of Health and Human Services, Centers for Disease Control and Prevention. Curr Intell Bull 2011;63.

[194] Heinrich U, Fuhst R, Rittinghausen S, et al. Chronic inhalation exposure of Wistar rats and two different strains of mice to diesel engine exhaust, carbon black, and titanium dioxide. Inhal Toxicol 1995;7:533—56.

[195] Furukawa F, Doi Y, Suguro M, et al. Lack of skin carcinogenicity of topically applied titanium dioxide nanoparticles in the mouse. Food Chem Toxicol 2011;49:744—9.

[196] Hansen T, Clermon G, Alves A, et al. Biological tolerance of different materials in bulk and nanoparticulate form in a rat model: sarcoma development by nanoparticles. J R Soc Interface 2006;3:767—75.

[197] Onuma K, Sato Y, Ogawara S, et al. Nano-scaled particles of titanium dioxide convert benign mouse fibrosarcoma cells into aggressive tumor cells. Am J Pathol 2009;175: 2171—83.

[198] Moon E-Y, Yi G-H, Kang J-S, et al. An increase in mouse tumor growth by an in vivo immunomodulating effect of titanium dioxide nanoparticles. J Immunotoxicol 2011; 8:56—67.

[199] ILSI Risk Science Institute Workshop Participants. The relevance of the rat lung response to particle overload for human risk assessment: a workshop consensus report. Inhal Toxicol 2000;12:1—17.

[200] Saber AT, Jensen KA, Jacobsen NR, et al. Inflammatory and genotoxic effects of nanoparticles designed for inclusion in paints and lacquers. Nanotoxicology 2012;6:453—71.

[201] Sakamoto Y, Nakae D, Fukumori N, et al. Induction of mesothelioma by a single intrascrotal administration of multi-wall carbon nanotube in intact male Fischer 344 rats. J Toxicol Sci 2009;34:65—76.

[202] Shvedova AA, Kisin ER, Mercer R, et al. Unusual inflammatory and fibrogenic pulmonary responses to single walled carbon nanotubes in mice. Am J Physiol Lung Cell Mol Physiol 2005;283:L698—708.

[203] Park EJ, Roh J, Kim SN, et al. A single intratracheal instillation of single-walled carbon nanotubes induced early lung fibrosis and subchronic tissue damage in mice. Arch Toxicol 2011;85:1121—31.

[204] Nagai H, Okazaki Y, Chew SH, et al. Diameter and rigidity of multiwalled carbon nanotubes are critical factors in mesothelial injury and carcinogenesis. Proc Natl Acad Sci USA 2011;108:E1330—8.

[205] Muller J, Delos M, Panin N, et al. Absence of carcinogenic response to multiwall carbon nanotubes in a 2-year bioassay in the peritoneal cavity of the rat. Toxicol Sci 2009;110:442—8.

[206] Xu J, Futakuchi M, Shimizu H, et al. Multi-walled carbon nanotubes translocate into the pleural cavity and induce visceral mesothelial proliferation in rats. Cancer Sci 2012;103:2045—50.

[207] Murphy FA, Schinwald A, Poland CA, Donaldson K. The mechanism of pleural inflammation by long carbon nanotubes: interaction of long fibres with macrophages stimulates them to amplify pro-inflammatory responses in mesothelial cells. Part Fibre Toxicol 2012;9:8.

[208] Schinwald A, Murphy FA, Prina-Mello A, et al. The threshold length for fiber-induced acute pleural inflammation: shedding light on the early events in asbestos-induced mesothelioma. Toxicol Sci 2012;128:461−70.

[209] van Broekhuizen PJC. Nano Matters: Building Blocks for a Precautionary Approach. PhD thesis:198. Available at: http://www.ivam.uva.nl/?nanomatters; 2012.

[210] van Broekhuizen P, van Broekhuizen F, Cornelissen R, Reijnders L. Workplace exposure to nanoparticles and the application of provisional nano reference values in times of uncertain risks. J Nanopart Res 2012;14:770.

[211] Fadeel B, Feliu N, Vogt C, et al. Bridge over troubled waters: understanding the synthetic and biological identities of engineered nanomaterials. WIREs Nanomed Nanobiotechnol 2013;5:111−29.

[212] Tantra R, Shard A. We need answers. Nat Nanotechnol 2013;8:71.

[213] Nel A, Xia T, Meng H, et al. Nanomaterial Toxicity Testing in the 21st Century: Use of a Predictive Toxicological Approach and High-Throughput Screening. Acc Chem Res 2012. [Epub ahead of print].

[214] Hartung T. Toxicology for the twenty-first century. Nature 2009;460(7252):208−12. http://dx.doi.org/10.1038/460208a.

From Source to Dose: Emission, Transport, Aerosol Dynamics and Dose Assessment for Workplace Aerosol Exposure

Martin Seipenbusch [1], Mingzhou Yu [2],
Christof Asbach [3], Uwe Rating [3],
Thomas A.J. Kuhlbusch [3], Göran Lidén [4]

[1] Karlsruhe Institute of Technology KIT, Karlsruhe, Germany
[2] Institute of Earth Environment, Chinese Academy of Sciences, Xi'an, China; China Jiliang University, Hangzhou, China
[3] Institute of Energy and Environmental Technology (IUTA), Air Quality & Sustainable Nanotechnology, Duisburg, Germany
[4] Stockholm University, Stockholm, Sweden

4.1 SOURCES OF NANOPARTICLES IN THE WORKPLACE (SEIPENBUSCH)

4.1.1 Sources of Nanoparticles in the Workplace

Nanoparticles can be emitted by a broad spectrum of potential sources varying in concentration, temporal release characteristics and particle size [1]. One way of classifying these emitters is by following the process chain of nanoparticles in production and application. A primary source then releases particles directly from a synthesis process or a welding operation for instance, wherever particles are formed immediately prior to emission. Nanoparticle release from powders during bagging or handling can be classified as a secondary source. In the application and use of nanoparticle-containing materials such as filled polymers a tertiary release of particle aerosols is eventually possible, e.g., during sanding or grinding.

4.1.2 Primary Nanoparticle Aerosol Sources

The release of primary nanoparticle aerosols may occur due to leakage from pressurized systems, potentially releasing large quantities of material. Examples are the 'rapid expansion of supercritical fluids' (RESS) process [2], where nanoparticles are generated in the expansion of supercritical CO_2, or hydrothermal synthesis in supercritical water [3]. In the event of a leak particles can be released into workplace air from a localized source at high concentrations. A similar scenario is possible when reactive precursors are released and get in contact with air, leading to the formation of oxide particles in the work environment itself. Next to leaks in reaction systems, failure of the downstream collection apparatus is a potential source, particularly for mass-produced particulate products like soot or silica. The particle concentrations can be medium to high, but tend to be localized to a lesser extent. All of the processes mentioned thus far typically lead to a continuous particle release over a longer period of time.

Apart from accidental release nanoparticle aerosols can reach the workplace environment by emission from open gas phase processes [4–6] or even liquid phase synthesis [6]. The source characteristics vary strongly in these cases, from a continuous release for open flames [4] to discontinuous, when closed containments are only temporarily opened for certain manual operations [6]. Also the particle size varies from free nanoparticles [4,6] to agglomerates [7,8], depending on the process and the material.

Other than the release of manufactured nanoparticles, incidental sources for nanoparticles, such as welding, soldering or plasma spraying can produce high levels of ultrafine particles. Typical size distributions of aerosols produced from welding [9] are bimodal with modal values around 30 nm for the first and close to 200 nm for the second, which points to two different principles of particle formation. The total number concentration for welding fumes is in the range of 10^7 cm^3 [10].

4.1.3 Secondary Nanoparticle Aerosol Sources

A secondary release of nanoparticles from powders can occur in several different ways. Reactive nanoparticles like nanoscale iron or other pure metals can oxidize when removed from the protective atmosphere of a collection apparatus or reactor and brought into contact with air. The release of nanoparticles due to a strong exothermic effect is then possible. Apart from this accidental release everyday procedures in the small-scale industrial production of nanomaterials or in research laboratories, such as cleaning or substrate removal in a CVD, flame or hot wall reactor, can potentially lead to particle release as well [11]. However, most experimental studies do not show a significant concentration of nanoparticles in workplace air for these procedures [12,13]. For the removal and handling of single-walled carbon nanotubes (SWCNT) from a CVD reactor, for instance, no increase in the number concentration was observed in the critical range between 10 nm and 1 μm [12]. Even for the production of commodity nanopowders on the large scale, such as carbon black, handling and bagging of powders apparently does not lead to the emission of free nanoparticles into the workplace environment. Kuhlbusch et al. [13] measured particle size distributions and mass concentrations in the bagging area of a carbon black production facility and found a strong increase in the particle mass below 10 μm particle size (PM10) but only a minor increase in the particle number in the size range below 100 nm. A release of particles thus occurred but merely of highly agglomerated structures with a median mobility diameter of about 400 nm. Similar findings were reported for other powder handling operations [14,15]. The measurement of the dustiness of nanopowders showed the generation of high concentrations of submicron particles, but not in the nano-regime [16,17].

The resuspension and deagglomeration of nano-powders requires a significant amount of energy [18,19], which is not provided in simple handling operations. Therefore, secondary sources will generally produce agglomerated aerosols well above 100 nm.

4.1.4 Tertiary Nanoparticle Sources

A tertiary nanoparticle aerosol source may originate from the use or processing of products or intermediates containing nanoparticles. Examples for this are spraying of particulate suspensions or the compounding of nanopowders with polymers [14]. Also in the production of nanostructured ceramic parts by ceramic injection moulding the release of nanoparticles is thinkable in the moulding and debinding steps. Machining, sanding and grinding of all materials containing nanoparticles are other potential tertiary sources for nanoparticles [20–24]. Since in all of these cases the particles are embedded in a solid or liquid matrix, the likelihood of the release of an aerosol in the nano-range stemming from the original particles is extremely low and was not found in the cited studies. Therefore, larger matrix-bound particles well above 100 nm can be expected.

4.2 AEROSOL DYNAMICS IN WORKPLACE ATMOSPHERES (SEIPENBUSCH)

4.2.1 Basic Mechanisms

The number concentration and size distribution of aerosols generally are subjected to constant change. A large number of mechanisms is responsible for this, leading to increasing or decreasing particle concentration and shifts in the particle size distribution to smaller or larger particles or even the alteration of the shape of the distribution. In this section the discussion will be limited to the mechanisms relevant in typical workplace environments.

Dilution due to ventilation is a major factor in reducing the concentration of aerosol particles in workplaces. Figure 4.1 shows the calculated reduction of a particle concentration in a completely mixed room at different air exchange rates. The highest rate ($20\ \mathrm{h}^{-1}$) leads to a reduction to half of the initial concentration after about two minutes. In real-world workplace environments, however, mixing of the air will not be quite as good as assumed in the mathematical description of the problem, therefore dilution will be substantially slower.

An important group of mechanisms leading to changes in particle number and size distribution are losses to surfaces and walls. Figure 4.2 summarizes several of these in a single diagram, characterized by the deposition velocity versus the particle diameter. Losses by *gravitational settling* become important for particle sizes larger than 1 μm and can be neglected in the nanoparticle regime. The curve is plotted for spheres of unit density. Agglomerates of nanoparticles, which can very well be in the range of several 100 nm in diameter, have a much lower density than the solid sphere, leading to a further reduction of the settling velocity due to gravitation.

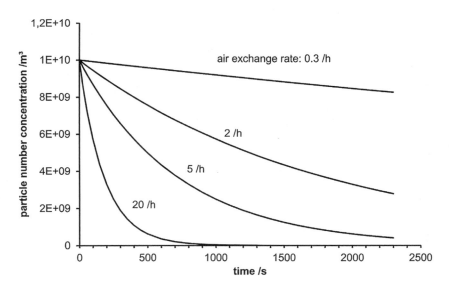

FIGURE 4.1 Dilution of an aerosol in a completely mixed chamber by ventilation at varied air exchange rate (calculated).

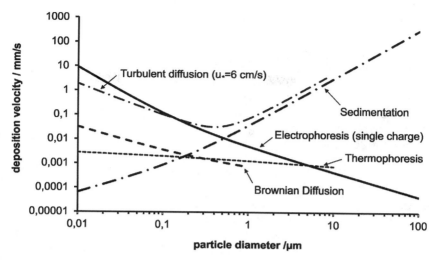

FIGURE 4.2   Deposition velocities of various surface loss mechanisms as functions of the particle size. Turbulent diffusion and settling according to [26], thermophoresis data from [27].

Thermophoretic losses require high temperature gradients and are only effective for particles smaller than 100 nm under standard conditions. Figure 4.2 shows the velocities for a temperature gradient of 100 K/m^{-1} at a temperature of 20°C. Since the thermophoretic transport velocity scales linearly with the temperature gradient, a ΔT of 100,000 K/m would be necessary to obtain values in the same order of magnitude as the dominant mechanisms [25], a value which can only be reached in close proximity to two differently tempered surfaces.

Electrostatic forces efficiently remove particles in the submicron range. The curve displayed in Figure 4.2 was calculated for an electrical field strength of 4500 V/m and a singly charged spherical particle. In a realistic workplace aerosol, however, most particles will not carry a charge and will therefore not be affected by electric fields. For very small particles the charging efficiency by ion attachment decreases further, which leads to a further reduction in the efficiency of the deposition mechanism.

Brownian diffusion is a relatively slow transport mechanism compared to electrophoresis, however it is non-negligible in the nano-regime. *Turbulent diffusion*, however, is a much faster process and clearly dominates the loss mechanisms in the absence of strong electric fields. Figure 4.2 shows the size dependence of the deposition velocity due to this mechanism. The model by Lai and Nazaroff [26] includes gravitational settling for large particles, therefore the decreasing deposition velocity for increasing particle size is countered by the acceleration of sedimentation for particles larger than approximately 500 nm. The efficiency of turbulent diffusion strongly depends on the turbulence of the system. In the model this parameter is expressed by the friction velocity u*, which characterizes the intensity of near-surface turbulent flow [26].

Apart from losses to the walls and other surfaces, *coagulation* of aerosol particles is a major modifier of the particle number concentration and the particle size. Coagulation generally occurs when particles collide due to relative motion, which results in adherence and agglomerate formation. Agglomerates are generally held together by van-der-Waals forces, which in the nano-scale are much larger than the weak shear or thermal forces present in a workplace atmosphere. Thus particle contacts established by coagulation are stable and will generally not be broken in this environment. The nature of the forces leading to the necessary relative motion of particles defines the type of agglomeration. If thermal relative motion causes the collision of aerosol particles the process is termed *Brownian coagulation.* If electrical or aerodynamic effects are the cause, coagulation is not a stochastic process as the latter and the term *kinematic coagulation* is used [28]. Brownian coagulation is however the dominant mechanism for common workplace aerosols, since a large number of charges and high flow velocities are uncommon. Coagulation theory mathematically describes the change of particle size and number with time employing the number concentration N, the particle diameter d_p and the diffusion coefficient D. The rate of change of the particle number is proportional to N^2 for a monodisperse aerosol and also proportional to the particle diameter and the diffusion coefficient [27]. Since the diffusion coefficient for particles is inversely dependent on the particle size, however, the coagulation between particles of different sizes in polydisperse aerosols is strongly accelerated. The large size of one partner increases the likelihood of a collision while the small mass of the second partner increases its diffusivity and therefore maximizes the relative velocity between the two particles [27,28]. Scavenging of small particles by larger ones is therefore an important mechanism in atmospheric aerosols but also plays a role in the dynamics of nanoparticle aerosols in workplace atmospheres with pre-existing background aerosols in the micron range.

4.2.2 The Influence of Background Particles on the Dynamics of a Nanoparticle Aerosol

In a series of laboratory experiments the interactions of nanoparticles with a background aerosol were explored [29]. A nanoparticle source continuously generated particles at a concentration of 5×10^6 cm^{-3}. The background aerosol was simulated by a mist of oil droplets at varied concentrations (8×10^4 cm^{-3}, 1×10^5 cm^{-3}, 3×10^5 cm^{-3}) at a flow rate of 5 L/min and a median particle size around 250 nm. The results are

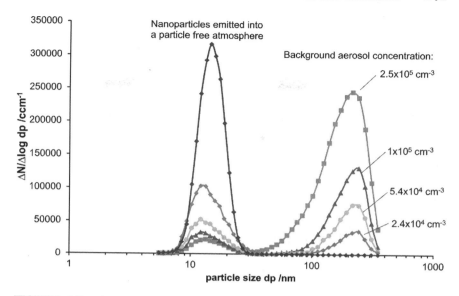

FIGURE 4.3 Size distributions measured for different concentrations of the background aerosol.

displayed in Figure 4.3; for comparison, the size distribution of a nano-particle aerosol in the absence of background particles is also shown. The variation of the number concentration of the background aerosol had a strong influence on the size distribution in the nano-regime. For the highest concentration, 2.5×10^5 cm^{-3}, the nanoparticle source is hardly visible in the size distribution. Under these conditions, the particles coagulate with the larger background particles at a rate approaching the source strength, and are thus effectively scavenged. With decreasing concentration of the oil mist aerosol the coagulation between nano-particles and background becomes less rapid due to the concentration dependence of the kinetics of this process. In consequence the nano-particle source becomes more and more prominent in the overall size distribution. For lower concentrations of the background aerosol the coagulation of the nanoparticles with each other becomes dominant.

4.3 MODELLING APPROACHES TO AEROSOL DYNAMICS AND TRANSPORT IN THE WORKPLACE (YU)

In addition to the traditional experimental methods, modelling approaches can be also used to predict the characteristics of aerosols in workplaces. Modelling approaches are able to predict the evolution of particle size distribution in time and space, and thus can be applied in the assessment of aerosol exposure and indoor air quality [30–35]. Based on the conservation of particle number concentration, the particle population balance equation, also known as the general dynamic equation (GDE), is constructed in terms of the particle size distribution which is considered to be the leading quantity characterizing nanoparticle aerosols [36]. With initial and boundary conditions, the temporal and spatial particle size distribution in the workplace can be obtained through solving the GDE equation with analytical and numerical methods [29,30]. In the presence of a flow, the combination of the GDE equation and the CFD equations is needed.

The key to the modelling approaches lies in the solution of the GDE equation, whose analytical solution cannot be achieved in the presence of internal processes including coagulation, condensation, deposition, gas-to-particle conversion, breakage, and the external process due to flow convection [36]. Almost all the aerosol processes in workplaces can be represented by the schematic in Figure 4.4 [37]. For nanoparticles, due to very small size, some processes such as sedimentation, agglomeration and breakage are negligible. Even in the absence of flow, the analytical solution of the GDE equation is not easily accomplished due to its complicated non-linear integral-differential mathematical form [38–40]. The difficulty arises mainly from the coagulation term which inherently has the structure of the Boltzmann equation. In the past, some efforts have been made to achieve the closure of the GDE equation. Three main numerical methods, the method of moments [39–44], the sectional method [45–48], and the stochastic particle method [49,50], have been proposed

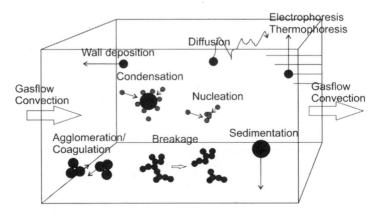

FIGURE 4.4 Schematic representation of aerosol processes within an elemental volume (including external and internal processes).

and evaluated. Both advantages and disadvantages of these three methods have been compared in the studies of Kraft [49]. If the coagulation kernel is simplified with the assumption of a homogeneously mixed aerosol, the analytical solution of the GDE equation can be achieved. The analytical method has been used to study nanoparticle dynamics at an experimental chamber [29].

Because of the relative simplicity of the implementation and the low computational cost, the method of moments has been extensively used to solve aerosol problems, and also can be applied in the assessment of nanoparticle dynamics in workplaces with high efficiency. The key to this method is to solve a set of moment equations related to the lower-order moments of aerosol size distribution using different numerical techniques. Five prominent techniques, i.e., making a prior assumption for the shape of the aerosol size distribution [40,51], approximating the integral moment by an n-point Gaussian quadrature [43,44], assuming the pth-order polynomical form for the moments [52], achieving closure with the interpolative method [42], and achieving closure with the Taylor-Expansion technique [38,39], have been proposed by different researchers. Due to the low requirement for computational resources, in the last 10 years, the combination of the method of moments and the CFD technique has been an emerging research field in which the task is to investigate and predict the temporal and spatial evolution of nanoparticles at different turbulent flows, including workplace aerosol flow.

The information of the particle size distribution is lost due to integral in the transfer from the population balance equation to the moment equations in the method of moments, and thus this method is unable to exactly trace the evolution of the particle size distribution even though today there have been some techniques for the reconstruction of the particle size distribution from the known moments [39,49]. The sectional method, which divides the population balance equation into a set of size classes, has an ability to overcome the limit of method of moments. This method was widely applied in the studies on the evolution of the particle size distribution in engineering applications due to the different dynamical processes including coagulation, condensation, gas-particle conversion, and the interaction between the continuum and the dispersed particles. In order to extract the size-resolved emission rate, the sectional method has been applied in the construction of the emission model for nanoparticle aerosol [30,35], which will be described in details below.

An alternative to the sectional method and the method of moments for solving the GDE equation is the stochastic particle method (or Monte Carlo method), which can also be used in the assessment of air quality and prediction of the characteristics of aerosols composed of nanoparticles. The application of this method, however, is limited because of the low efficiency, even though it is able to capture the evolution of the particle size distribution. Usually, in the construction of the GDE equation, the particle size distribution is defined only as a function of particle volume (or diameter). If the particle morphology and chemical composition are also focused, then the particle size distribution should be a function of three variables, i.e., volume, morphology parameter (e.g., mass fractal dimension) and chemical composition parameter (e.g., volume or mass percent fraction). The addition of variables in the GDE equation greatly increases the difficulty of solving the GDE equation, regardless of the numerical method or the analytical method.

Although the GDE equation has to be combined with the CFD equation to solve nanoparticle aerosol problems in flows, the solution for this kind of two-phase equation is easier than for relative larger particle-laden two-phase systems (Stokes number is approximate to or more than 1). This is because, relative to the Kolmogorov length scale $(v^3/\varepsilon)^{1/4}$, v is the kinetic viscosity of the continuum and ε is the dissipation rate pure unit mass) and in the Kolmogorov time scale $((v/\varepsilon)^{1/2})$, there are generally very small size scales and inertial response times for nanoparticles. At this case, the particle Stokes number is sufficiently small '( )' implying the particles follow the local fluid motion precisely and their velocity slip can be neglected [53,54]. Consequently, the study on nanoparticle dynamics due to external and internal processes can be decoupled [55], which is usually named as the one-way coupling solution in the research field of multiphase flow. The one-way coupling solution is greatly valid in almost all nanoparticle-laden multiphase systems, especially in the dilution condition where the particle volume fraction is below 0.1% [56]. The two-way coupling, which concerns not only the effect of the continuum on the dispersed phases but also the inverse, is needed if the high gas-particle conversion occurs where the transfer of mass and energy between them cannot be ignored.

4.3.1 General Remarks on CFD Modelling of Nanoparticle Aerosols

Computational fluid dynamics (CFD) is considered to be an effective tool in predicting airflow, particle transport, deposition and particle number concentration in aerosol studies, and today it has been extensively used in nanoparticle instrumentation (sampling, sensing, dilution, focusing, mixing, etc.), evolution of nanoparticle dynamics in specific environments (human respiratory track, aerosol transport/delivery system, energy systems, workplaces, etc.), and nanoparticle synthesis (combustion flow, high-temperature rector) [37]. Compared to its counterpart, i.e., experimental study, it has the following advantages [57]: 1) results are more complete because the particle number concentration and size distribution, as well as airflow pattern including velocity, turbulence characteristics, temperature and humidity, can be simultaneously obtained in the spatial—temporal space; 2) results are obtained more rapidly and easily, and thus much more time and money are saved; 3) problems related to contaminant toxicity, explosivity and flammability are avoided; and 4) it has the ability to provide some information that can hardly or cannot be obtained from experiments, such as the historical details of particles undergoing transport, coagulation and condensation in workplaces, the continuous change of particle dynamics with time at any location, and the deposition of size-resolved nanoparticles within the concentration boundary.

Depending on the particle volume fraction and its response time to a time characteristic of the fluid motion in workplaces, the gas particle flows can be classified as being either dilute or dense, which corresponds to two solutions of numerical studies of multiphase flow, one-way coupling and two-way coupling. Due to the low-volume fraction and small Stokes number of nanoparticles in workplaces, one-way coupling solution is suited, meaning in the numerical models the presence of particles has no effect or a negligible effect on the properties of the carrier phase [37]. This is typically valid due to the fact that even for the particulate system with diameter 100 nm and number concentration $1.0^8/cm^3$, the volume fraction is only 5.2×10^{-6}. In fact, the particle number concentration is far below this value in almost all workplaces.

Based on the solution for handling the dispersed phase, there are two CFD modelling approaches for nanoparticle aerosol problems in workplaces: namely the Eulerian-Eulerian and the Eulerian-Lagrangian approaches [58]. Both approaches consider the carrier phase as a continuum in that the flow details of the carrier, such as velocity and pressure, are obtained by solving Navier-Stokes equations. For the dispersed phase, however, in the Eulerian-Eulerian approach the continuum assumption is used, while in the Eulerian-Lagrangian approach the dispersed phase is treated as being some separated particles each having its own trajectory. Today, both approaches have their own advantages and disadvantages in nanoparticle aerosol studies in workplaces, and they are used in different fields in terms of different requirements. In the performing of CFD calculation, the selection of methods for solving Navier-Stokes equations is very important because almost all the computational costs lie in the iterative calculation of Navier-Stokes equations. Generally, the Navier-Stokes equations can be calculated by the Reynolds-Average-Navier-Stokes (RANS) turbulent model, the direct numerical simulation (DNS), and the large eddy simulation (LES). The RANS turbulent model is more widely used in the nanoparticle aerosol transport and deposition due to its high efficiency as compared to two other counterparts, but limited to the steady flow condition where only time-averaged results are provided [58—61]. Although the DNS has an ability to give the most comprehensive information for both airflow and particles with the most accuracy, it is now still limited in some basic researches due to too much computational cost. The LES is an ideal choice for unsteady airflow problems with reasonable accuracy and computational cost, and it is expected to play more important roles in nanoparticle dynamics and transport in workplaces in the future.

In workplaces, the coagulation process can be ignored when the nanoparticle number concentration is less than $10^4/cm^3$. In this case, if other internal processes including gas-to-particle conversion and condensation are also not considered, both the Eulerian-Lagrangian and Eulerian-Eulerian approaches applied in the micro-scale particle-laden multiphase flow research can be directly extended to be here. In these researches, the information of the spatial—temporal evolution of particle dynamics can be easily obtained by tracing each particle motion in the framework of the Eulerian-Lagrangian approach (in this approach, each particle may represent a number of particles having the same conditions, e.g., size, velocity, and initial condition) and by solving particle concentration convection-diffusion equations in the framework of the Eulerian-Eulerian approach. In workplaces where the particle number concentration is more than $10^4/cm^3$, however, the primary quantity of interest is the particle size distribution and its variance in space and time. The reliable and ideal numerical model should provide the spatial—temporal evolution of this quantity when nanoparticles undergo both internal and external processes. Here, internal processes commonly refer to particle coagulation, condensation and gas-to-particle conversion, while external processes include transport, diffusion, deposition, and particle motion induced by different forces, such as electrostatic force and thermophoresis force. Due to the fact that internal processes, including coagulation and gas-to-particle conversion, are not easily included in the kinetic calculation for tracing particles, and there are also too many particles which need to be tracked, the Eulerian-Lagrangian approach is not recommended in workplaces [62]. Based on the mean-field theory, both internal and external processes can be described by a population balance equation, also known as the general dynamic equation. This equation has all the characteristics to make it coupled with the Navier-Stokes equations within the framework of the Eulerian-Eulerian approach. The simultaneous solution of the Navier-Stokes equations and the general dynamic equation with initial and boundary conditions can provide a complete description of the size distribution at arbitrary workplace geometries and gas flows [37]. Due to the low efficiency, when coupled with the Navier-Stokes equations, the general equation needs to be further disposed by the sectional method, the moment method or the Monte Carlo method, which is discussed in Section 4.5.

4.4 APPLICATIONS OF AEROSOL DYNAMICS AND FLOW MODELLING TO WORKPLACE EXPOSURE (MINGZHOU YU)

4.4.1 Modelling Approaches for Source Identification and Isolation from a Background Aerosol Particle Transport

In recent years, numerous efforts have been focused on the examination and identification of indoor aerosol sources and especially their strength in workplaces [30,33,35,63−66]. In order to accomplish this, the only selection is to properly extract the size-resolved or total particle emission rate from the measured data. However, this is not easy to achieve because in the measured data the newly emitted particles and the evolving particles cannot be separately disposed, even if coagulation, deposition and ventilation are negligible [32,34,35,64]. According to the type of data provided by measurement instrument, numerous model approaches have been proposed to solve this problem.

In these model studies, a time series of measured particle concentration over the whole size range or total particle number (or mass) concentration usually provided by the scanning mobility particle sizer technique and the optical particle counter technique are needed. Depending on whether the estimated result is based on size-resolved or not, these models are divided into two categories, size-resolved model and size-unresolved model. In addition, these models are also divided into another two categories, a time-varying model and a time-averaged model, depending on whether the the estimated result is time-varying or time-averaged.

The difference of ENP and macro-particle aerosol lies in that ENP has very large number concentration and surface area concentration, but contributes very little to mass concentration. This novel characterization of ENP poses great challenges for the identification of source emission rate. The mass-dependent size-unresolved model usually used in the indoor air assessment is no longer applied in this field; background particles especially dominate the mass concentration. In order to properly characterize the strength of ENP emission, it is necessary to provide particle size distribution and particle number concentration using size-resolved models. In this section, different models in terms of size-resolved and size-unresolved methods are outlined, but the emphasis lies on the size-resolved model since the ENP is the main investigated object in this section. It is noted here that all the models below follow the same assumption that the aerosol is well-mixed, meaning the transport mechanism is not concerned.

4.4.1.1 Size-Resolved Emission Model

The size-resolved emission rate is commonly found by minimizing the difference between the measured and the simulated data with a complicated iterative optimization procedure. In these models, if it is time-varying, the measured size-resolved data are usually the initial condition of the GDE equation, and thus the accuracy of the model is strongly dependent on the time resolution of measurement instrument [30,35,66]. This kind of aerosol model has an ability to produce the time-varying emission rate with high resolution in the size space, but limited to low efficiency due to the complicated calculation.

According to the type of measured data provided by the measurement instrument, the particle size distribution is usually defined as a function of particle mass or particle number, and correspondingly there are two different size-resolved models, i.e., number-dependent and mass-dependent. Today, the widely used mass-dependent model is a method which infers size-specific mass emission factors for environmental tobacco smoke [64]. In this model, two unknown model parameters, i.e., emission rate and deposition rate, need to be determined in the procedure of iterative calculation, which leads to the very complicated iterative procedure in the calculation.

Based on the atmospheric aerosol dynamic model UHMA [67], Hussein et al. [35] proposed a semi-empirical model to estimate the size-resolved number emission rate due to indoor sources of aerosol particles. This model has been widely used in indoor aerosol studies on particles released from human activity and printer. In this model, the emission rate is achieved by several steps based on the difference between the measurement and the simulated data. It is noted here that the measurements represent all the evolving aerosol particles and the newly emitted particles, while the simulated concentrations represent the amount of aerosol particles after being engaged in the processes handled in the aerosol models. The aerosol concentration is firstly calculated by ignoring the source, and then the difference between the measurement and simulation, $J_{k,i}$, is obtained, $(S_{k,\text{emission},i})$.

$$J_{k,i} = N_{k,i,\text{measured}} - N_{k,i,\text{simulated}} \qquad (4.1)$$

Here, k represents the size section for the particle size distribution and i represents the time step in measurement and simulation. Since all particles of the aerosol source are included in the term $J_{k,i}$, the change rate of particle number concentrations that remain suspended is

$$\frac{dJ_{k,i}}{dt} = S_{k,\text{emission},i} + S_{k,\text{coag},i} + S_{k,\text{cond},i} - S_{k,\text{deposition},i} - S_{k,\text{venti},i} \qquad (4.2)$$

On the right side of Eq. (4.2), the five terms represent the change rate due to emission, coagulation , condensation, deposition and ventilation. In the aerosol model studies, the last four terms on the left side of Eq. (4.2) can be easily determined in advance, and thus the emission rate term ($S_{k,emission,i}$) can be achieved by iterating procedure until convergent solutions are obtained for the emission rate terms.

In the current studies, a new aerosol model for the size-resolved emission rate was proposed which is also based on the iterative calculation in terms of the difference between measured data and simulated data [30]. In this model, the high and low limit of emission rate is firstly determined by the theoretical analysis, and then the emission rate is found by the iterative procedure through solving the GDE equation with high demanding criterion,

$$\frac{\sum_{k=1}^{MAX}|\overline{\overline{N_k}}\,(t+\Delta t) - N_k(t+\Delta t)|}{\sum_{K=1}^{MAX}N_k(t+\Delta t)} < 10^{-10} \tag{4.3}$$

where, $\overline{\overline{N_k}}\,(t+\Delta t)$ is the calculated data and $N_k\,(t+\Delta t)$ is the measured data at time $t+\Delta t$. In order to interpret the interaction between particles due to Brownian coagulation, the sectional method was introduced into this model. The model took into account coagulation, particle removal by deposition and ventilation, and emission. In principle, the condensation can be included in this model without additional trouble. If the high and low limits are in advance determined by theoretical analysis, then the procedure for finding the real emission rate takes the following steps:

i Find the high and low limits of emission rate using the above analysis. For the first iteration, the high limit (l_H) is $\widetilde{S}_k$, and the low limit (l_L) is $\overline{S}_k$.

ii Introduce the emission rate, $S_k(try)$ (the arithmetic mean of the high limit and the low limit) into Eq. (4.3), and then get its numerical result using the 4th-order Runge-Kutta method with fixed time step. It should be noted that the initial calculated particle number concentration at time t is measured data $N_k(t)$ and the final calculated particle number concentration using the numerical method is $\overline{\overline{N_k}}\,(t+\Delta t)$.

iii Compare the calculated particle number concentration $\overline{\overline{N_k}}\,(t+\Delta t)$ and the measured concentration $N_k(t+\Delta t)$ with

$$\frac{\sum_{k=1}^{MAX}|\overline{\overline{N_k}}(t+\Delta t) - N_k(t+\Delta t)|}{\sum_{K=1}^{MAX}N_k(t+\Delta t)} < 10^{-10} \tag{4.4}$$

and if

(1) $N_k(t+\Delta t) - \overline{\overline{N_k}}\,(t+\Delta t) > 0$, the real emission rate is in the range $((l_H + l_L)/2, l_H)$, and if

(2) $N_k(t + \Delta t) - \overline{\overline{N_k}}(t + \Delta t) < 0$, the real emission rate is in the range $(l_L, (l_H + l_L)/2)$; after the high and low limit are fixed, the iteration goes to step (2);

(3) when the requirement of Eq. (4.4) is met, the real emission rate is found,

$S_k(real) = (l_H + l_L)/2$, and the iteration is stopped.

Instead of seeking the time-varying size-resolved emission rate, Géhin et al. [68] proposed a mathematical model to get the time-averaged size-resolved emission rate (mean equivalent emission rate) based on the particle population balance equation. It needs less time to produce estimated results, but it is limited to cases where the emission rate is not changed largely during emitting. In this model, if the particle number concentrations at each section at the beginning and at the end of the source emission are obtained in advance by measuring, then the mean equivalent emission rate ($\overline{s_k}$) is represented,

$$\frac{\partial \overline{s_k}}{\partial \log(d)} = \frac{V}{t_2 - t_1} \times \left(\left(N_{k,t_2} - N_{k,t_1} \right) + \lambda \times \int_{t_1}^{t_2} N_{k,t} dt \right) \qquad (4.5)$$

Here, d is the particle diameter at section k, V is the chamber volume, N_{k,t_1} and N_{k,t_2} is the particle population balance intensity at the beginning time t_1 and the end time t_2, and λ is the particle loss rate due to deposition and ventilation. It is noted here that the effect of coagulation and condensation on particle number size distribution is included in the equivalent emission rate, which avoids the integral of coagulation and condensation terms which is very difficult to solve in mathematics, but leads to a discrepancy between the real time-averaged emission rate and the equivalent emission rate, especially in the case where the coagulation and condensation cannot be ignored. This model has been used to identify the source emission rate for the human activities indoors.

4.4.1.2 Size-Unresolved Model

Although the above size-resolved model has an ability to produce the details of particle emission rate in the size space and time space with high resolution, it is limited in efficiency due to the complicated calculation. In the present studies on identifying and characterizing aerosol sources, an alternative method is also widely used which estimates the total mass concentration (or total number concentration) rather than size-resolved quantities [32,33,63,69]. In this kind of model, the construction of the model equation is based on the mass conservation or the total number conservation, and the coagulation process is usually ignored, and thus it is limited in the case where particle number concentration is not high

($<10^4/cm^3$) [70]. Although the mass-dependent size-unresolved model is not recommended to be used in the identification of nanoparticle aerosol sources, it is also shown in this section since it has been widely used in the estimation of source indoors.

By considering the transfer of aerosols between indoors and outdoors, the size-unresolved model usually takes the following form [32,63]:

$$\frac{dC_{in}(t)}{dt} = a\alpha PC_{out}(t) - (\alpha + k)C_{in}(t) + \frac{S(t)}{V} \tag{4.6}$$

Here, C_{in} and C_{out} are the indoor and outdoor particle concentrations (mass or total number), P is the penetration efficiency, α is the air exchange rate, k is the deposition rate, S is the indoor particle generation rate, t is time, and V is the efficient volume of the house.

In the model of Reference [63], the particle concentration C is considered to be the mass concentration, and the emission strength obtained is the time-averaged mass concentration. In this model, the mass balance equation is integrated from time t = 0 to T, and then the averaged emission is obtained by dividing T, where T is an arbitrary time greater than the source emission time t_s. Due to the fact that the emission happens within time t_s rather than T, the final emission rate needs further to be modified based on the ratio of T and t_s.

Using Eq. (4.6), He et al. identified the indoor source due to human activities and laser printer release [32,33]. In their studies, the particle number concentration rather than the mass concentration was identified and characterized. In order to obtain the ideal averaged source strength, they used the time at the peak concentration instead of the arbitrary time to integrate Eq. (4.6), and thus the time-averaged source emission is related to the initial and peak number concentrations, and the time difference between initial and peak concentration.

In the investigation on the ultrafine particle emissions from laser printers using emission test chambers at Fraunhofer Wilhelm-Klauditz-Institute, Germany, Schripp et al. [69] used a similar equation to Eq. (4.6) to identify the particle unit specific emission rate (SER) . In this model, the time-resolved rather than time-averaged unit specific emission was obtained by considering particle loss due to deposition and ventilation. In the case only the release from the chamber to outside is considered, the evolution of total particle number concentration at the chamber takes the following equation:

$$\frac{dC_{dp}}{dt} = L \cdot SER(t) - (\alpha + k) \cdot C_{dp}(t) \tag{4.7}$$

Here, C_{dp} represents the total particle number concentration in the chamber, L is the loading factor (1 printer m^{-3}), α is the air exchange rate, and k is the deposition rate. By integrating Eq. (4.7) with the assumption that SER(t) is a constant at the integral time Δt, the equation which extracts SER(t) from the measurement is:

$$SER(t) = \frac{C_{dp}(t) - C_{dp}(t - \Delta t) \cdot e^{-k \cdot \Delta t}}{L} \qquad (4.8)$$

Eq. (4.8) can be used to produce the time-resolved emission rate along with the measured particle size distribution. Similar to the above time-averaged models, this model is still limited in the case where particle number concentration is not high, usually less than $10^4/cm^3$, because coagulation process is ignored. In fact, if the coagulation process is added into Eqs. (4.7) and (4.8), the integral is not easy to achieve due to the complicated mathematical form of coagulation kernel. The accuracy of this model is strongly dependent on the time resolution of measured data which is similar to the size-resolved model.

4.5 MODELLING OF THE EVOLUTION OF A NANOPARTICLE AEROSOL IN A SIMULATED WORKPLACE (ASBACH, RATING, KUHLBUSCH)

Exposure to engineered nanoparticles in the workplace has thus far mainly been assessed by using stationary measurement equipment [1,15], such as mobility particle sizers, particle counters or surface area monitors. Some papers have recently been published that describe mapping of submicron and nanoscale particles using mobile equipment in order to obtain higher spatial resolution of the measurements [71−73]. Only a few personal sampling devices exist that are able to determine personal exposure specifically to nanoparticles. All the aforementioned methods have in common that they can only deliver results with limited or no temporal and/or spatial resolution. Computational simulations of particle dispersion in a workplace, however, can provide information on temporal and spatial variations of aerosol or particle properties [29]. Furthermore, modelling can enable the study of the effects of single particle dynamic processes, such as coagulation or Brownian diffusion on the variation of aerosol and particle parameters. Another advantage of modelling is that boundary conditions, such as wall temperatures or ventilation scenarios, can be varied in order to study their effect on particle dispersion and conse-quently possible exposure to particles in a workplace. One major advantage of such simulations is that worst case scenarios can be assumed and their effects studied in detail, eventually leading to enhanced risk management procedures. In principle any workplace can be simulated, but the more complex the simulated geometry, the higher

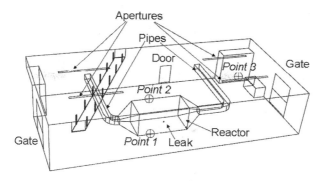

FIGURE 4.5 Modelled geometry, including a reactor with pipes; a leak is assumed to emit particles from the reactor.

is the tendency of the simulations to not converge. In the following, an example, developed in the German project NanoCare [74], is given for the dispersion of nanoparticles from a leak in a reactor. A virtual workplace environment was assumed for this scenario. The simulated geometry is shown in Figure 4.5 and includes a reactor with upstream and downstream piping. Nanoparticles are assumed to leak from the reactor into the workplace, which is naturally ventilated through doors and gates as well as apertures in the ceiling, as commonly found in industrial settings. In addition the geometry includes several pieces of furniture as flow obstacles.

Airflows were simulated using the CFD software FLUENT, whereas the Fine Particle Model (FPM) was used to include aerosol dynamic effects on particle dispersion. The FPM allows for a detailed analysis of the parameters affecting particle dispersion and aerosol dynamic processes. Three exemplary points were chosen in the modelled workplace for further analysis. These locations were evaluated 1 m and 1.5 m above ground, mimicking the breathing height of a sitting and standing worker, respectively. For the sake of clarity, only results for 1 m above ground are shown here. In a basic case, all temperatures in the room were assumed to be 300 K; gates and doors have an open slit of 10 cm height at the bottom. Apertures in the ceiling were open for ventilation. It was assumed that the reactor is operated at a slight overpressure of 50 Pa. The size distribution emitted through the leak was assumed to be lognormal with a median diameter of 50 nm and a concentration of 10^{16} m^{-3}. Figure 4.6 shows the distribution of the flow in the modelled workplace (left) as well as the change of the mean particle size (right). As can be seen in the figure, the near field of the leak was simulated as a nest with a very high number of grid cells. The simulation procedure was an iterative process. Initially the dispersion of the flow and particles was simulated with rather large grid

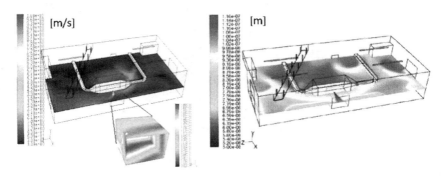

FIGURE 4.6   Flow distribution in the modelled workplace (left) and change of mean particle size (right) in the basic case; 1 m above ground.

cells in the entire workplace. After the first run, all relevant data from the interface to the nest were imported into the simulation file for the nest and repeated with the high resolution. After convergence of the nest simulations, the interface data were again imported to the larger geometry file and used for a refined simulation of the dispersion in the entire workplace.

The right contour plot in Figure 4.6 shows that the particle size changes during dispersion due to coagulation, because of the very high particle concentrations in the reactor. It can be seen that the changes are small in the vicinity of the leak but get noticeably higher at distances farther away, because these particles had more time to coagulate.

Parameters affecting particle size distribution and dispersion, including emitted size distributions, ventilation scenarios, background aerosol, and temperatures of floor, reactor wall and leak flow were systematically varied. A number of variations, listed in Table 4.1, were simulated in order to study the effect of different processes on particle dispersion and size distribution.

The effect of the aerosol temperature leaking from the reactor on the particle dispersion in the near field is shown in Figure 4.7. With increasing temperature, the flow drifts upwards, due to temperature-induced buoyancy effects. It is obvious that such differences in the near field can affect the dispersion in the room and hence the consequent worker exposure. The resulting size distributions in points 3 in 1 m above ground are shown in Figure 4.8. The effect on the size distribution is highest in point 1, but is getting lower for the points farther away from the leak. The higher deviation in point 1 is caused by the fact that the leak occurs approximately 1 m above ground, i.e., the same height where the three points were evaluated. While at 300 K the aerosol disperses more or less horizontally (see Figure 4.7), it drifts

TABLE 4.1 Variations of Simulated Parameters from the Basic Case

Pos.	Variation	Values
1	Flow	Right gate fully open
2	Emitted number concentration	$10^{14}/m^3$
3	Emitted particle diameter	20 nm or 90 nm
4	Temperature of leak flow	300 K, 350 K or 450 K
5	Wall temperature of reactor	350 K
6	Floor temperature	290 K
7	Background aerosol	200 nm, $5*10^9/m^3$, $\sigma_g = 2,0$ 200 nm, $2*10^{10}/m^3$, $\sigma_g = 2,0$
8	Flow velocity through door and gate apertures	2 cm/s

vertically at the higher temperature. As a result, the highest concentrations are found higher above ground at higher temperature. Figure 4.7 also shows that the temperature of the released aerosol has a much lower effect on the particle size distribution in the far field (points 2 and 3).

Another exemplary case for the change of particle size distributions in points 1 to 3 is shown in Figure 4.9, where the surface temperature of the

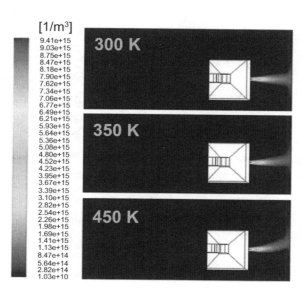

FIGURE 4.7 Effect of leak flow temperature on particle dispersion in the near field.

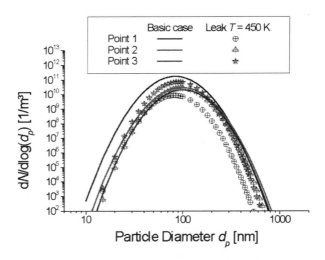

FIGURE 4.8 Change of particle size distribution at three points (see Figure 4.5) when the released aerosol flow temperature was 300 K (basic case) and 450 K.

reactor walls was increased from 300 K (basic case) to 350 K (case 5 in Table 4.1). It can be seen that the effect of a higher reactor wall temperature is opposite to the effect of higher aerosol temperature, i.e., it only negligibly affects the size distribution in point 1, where the size distribution still mainly represents the diluted emitted size distribution. The effect on size distributions is much more pronounced in points 2 and 3,

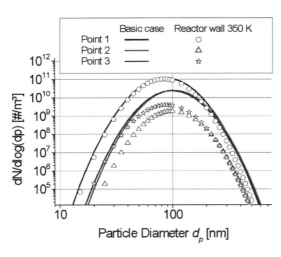

FIGURE 4.9 Change of particle size distribution at three points (see Figure 4.5) when the reactor wall temperature was increased from 300 K (basic case) to 350 K.

because particles have travelled over a longer distance and time before they reach these locations, and therefore processes leading to slower changes, such as coagulation and thermophoresis, are more effective. The effect on particle concentration is obviously higher in point 2 compared with point 3. This is caused by the temperature gradient between the warmer reactor wall (350 K) compared with the surrounding air (300 K), which leads to a thermophoretic repulsion of particles away from the reactor.

Similar analyses were conducted for all variations listed in Table 4.1. It was shown that especially temperature changes have significant impact on the particle dispersion and deposition in the modelled workplace, because of thermophoresis and buoyancy effects. In contrast to other studies in the literature [29], the inclusion of a background aerosol did not show enhanced coagulation. The reason may be the assumed rather low background concentration compared with the concentration emitted through the leak.

The results show that CFD simulations can be a very efficient tool for studying dispersion and changes of particles during airborne transport in workplaces. Modelling can furthermore help in identifying hotspots. Hence such modelling can also assist exposure measurement or monitoring by providing guidelines for a suitable positioning of measurement instruments and monitors.

Acknowledgement

This work was funded by the German Federal Ministry for Education and Research (BMBF) as part of the NanoCare project.

4.6 DOSE ASSESSMENT (LIDÉN)

4.6.1 The Human Respiratory Tract

The entrance to the airways consists of two paths, the nose and the mouth. The part of the airways outside the thorax is termed the head airways or the extrathoracic region. Along the nasal path on the way to the larynx, the air passes the nostrils, the anterior and posterior parts of the nose and the nasopharynx. If inhaled through the mouth, along the way to the larynx, the air passes the lips, the oral cavity and the oropharynx. In the head airways the inhaled air is heated to approximately 37°C and humidified to approximately 99% relative humidity. The anterior and posterior parts of the nose, the back of the oral cavity and the pharynx act as an effective filter for large (supermicron) particles and a less good filter for ultrafine particles (nanoparticles). Clearance of deposited particles is fast, on the scale of minutes, either out of the nostrils or backwards towards the esophagus and swallowed into the gastrointestinal tract.

The tracheo-bronchial airways consist of the trachea, which divides into the main bronchi and approximately 16 further divisions (generations) until air sacs (alveoli) occur in the wall of the terminal bronchioles. These airways are coated with cilia and covered by a mucus. The mucociliary escalator transports particles upwards to be swallowed. Clearance of deposited particles is fast, on the order of days, but the additional existence of a slower mode is not excluded. The trachea, bronchi and bronchioles are also termed the (proximal and distal) tracheo-bronchial region.

Further down into the lung, the terminal bronchioles are replaced with respiratory bronchioles with the coverage of the walls by alveoli increasing for each generation. This region is termed the gas-exchange region. From the alveoli oxygen is diffusing through the cell membranes into the bloodstream and carbon dioxide diffuses in the opposite direction. An alveolus is approximately 250 μm in diameter and a human lung has approximately 700 million alveoli, which add up to approximately 70 m^2. The surface of the alveoli is covered by a thin layer of a liquid (lung-lining liquid) that reduces the surface tension, otherwise we would not have been able to expand the size of the lung (and the alveoli) upon inhaling. Many types of deposited particles undergo partial (or possibly complete) dissolution and subsequent absorption by adjacent cells. Macrophages in the alveoli search for deposited particles and phagocytes them if possible. Other clearance routes are to the lymphatic system and the bloodstream. Clearance from the gas-exchange region is much slower and is of the order of months or years, depending on substance. At the end of their life the macrophages move to the beginning of the mucociliary escalator and are finally deposited into the gastrointestinal tract.

Deposition of particles in the extrathoracic and the tracheo-bronchial regions acts as a defensive filter for the deepest region of the lungs, the alveoli, where only a thin membrane of approximately 1 μm separates the air from the blood.

The extrathoracic and the tracheo-bronchial regions are more or less emptied and filled with fresh inhaled air with each breath. The gas-exchange region contains a residual volume of air that cannot be exhaled during exhalation. During inhalation the incoming air is mixed with this residual air. This leads to much longer residence times for airborne particles in the gas-exchange region (and hence also higher probabilities of deposition for the particles that manage to reach this region) than in the regions upstream.

4.6.2 Physical Deposition Mechanisms

Once a particle is inhaled it is carried by the inhaled air through the various compartments of the respiratory tract. Deposition may occur both during inhalation as well as during exhalation. There are four mechanical mechanisms and one electrical that may cause a particle to deposit, viz., impaction, sedimentation, interception, diffusion and electrical image charge.

Impaction occurs when a particle carried by an air stream encounters an object partially blocking the path, for example, the next bronchial airway generation. Due to its inertia, the particle will deviate from the streamlines and the deviation increases with particle size, density and speed. The relevant measurement of the particle size for impaction is termed the *aerodynamic (equivalent) diameter*. Impaction occurs mainly in the extra-thoracic and the tracheo-bronchial regions.

Sedimentation is caused by the gravitation and occurs mainly in horizontal small-sized airways with relatively high residence times. Also here *aerodynamic diameter* is the relevant metric of the particle size.

Interception is caused by the edges of the particle touching and henceforth sticking to the internal walls of the respiratory tract, i.e., it is no longer airborne. This is mainly important for a special type of particles which have one dimension much longer than the others, and for which the deposition by the other mechanisms does not depend on this much larger dimension, namely so-called fibers and ropes. The relevant metric of the particle size for interception can be termed the *geometric (equivalent) diameter*, and for fibers/ropes the length can be used.

Diffusion is caused by the random movement of the air molecules continuously hitting the particles from all sides. The movement thus imparted to a particle decreases with increasing particle size. The relevant metric of the particle size for diffusion is termed the *thermodynamic (equivalent) diameter*.

The meaning of the term *equivalent* diameter is that the particle under consideration has the identical property (relevant to the specific equivalent diameter chosen) as a spherical particle, i.e., a particle having a uniquely defined size. Only for aerodynamic diameters is the density of the spherical particle specified, viz., 1 g/cm^3. Any given particle may therefore be characterized with three different equivalent diameters for the four mechanical deposition/separation mechanisms.

Deposition mechanisms measured according to aerodynamic and geometric diameters increase in efficiency with particle size, and for smaller particles the deposition is higher further down into the respiratory tract. Deposition mechanisms measured according to thermodynamic diameter decrease in efficiency with particle size, and for larger particles the deposition is higher further down into the respiratory tract.

Electrical image forces appear as the internal surfaces of the airways and alveoli are covered with an ion-containing fluid. Electrically, these surfaces will act as if they had an identical charge of opposite sign as the particle charge behind the external side of the surface. This will attract the particle towards the surface. For this to be important, the particle needs to carry high charges, which in most cases also only occur for fibers and ropes [75].

4.6.3 Inhaled Particles

The efficiency by which airborne particles are inhaled depends on their aerodynamic size. However, for nanoparticles this efficiency is effectively 100% and for respirable particles (up to an aerodynamic diameter of 4 μm) it drops slightly to approximately 90%.

Inhaled particles may change their size upon inhalation by condensation of water vapor (due to the high relative humidity) or evaporation of organics (due to the higher temperature inside the body relative to the workplace or ambient environment, the saturation vapor pressure will be increased for the compounds of the particle). As most engineered and manufactured nanoparticles are inorganic (or made from almost pure carbon, like carbon nanotubes, CNT) with high boiling temperatures, the latter mechanism can be disregarded. A third mechanism is coagulation, but this is only fast enough for very high particle concentrations ($>1 \cdot 10^{12}$ m^{-3}), which only occur in welding plumes proper and similar locations extremely close to the generation source.

Particles containing soluble salts, oxides or other compounds consisting of ions, will grow in size upon inhalation due to water condensation. The amount of water condensed depends upon the number of ions each compound dissolves into, and the ratios of molecular weight and dry density to the corresponding values for water. For particles less than 400 nm the condensation of water is so fast that it can be considered to be instantaneous, whereas for larger particles it is a slower process that is not finished until the particles have travelled a significant distance into the trachea and hence is much more difficult to account for.

Compared to non-soluble particles, soluble particles of the same initial size that deposit according to their aerodynamic size will have a higher likelihood of deposition and/or deposit higher up in the respiratory tract. On the other hand, soluble particles that deposit according to their thermodynamic size will be less likely to deposit and/or deposit further down in the respiratory tract than non-soluble particles of the same size as the initial sizes of the soluble particles.

4.6.4 The Basis of the Knowledge on the Probability of Deposition of Airborne Particles

Our knowledge of the deposition of airborne particles in the human respiratory tract originates from both experiments and numerical calculations.

Experiments to determine the regional deposition, i.e., the deposition for different compartments of the respiratory tract, have mainly been carried out with monodisperse particles labelled with some radioactive material [76] but also with bolus deposition [77]. Experiments have also been carried out with casts of the first generations of the bronchi. Presently, with the widespread availability of particle size spectrometers, the total deposition has been measured in several environments, using polydisperse test aerosols, where previously these kind of experiments would not have been conceived as possible [78].

Numerical calculations use a variety of models for the respiratory tract, for example, concerning a deterministic or random sequence of diminishing size of each new generation of bronchi, fluid dynamics, deposition mechanisms, etc. (see 79 and 80).

4.6.5 Factors Influencing the Rate/Probability of Deposition

Major external determinants are the inhaled volume and the worker's breathing pattern. With increased workload, larger volumes are inhaled/exhaled during shorter periods of time. Per unit time, this gives higher air speeds and shorter residence times, which in turn increases deposition by impaction but reduces deposition by sedimentation and diffusion. Additionally, the total amount of the inhaled air is increased, which increases the total amount deposited in the respiratory tract. The amount of air inhaled per minute increases substantially when the work carried out changes from sitting to light, moderate and high workload. Average values for Caucasian males for these workloads are in the range 9 L to 50 L, and for female Caucasians 3–5 L less.

The body size of a worker has a small but definite effect. What is of main interest is not the total size/weight (which, however, will influence the amount of air inhaled), but the diameters and lengths of the airways. A larger body size will cause decreases in the deposition by impaction and sedimentation in all regions and by diffusion in the tracheo-bronchial and gas-exchange regions, whereas it will cause an increase in the deposition by diffusion in the extrathoracic region.

Most people inhale practically all air through the nose except during heavy work, when a substantial fraction is also inhaled by the mouth. Some people inhale partially through the mouth whatever the workload. The main difference related to particle deposition is that the nasal pathway is a much better filter for small and large particles than the oral pathway, and therefore the deposition in the tracheo-bronchial and gas-exchange regions is higher for oral breathing.

4.6.6 Subsequent Fate of Deposited Nanoparticles

Nanoparticles that deposit in the anterior part of the nose may be transported to the site of the olfactory nerve, and then move along this nerve up to the olfactory bulb inside the brain. This has been demonstrated for rats [81] and non-human primates [82]. The question on the extent to which material translocated to the olfactory bulb may be further distributed inside the brain is still unresolved.

Ultrafine particles and nanoparticles deposited in the gas-exchange region are covered by a layer of macro molecules (proteins and lipids) from the lung lining fluid. This causes the size of the combined particle and surface layer of macro molecules to grow. Other clearance routes are to the lymphatic vessels or the bloodstream. Some particles can be translocated into the lung lining cells.

4.6.7 Sampling Conventions for Deposited Aerosol Fraction

Sampling/monitoring related to occupational health should be health-related. If a certain (negative) health effect is associated with particles depositing in a specific region/compartment of the human respiratory tract, it is only relevant to sample/monitor the size fraction consisting of all airborne particles at the workplace depositing in the region/compartment in question. Such *sampling conventions* have been specified by EN ISO13138 [83]. A sampling convention for a specific compartment is a particle size-selective sampling efficiency curve that samplers/monitors for airborne particles depositing in the compartment in question should try to emulate.

The concept of sampling conventions stems from the sampling conventions for respirable particles defined by BMRC [84], the Johannesburg convention [85] and the (US) Atomic Energy Commission [86], originally for the size fraction of airborne coal dust responsible for coal workers' pneumoconiosis, i.e., the aerosol fraction depositing deepest into the lung. The sampling convention for the respirable fraction was designed to cut away the large super-micron particles that could by no means reach deep into the lung, by using the aerosol technology available in the fifties and sixties.

The model of the deposition of airborne particles in the human respiratory tract by the International Commission for Radiological Protection (ICRP) is widely used for dose estimations [76]. The ICRP model can be bought as a computer program running under MS-DOS, but the model can (easily) be programmed in any computer language as the ICRP report gives a detailed description and explicitly presents all equations. In order to use the model for a specific worker in a specific work situation, one needs to specify the body size, the breathing pattern and the particle size distribution s/he is exposed to. In practical life, the physiological parameters can be difficult to determine and hence average values given by the ICRP often are used based on the workload and sex of the worker.

CEN and ISO have published a common standard on sampling conventions for the aerosol fractions that deposit in the individual compartments of the respiratory tract [87]. In this standard the compartmentalization of ICRP is used, viz., the head airway region is split into two compartments, *ET1* and *ET2*, and the tracheo-bronchial region is also split into two compartments, *BB* and *bb*. ET1 consists of the anterior part of the nose and ET2 consists of the rest of the extra-thoracic region. BB consists of the trachea and the bronchi, whereas bb consists of the bronchioles and terminal bronchioles. The ICRP model calculates the deposition efficiency for all five compartments for selected workload (breathing pattern), sex and breathing mode as a function of initial particle size, i.e., before inhalation, separately for particles deposited by aerodynamic forces and by diffusion. However, the effect of sex is small and can be disregarded. For each particle size (aerodynamic *and* thermodynamic sizes, respectively), the deposition efficiency for a specific compartment is the ratio of the number of particles depositing in the compartment according to the corresponding mechanism(s) to the number of particles being inhaled. It can therefore be regarded as the probability that a particle of a given size will deposit in the compartment in question. The ICRP model does not account for deposition by electrostatic charge or by interception.

Based on the predicted deposition efficiencies for three workloads, both sexes and two breathing modes (normal and oral breathing), the average deposition efficiency curve per compartment is defined as a sampling convention. Each sampling convention is given both for deposition based on diffusion (thermodynamic diameter) and impaction/sedimentation (aerodynamic diameter), and they are functions of the respective (equivalent) diameter.

Sampling conventions for five compartments in both thermodynamic and aerodynamic diameters comprise ten different sampling conventions. However, three sampling conventions can be expressed as a multiple of another sampling convention. Therefore, there are seven independent sampling conventions. Each is given as a mathematical function with a few parameters. Table 4.2 presents the mathematical representations and parameters for the independent sampling conventions and Table 4.3 shows how to calculate the dependent sampling conventions. Bartley and

TABLE 4.2 Deposition Sampling Conventions Relative to Inhaled Aerosol Represented in Mathematical Functions of Aerodynamic or Thermodynamic Diameter

Deposition Regimen	Sampling Convention	Mathematical Representation	c_1 [–]	c_2 [–]	d_c [nm]	σ [–]
Thermodynamic	$D_{tET1}(d_{th})$	$c_1\left(\dfrac{d_{th}}{d_c}\right)^{c_2}$	0.50	-0.7	0.55	N/A
	$D_{tbb}(d_{th})$	$\left(c_1 + c_2\left[\dfrac{d_{th}-d_c}{d_c}\right]^2\right)\exp\left(-\dfrac{1}{2}\left(\dfrac{\ln d_{th}/d_c}{\ln\sigma}\right)^2\right)$	0.31	0.0202	4.1	2.57
	$D_{tGE}(d_{th})$	$\left(c_1 + c_2\left[\dfrac{d_{th}-d_c}{d_c}\right]^2\right)\exp\left(-\dfrac{1}{2}\left(\dfrac{\ln d_{th}/d_c}{\ln\sigma}\right)^2\right)$	0.48	0.0225	15	2.54
Aerodynamic	$D_{aET1}(d_{ae})$	$c_1\Phi\left(\dfrac{\ln(d_{ae}/d_c)}{\ln\sigma}\right)$	0.325	N/A	$2.7\cdot10^3$	2.5
	$D_{aBB}(d_{ae})$	$c_1\exp\left(-\dfrac{1}{2}\left(\dfrac{\ln d_{ae}/d_c}{\ln\sigma}\right)^2\right)$	0.12	N/A	$5.2\cdot10^3$	2.0
	$D_{abb}(d_{ae})$	$c_1\exp\left(-\dfrac{1}{2}\left(\dfrac{\ln d_{ae}/d_c}{\ln\sigma}\right)^2\right)$	0.05	N/A	$3.4\cdot10^3$	2.0
	$D_{aGE}(d_{ae})$	$c_1\exp\left(-\dfrac{1}{2}\left(\dfrac{\ln d_{ae}/d_c}{\ln\sigma}\right)^2\right)$	0.19	N/A	$2.3\cdot10^3$	2.0

Note 1. $\Phi(x)$ is the cumulative normal distribution function.

Note 2. The parameters c_1 and c_2 for $D_{tET1}(d_{th})$ and c_2 for $D_{tbb}(d_{th})$ and $D_{tGE}(d_{th})$, respectively, have other numerical values than in EN ISO13138:2012, because they are differently defined, but the calculated values of respective sampling convention are identical.

TABLE 4.3 Dependent Deposition Sampling Conventions

Deposition Regimen	Sampling Convention	Mathematical Representation
Thermodynamic	$D_{tET2}(d_{th})$	$D_{tET2}(d_{th}) = 1.67\,D_{tET1}(d_{th})$
	$D_{tBB}(d_{th})$	$D_{tBB}(d_{th}) = 0.70\,D_{tET1}(d_{th})$
Aerodynamic	$D_{aET2}(d_{ae})$	$D_{aET2}(d_{ae}) = 2.08\,D_{aET1}(d_{ae})$

Vincent [88] and Bartley [89] give information on how the sampling conventions were designed.

Figure 4.10 presents the seven independent sampling conventions for deposition based on diffusion and aerodynamics, respectively, in the corresponding compartments of the respiratory tract. For each compartment, the deposition due to diffusion is negligible when aerodynamic deposition dominates, and vice versa. The vertical bar is located in the size range where, for each compartment, the sampling convention as a function of thermodynamic diameter approximately equals that of

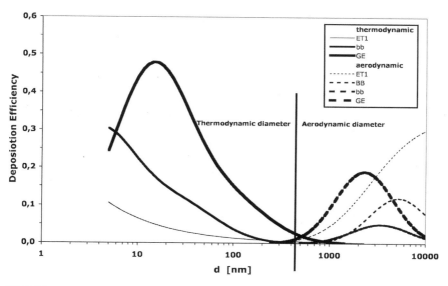

FIGURE 4.10 Sampling conventions for deposition efficiencies in the ET1, BB, bb and GE compartments as a function of thermodynamic diameter and aerodynamic diameter, respectively. Not shown are the dependent sampling conventions for BB thermodynamic deposition and ET2 for both thermodynamic and aerodynamic deposition. The vertical bar is located in the size range where, for each compartment, the sampling convention as a function of thermodynamic diameter approximately equals that of aerodynamic diameter.

aerodynamic diameter. Depending on the compartment, this occurs in the range 370–750 nm, and the value of the two corresponding sampling conventions is much less than 0.01 except for the gas-exchange region where the value is 0.02. For all practical purposes any cross-contamination between aerodynamic and diffusive deposition can be disregarded.

It is expected that the sampling conventions can be used in three ways: 1) to develop a sampler with seven independent collection substrates, each of which has a deposition efficiency equal to one of the independent deposition sampling conventions; 2) to apply to measured size distributions as functions of both thermodynamic and aerodynamic particle size distributions in order to estimate the deposited fraction in each compartment; or 3) to develop a sampler for one (or several) of the deposition sampling conventions.

The standard EN ISO13138:2012 describes how the information obtained by sampling or otherwise determining all fractions corresponding to the seven independent deposition sampling conventions can be used to determine also the fractions for the three workloads (sitting, light and heavy exercise) and two breathing patterns (normal and oral) for which the ICRP report presents the deposited fractions. This is carried out by applying a set of correction factors that depend on the fraction of interest. More information is found in the work by Bartley and Vincent [88].

In order to estimate the dose of a specific worker in a specific compartment of his/her respiratory tract, one first needs to determine (or estimate) the worker's breathing pattern (normal or oral) and volume inhaled per minute (Q) and then estimate which workload (sitting, light or heavy work) best characterizes the worker's job. Based on the measured and corrected deposited fractional concentration for the specific compartment ($C_{dep,comp}$), the dose (m_{dose}) can be calculated as

$$m_{dose} = C_{dep,comp} t Q$$

where t is the length of the workday, or the time over which exposure occurs.

References

[1] Kuhlbusch TAJ, Asbach C, Fissan H, et al. Nanoparticle exposure at nanotechnology workplaces: a review. Part Fibre Toxicol 2011;8:22.
[2] Türk M, Lietzow R. Formation and stabilization of submicron particles via rapid expansion processes. J Supercrit Fluids 2008;45(3):346–55.

[3] Lu J, Minami K, Takami S, et al. Supercritical hydrothermal synthesis and in situ organic modification of indium tin oxide nanoparticles using continuous-flow reaction system. ACS Appl Mater Interfaces 2012;4(1):351—4.

[4] Demou E, Stark W, Hellweg S. Particle emission and exposure during nanoparticle synthesis in research laboratories. Ann Occup Hyg 2009;53(8):829—38.

[5] Demou E, Peter P, Hellweg S. Exposure to manufactured nanostructured particles in an industrial pilot plant. Ann Occup Hyg 2008;52(8):695—706.

[6] Park J, Kwak BK, Bae E, et al. Characterization of exposure to silver nanoparticles in a manufacturing facility. J Nanopart Res 2009;11(7):1705—12.

[7] Peters TM, Elzey S, Johnson R, et al. Airborne monitoring to distinguish engineered nanomaterials from incidental particles for environmental health and safety. J Occup Environ Hyg 2009;6(2):73—81.

[8] Evans DE, Ku BK, Birch ME, Dunn KH. Aerosol monitoring during carbon nanofiber production: mobile direct-reading sampling. Ann Occup Hyg 2010;54(5):514—31.

[9] Wake D, Mark D. Ultrafine aerosols in the workplace. Ann Occup Hyg 2002;46(1): 235—8.

[10] Holzinger K, Pohlmann G. Comparative investigations on selected physical and chemical characteristics of welding fumes emitted by different processes—first results. Copenhagen, Denmark, May 2005: International Conference: Health and Safety in Welding and Allied Processes; 2005.

[11] Wang J, Asbach C, Fissan H, et al. How can nanobiotechnology oversight advance science and industry: examples from environmental, health, and safety studies of nanoparticles (nano-EHS). J Nanopart Res 2011;13(4):1373—87.

[12] Maynard AD, Baron PA, Foley M, et al. Exposure to carbon nanotube material: aerosol release during the handling of unrefined single-walled carbon nanotube material. J Toxicol Environ Health Part A 2004;67(1):87—107.

[13] Kuhlbusch TAJ, Neumann S, Fissan H. Number size distribution, mass concentration, and particle composition of PM1, PM2.5, and PM10 in bag filling areas of carbon black production. J Occup Environ Hyg 2004;1(10):660—71.

[14] Tsai S, Ashter A, Ada E, et al. Airborne nanoparticle release associated with the compounding of nanocomposites using nanoalumina as fillers. Aerosol Air Qual Res 2008;8(2):160—77.

[15] Brouwer D, Duuren-Stuurman B, Berges M, et al. From workplace air measurement results toward estimates of exposure? Development of a strategy to assess exposure to manufactured nano-objects. J Nanopart Res 2009;11(8):1867—81.

[16] Ibaseta N, Biscans B. Ultrafine Aerosol Emission from the Free Fall of TiO_2 and SiO_2 Nanopowders. Kona 2007;25:190—204.

[17] Tsai C-J, Wu C-H, Leu M-L, et al. Dustiness test of nanopowders using a standard rotating drum with a modified sampling train. J Nanopart Res 2008;11(1):121—31.

[18] Stahlmecke B, Wagener S, Asbach C, et al. Investigation of airborne nanopowder agglomerate stability in an orifice under various differential pressure conditions. J Nanopart Res 2009;11(7):1625—35.

[19] Seipenbusch M, Rothenbacher S, Kirchhoff M, et al. Interparticle forces in silica nanoparticle agglomerates. J Nanopart Res 2009;12(6):2037—44.

[20] Bello D, Wardle BL, Yamamoto N, et al. Exposure to nanoscale particles and fibers during machining of hybrid advanced composites containing carbon nanotubes. J Nanopart Res 2008;11(1):231—49.

[21] Koponen IK, Jensen KA, Schneider T. Comparison of dust released from sanding conventional and nanoparticle-doped wall and wood coatings. J Exposure Sci Environ Epidemiol 2011;21:408—18.

[22] Vorbau M, Hillemann L, Stintz M. Method for the characterization of the abrasion induced nanoparticle release into air from surface coatings. J Aerosol Sci 2009;40(3): 209–17.
[23] Göhler D, Stintz M, Hillemann L, Vorbau M. Characterization of nanoparticle release from surface coatings by the simulation of a sanding process. Ann Occup Hyg 2010; 54(6):615–24.
[24] Wohlleben W, Brill S, Meier MW, et al. On the Lifecycle of Nanocomposites: comparing released fragments and their in-vivo hazards from three release mechanisms and four nanocomposites. Small 2011;7(16):2384–95.
[25] Messerer A, Niessner R. Thermophoretic deposition of soot aerosol particles under experimental conditions relevant for modern diesel engine exhaust gas systems. J Aerosol Sci 2003;34(8):1009–21.
[26] Lai AK, Nazaroff WW. Modelling indoor particle deposition from turbulent flow onto smooth surfaces. J Aerosol Sci 2000;31(4):463–76.
[27] Hinds WC. Aerosol Technology. New York: John Wiley & Sons; 1999.
[28] Friedlander SK. Smoke, dust and haze: Fundamentals of aerosol behavior. New York: Wiley-Interscience; 1977.
[29] Seipenbusch M, Binder A, Kasper G. Temporal evolution of nanoparticle aerosols in workplace exposure. Ann Occup Hyg 2008;52:707–16.
[30] Koivisto AJ, Yu M, Hämeri K, Seipenbusch M. Size resolved particle emission rates from an evolving indoor aerosol system. J Aerosol Sci 2012;47:58–69.
[31] Hussein T, Glytsos T, Ondráček J, et al. Particle size characterization and emission rates during indoor activities in a house. Atmos Environ 2006;40(23):4285–307.
[32] He C, Morawska L, Hitchins J, Gilbert D. Contribution from indoor sources to particle number and mass concentrations in residential houses. Atmos Environ 2004;38(21): 3405–15.
[33] He C, Morawska L, Taplin L. Particle emission characteristics of office printers. Environ Sci Technol 2007;41(17):6039–45.
[34] Hussein T, Hameri K, Heikkinen M, Kulmala M. Indoor and outdoor particle size characterization at a family house in Espoo–Finland. Atmos Environ 2005;39(20): 3697–709.
[35] Hussein T, Korhonen H, Herrmann E, et al. Emission rates due to indoor activities: indoor aerosol model development, evaluation, and applications. Aerosol Sci Technol 2005;39(11):1111–27.
[36] Friedlander SK. Smoke, dust and haze: fundamentals of aerosol dynamics. 2nd ed. Oxford University Press; 2000.
[37] Crowe CT. Multiphase flow handbook. 2nd ed. FL: CRC Press; 2006.
[38] Yu M, Lin J. Taylor-expansion moment method for agglomerate coagulation due to Brownian motion in the entire size regime. J Aerosol Sci 2009;40(6):549–62.
[39] Yu M, Lin J, Chan T. A new moment method for solving the coagulation equation for particles in brownian motion. Aerosol Sci Technol 2008;42(9):705–13.
[40] Lee KW, Chen H, Gieseke JA. Log-normally preserving size distribution for brownian coagulation in the free-molecule regime. Aerosol Sci Technol 1984;3(1):53–62.
[41] Hulburt HM, Katz S. Some problems in particle technology: A statistical mechanical formulation. Chem Eng Sci 1964;19(8):555–74.
[42] Frenklach M. Method of moments with interpolative closure. Chem Eng Sci 2002; 57(12):2229–39.
[43] McGraw R. Description of aerosol dynamics by the quadrature method of moments. Aerosol Sci Technol 1997;27(2):255–65.

[44] Marchisio DL, Vigil RD, Fox RO. Quadrature method of moments for aggregation-breakage processes. J Colloid Interface Sci 2003;258(2):322—34.

[45] Gelbard F, Seinfeld J. Simulation of multicomponent aerosol dynamics. J Colloid Interface Sci 1980;78(2):485—501.

[46] Talukdar SS, Swihart MT. Aerosol dynamics modelling of silicon nanoparticle formation during silane pyrolysis: a comparison of three solution methods. J Aerosol Sci 2004;35(7):889—908.

[47] Kostoglou M. Extended cell average technique for the solution of coagulation equation. J Colloid Interface Sci 2007;306(1):72—81.

[48] Landgrebe JD, Pratsinis SE. A discrete-sectional model for particulate production by gas-phase chemical reaction and aerosol coagulation in the free-molecular regime. J Colloid Interface Sci 1990;139(1):63—86.

[49] Kraft M. Modelling of Particulate Processes. KONA 2005;23:18—35.

[50] Morgan N, Wells C, Goodson M, et al. A new numerical approach for the simulation of the growth of inorganic nanoparticles. J Comput Phys 2006;211(2):638—58.

[51] Pratsinis S. Simultaneous nucleation, condensation, and coagulation in aerosol reactors. J Colloid Interface Sci 1988;124(2):416—27.

[52] Barrett JC, Jheeta JS. Improving the accuracy of the moments method for solving the aerosol general dynamic equation. J Aerosol Sci 1996;27(8):1135—42.

[53] Wang L-P, Wexler AS, Zhou Y. On the collision rate of small particles in isotropic turbulence. I. Zero-inertia case. Phys Fluids 1998;10(1):266—76.

[54] Yu M, Lin J. Nanoparticle-laden flows via moment method: A review. Int J Multiphase Flow 2010;36(2):144—51.

[55] Barthelmes G, Pratsinis S, Buggisch H. Particle size distributions and viscosity of suspensions undergoing shear-induced coagulation and fragmentation. Chem Eng Sci 2003;58(13):2893—902.

[56] Heine MC, Pratsinis SE. Brownian coagulation at high concentration. Langmuir 2007; 23(19):9882—90.

[57] Friedlander SK, Pui DYH. Emerging Issues in Nanoparticle Aerosol Science and Technology. J Nanopart Res 2004;6(2):313—20.

[58] Nerisson P, Simonin O, Ricciardi L, et al. Improved CFD transport and boundary conditions models for low-inertia particles. Comput Fluids 2011;40(1):79—91.

[59] Zaichik LI, Drobyshevsky NI, Filippov AS, et al. A diffusion-inertia model for predicting dispersion and deposition of low-inertia particles in turbulent flows. Int J Heat Mass Transfer 2010;53(1-3):154—62.

[60] Liao C-M, Chiang Y-H, Chio C-P. Assessing the airborne titanium dioxide nanoparticle-related exposure hazard at workplace. J Hazard Mater 2009;162(1):57—65.

[61] Lai ACK, Chen FZ. Comparison of a new Eulerian model with a modified Lagrangian approach for particle distribution and deposition indoors. Atmos Environ 2007;41(25): 5249—56.

[62] Tang Y, Guo B. Computational fluid dynamics simulation of aerosol transport and deposition. Front Environ Sci Eng China 2011;5(3):362—77.

[63] Ferro AR, Kopperud RJ, Hildemann LM. Source strengths for indoor human activities that resuspend particulate matter. Environ Sci Technol 2004;38(6):1759—64.

[64] Klepeis NE, Apte MG, Gundel LA, et al. Determining size-specific emission factors for environmental tobacco smoke particles. Aerosol Sci Technol 2003;37(10):780—90.

[65] Glytsos T, Ondráček J, Džumbová L, et al. Characterization of particulate matter concentrations during controlled indoor activities. Atmos Environ 2010; 44(12):1539—49.

[66] Koivisto AJ, Hussein T, Niemelä R, et al. Impact of particle emissions of new laser printers on modeled office room. Atmos Environ 2010;44(17):2140−6.

[67] Korhonen H, Lehtinen KEJ, Kulmala M. Multicomponent aerosol dynamics model UHMA: model development and validation. Atmos Chem Phys Discussions 2004; 4:471−506.

[68] Géhin E, Ramalho O, Kirchner S. Size distribution and emission rate measurement of fine and ultrafine particle from indoor human activities. Atmos Environ 2008;42(35): 8341−52.

[69] Schripp T, Wensing M, Uhde E, et al. Evaluation of ultrafine particle emissions from laser printers using emission test chambers. Environ Sci Technol 2008;42(12):4338−43.

[70] Hussein T, Hruška A, Dohányosová P, et al. Deposition rates on smooth surfaces and coagulation of aerosol particles inside a test chamber. Atmos Environ 2009;43(4): 905−14.

[71] Peters TM, Heitbrink WA, Evans DE, et al. The mapping of fine and ultrafine particle concentrations in an engine machining and assembly facility. Ann Occup Hyg 2006; 50(3):249−57.

[72] Evans DE, Heitbrink WA, Slavin TJ, Peters TM. Ultrafine and respirable particles in an automotive grey iron foundry. Ann Occup Hyg 2008;52(1):9−21.

[73] Heitbrink WA, Evans DE, Peters TM, Slavin TJ. Characterization and mapping of very fine particles in an engine machining and assembly facility. J Occup Environ Hyg 2007;4(5):341−51.

[74] Asbach C, Dahmann D, Kaminski H, et al. NanoCare — health related aspects of nanomaterials. In: Kuhlbusch TAJ, Krug H, Nau K, editors. Frankfurt: Dechema; 2009. p. 74−88.

[75] Yu CP. Theories of electrostatic lung deposition of inhaled aerosols. Ann Occup Hyg 1985;29(2):219−27.

[76] ICRP. Human Respiratory Tract Model for Radiological Protection, 1st Edition. Ann ICRP 1994;24(1-3):1−482.

[77] Kim CS, Hu SC. Regional deposition of inhaled particles in human lungs: comparison between men and women. J Appl Physiol 1998;84(6):1834−44.

[78] Löndahl J. Exerimental determination of the deposition of aerosol particles in the human respiratory tract. Doctoral dissertation. Sweden: Lund University; 2009.

[79] Werner H. Modelling inhaled particle deposition in the human lung—A review. J Aerosol Sci 2011;42(10):693−724.

[80] Wang C-S. A brief history of respiratory modelling. In: Ensor DS, editor. Aerosol Science and Technology: History and Reviews 2011. p. 391−409. Chapter 15.

[81] Elder A, Gelein R, Silva V, et al. Translocation of inhaled ultrafine manganese oxide particles to the central nervous system. Environ Health Perspect 2006;114(8):1172−8.

[82] Dorman DC, Struve MF, Marshall MW, et al. Tissue manganese concentrations in young male rhesus monkeys following subchronic manganese sulfate inhalation. Toxicol Sci 2006;92(1):201−10.

[83] International Organization for Standardization. ISO 13138:2012. Air quality—Sampling conventions for estimating ultrafine and fine aerosol particle deposition in the human respiratory system. Geneva, 2012. Switzerland.

[84] In: Minutes of a joint meeting of panels 1, 2 and 3 held on 4 March 1952. In: BMRC, editor. Recommendations of the BMRC Panels relating to Selective Sampling. London: British Medical Research Council; 1952.

[85] Orenstein AJ. Recommendations adopted by the pneumoconiosis conference. In: second pneumoconiosis conference 1959. p. 617−21. Johannesburg.

[86] Hatch T, Gross P. Pulmonary Deposition and Retention of Inhaled Aerosols. New York: Academic Press; 1964.

[87] Comité Européen de Normalisation. EN 13138:2012 Air quality—Sampling conventions for estimating ultrafine and fine aerosol particle deposition in the human respiratory system. Belgium, Brussels; 2012.

[88] Bartley DL, Vincent JH. Sampling conventions for estimating ultrafine and fine aerosol particle deposition in the human respiratory tract. Ann Occup Hyg 2011;55(7):696—709.

[89] Bartley DL. Do we really need SEVEN new aerosol particle sampling conventions? Ann Occup Hyg 2011;55(7):692—5.

Monitoring and Sampling Strategy for (Manufactured) Nano Objects, Agglomerates and Aggregates (NOAA): Potential Added Value of the NANODEVICE Project

Derk H. Brouwer [1], Göran Lidén [2],
Christof Asbach [3], Markus G.M. Berges [4],
Martie van Tongeren [5]

[1] TNO, Quality of Life, Research & Development, Zeist, The Netherlands
[2] Stockholm University, Stockholm, Sweden
[3] Institute of Energy and Environmental Technology (IUTA),
Air Quality & Sustainable Nanotechnology, Duisburg, Germany
[4] Institute of Occupational Safety and Health of the German Social Accident
Insurance (DGUV-IFA), Sankt Augustin, Germany
[5] Centre for Human Exposure Science (CHES), Institute of Occupational
Medicine, Edinburgh, United Kingdom

5.1 INTRODUCTION

The production of nanomaterials and nano-enabled products, being the first stages of the product value chain, will involve workers who are potentially exposed to (manufactured) nano-objects, agglomerates and aggregates (NOAA). To determine actual workplace exposure, studies have been conducted [1,2], however, the methods and strategies applied

Handbook of Nanosafety
http://dx.doi.org/10.1016/B978-0-12-416604-2.00005-6 **173**

to measure concentrations of NOAA in workplace air show substantial differences [3,4].

The aim of this chapter is to outline the major elements of a measurement strategy to assess exposure of workers to NOAA. After the general introduction the major principles of monitoring and sampling will be discussed, followed by considerations on measurement strategy. The last part of the chapter discusses the potential contribution to measurement strategies of the pre-prototype devices that have been developed in the NANODEVICE project.

In general, a measurement strategy for workplace exposure assessment can be described as a framework for the selection of relevant considerations associated with a measurement campaign, e.g., what substance will be measured, at which location, when, for how long, how many individuals to sample, how many samples per individual to collect and by what methods, in particular, what exposure metrics will be measured. It is agreed that these considerations should also be addressed in the case of NOAA [1,3]. The objectives of a particular measurement campaign can vary widely and may include exposure exploration and characterization, risk assessment, epidemiology, evaluation of the effectiveness of exposure control measures, testing/comparing of (new) measurement methods and compliance with any occupational exposure limit or benchmark level. The actual measurement strategy should be designed to be consistent with the study objectives. For example, exposure characterization studies that attempt to identify exposure pathways (transport processes of the contaminant from source to the receptor (the worker)) and exposure-modifying factors/determinants should ideally include both measurements at the source as well as at the individual level. Exposure assessments for use in compliance assessment, epidemiologic studies or risk management should ideally focus on the individual worker, using personal breathing zone samples collected over a full work shift. In contrast, studies of the efficacy of a control measure may be carried out using static measurements at or near the source.

For measuring aerosols, and more specifically manufactured (nano-objects, agglomerates and aggregates ((m)NOAA) aerosols, the exposure assessment community is presently faced with some specific challenges.

- Firstly, no consensus exists on the relevant aerosol sizes fractions. ISO Technical Committee 229 (TC229) and other TCs have adopted the term 'nano-objects, agglomerates and aggregates' (NOAA) to demonstrate the (health) relevance of not only nano-objects (i.e., objects with one, two or three dimensions below 100 nm), but also their aggregates and agglomerates. Evidence so far suggests that exposure in the workplace is predominantly to airborne aggregates and agglomerates of nano-objects [4]. As described in Chapter 4, the gas exchange area with diffusion-driven deposition (aerosol sizes up to about 400 nm thermodynamic diameter) would be a relevant cutoff point here.

- Secondly, there is no agreement on (health-) relevant exposure/dose metric. As a consequence, exposure should ideally be expressed using multiple metrics, e.g., size-resolved particle number concentration over time, surface area concentration, mass concentration. Surface area concentration and mass concentration can be estimated from size-resolved particle number concentration with assumptions on the spherical shape of the aerosol and the specific density of the aerosol. One complication is that both shape and density of the airborne particle may differ from its powder form.

- Thirdly, the size-resolved particle number concentration in workplace air will vary both temporally and spatially. Due to aerosol dynamics (homogeneous and heterogeneous nucleation condensation, coagulation and deposition loss), mixing and interaction with other sources and transport through different influencing environments (e.g., turbulence, thermal, electric and magnetic fields), the size distribution in the breathing zone of a person is likely to be different compared with the size distribution at the source of release or emission due to aerosol dynamics (homogeneous and heterogeneous nucleation and condensation, coagulation) and losses through deposition and diffusion onto surfaces. These processes will be affected by environmental conditions (e.g, turbulence, thermal, electric and magnetic fields) and presence of aerosols from other sources.

- Fourthly, in order to estimate exposure to airborne (m)NOAAs, it is critical that these aerosols are distinguished from particles from background sources. Currently, no commercially available real-time device exists that is able to differentiate (m)NOAA from other, natural or process-generated nano aerosols. Consequently, as well as real-time devices, measurement strategies for (m)NOAA generally also include aggregated sampling methods followed by off-line chemical or microscopy analysis to characterize (m)NOAA.

- Fifthly, most currently available samplers meet the conventions that mimic the size-dependent penetration efficiency into the respiratory tract, e.g., the inhalable fraction, the thoracic fraction and the respirable fraction. However, no devices or samplers are currently available that meet conventions that indicate the size-dependent (location and efficiency of) lung deposition. An additional sampling bias is that results of the exposure measurements may not be representative of the actual deposited dose. Hygroscopic particles will grow in the saturated air in the airways, and the deposition in the respiratory tract will depend on the 'wet size' instead of the commonly measured 'dry size' [5]. This is an issue for estimating dose of all airborne particles, not just for those in the nano-size range; however, it might be more critical for the latter subgroup. Bartley and Vincent and EN ISO 13138:2012 [6] have proposed new deposition conventions, which enable a more accurate calculation of the deposited dose in the respiratory tract of airborne particles, including ultrafine and nano particles. The deposition, both in terms of efficiency and location in the respiratory tract, is influenced by the breathing patterns and the anatomy of the airways. Breathing patterns, e.g., inhaled tidal volume, breathing frequency, duty cycle, (pause between inhalation and exhalation), residual expiratory volume and distribution between nasal and oral inhalation will depend on the physical work load required for the task (scenario-specificity). The anatomy of the respiratory tract is composed of individual factors. Recently, it has been explored whether so-called physiologic sampling pumps (PSPs) would mimic parts of the variation of the breathing patterns over time by adjusting their sampling rates in proportion to wearers' minute ventilation rates and consequently would reduce the inherent uncertainty in exposure (and risk) assessment compared to constant-flow sampling approaches [7]. However, this will only be valid for vapors and gases since the flow rate of a sampler pump will determine the aspiration efficiency of particulate aerosols in combination with the sampler geometry.

5.2 MEASUREMENT PRINCIPLES AND INSTRUMENTATION

An important element in developing a sampling strategy is the selection of appropriate measurement method(s). In the present section the general principles of monitoring and detection, and sampling, as well as the metrics will be discussed with emphasis on nano-sized aerosols. Moreover, most commonly used types of instruments or devices and sampling substrates will be addressed.

In general a distinction can be made between devices that monitor (online) a chemical substance or aerosol by 'near or quasi' real-time detection and devices that sample (time-aggregated) chemical substances or aerosols on a substrate, followed by off-line analysis. Devices of the first subgroup are usually calibrated (often using a standard aerosol) to give an estimate of a real-time concentration, whereas the devices of the second subgroup, with the results of the analysis, will give an estimate of the time-weighted average concentration over the measurement period.

5.2.1 Monitoring and Detection Principles

Optical detection techniques evaluate the intensity of light scattered by airborne particles to determine the particles' sizes and concentration. This method works well as long as the particles are larger than approximately half the wavelength of the light used by the instrument. Optical detection is not appropriate for particles with a diameter smaller than approximately 200—300 nm. Below this particle size other particle detection principles have to be employed.

In general, the methods can be distinguished into those that determine size-integrated particle quantities, e.g., the total particle number concentration, and those that measure particle-size-resolved concentrations. While the prior group of instruments is usually relatively simple, the latter type requires an additional step to separate or distinguish particle size.

5.2.1.1 Principles for Measuring Size-Integrated Concentrations

The most commonly used instruments for measuring particle number concentrations down to the low nanometer size range are condensation particle counters (CPC, also referred to as condensation nuclei counters or Aitken nuclei counters). In a CPC, the particles are exposed to an atmosphere of a supersaturated working fluid, usually butanol [8] or water [9]. The vapor condenses onto the particles' surfaces and causes the particles to grow to optically detectable sizes. Consequently the enlarged particles are detected by light scattering in a measurement cell. The measurement principle dates back to the late 19th century [10], when a condensation chamber was used to visually detect dust particles in the atmosphere [11]. In modern CPCs, particles at low concentrations are detected in a single particle counting mode, i.e., light scattering events of single particles are counted. Depending on the CPC model, the upper concentration limit of the single particle count mode is commonly between $10^4/cm^3$ and $10^6/cm^3$. At higher concentrations, some CPC models switch to a photometric mode, i.e., they measure the scattered light from an ensemble of particles in the measurement cell to determine the number of particles in the ensemble. Common CPCs offer time resolutions down to 1 s. While the accuracy of the single particle count mode is usually very good [12], the photometric mode is known to be less accurate. In the case where water vapor is used as the condensation material, the growth rate of hygroscopic or hydrophilic particles is significantly higher than for hydrophobic particles, which could result in a material dependence of the accuracy of the photometric measurement.

Depending on the CPC model, the lowest detectable particle size varies from 2.5 nm with ultrafine condensation particle counters (UCPC) to 10 nm with regular CPCs. Besides static, mains-operated CPCs, a small variety of battery-operated, handheld CPCs exist for which the lowest detectable size range is somewhat higher, at approximately 10–20 nm. A comprehensive overview of CPCs is given in McMurry [13].

Recently, unipolar diffusion charging, followed by a step to manipulate the particle size distribution and eventual downstream measurement of the particle-induced current(s), has been introduced as a method of measuring nano-scale particle number concentrations. Instruments that apply this technique, commonly referred to as diffusion chargers, are portable and powered by a battery. Two different versions of diffusion chargers exist for measuring airborne particle number concentrations. Both first use a unipolar diffusion charger to charge the particles to a known, particle-size-dependent charge level. In one version, the particle size distribution is subsequently mechanically manipulated by passing the particles through a stack of diffusion screens [14]. Small particles preferentially deposit in this diffusion screen stage, while the particle fraction that is not deposited in this first stage is captured by a downstream high-efficiency filter. Both the diffusion and the filter stages are connected to electrometers to simultaneously measure the currents made up from the electrical charges carried by the depositing particles. The particle number concentration and the mean particle size are inferred from the two independent current measurements and provided with a time resolution down to 1 s. The second type of diffusion charger instruments use an electrical manipulation of the size distribution instead of mechanical manipulation [15]. Downstream of the charger, the charged particles are passed through a low-efficiency electrostatic precipitator (ESP). A square-wave electric field is applied to the precipitator. The lower field strength of the square wave is chosen such that only ions are removed from the sample flow, while with the higher field strength, preferentially small, highly mobile particles are removed. Downstream of the ESP, the particles are captured on a filter and the two independent currents during low and high ESP voltage, respectively, are measured sequentially. Number concentration and mean particle size are then determined from the two current values with a time resolution down to 10 s. Diffusion chargers have an advantage over CPCs, as the latter instruments require fluids and hence need to be carefully maintained in a horizontal position. When mains-operated, diffusion chargers can be used for weeks without special attention and are therefore also suitable as permanent monitors. Recent studies have suggested that the number concentrations obtained using diffusion chargers are comparable with those from CPCs (i.e., within a range of ±30%) [16–18]. The accuracy of the mean particle size estimate was also reported to be around ±30%. It should be noted here that the calculation of the particle number concentration and mean size assume that the airborne particles are spherical (i.e., not agglomerates).

It has been shown that following unipolar diffusion charging, the resulting charge level of an aerosol in some cases proportional is to the fraction of the particle surface area concentration that would deposit in either the alveolar (gas-exchange) or the thoracic region of the human lungs. The lung deposition rates were estimated based on a model developed by the International Commission for Radiological Protection [19], assuming the breathing pattern of a 'reference worker' under light exercise. The reference worker was assumed to breathe only through the nose with 2200 cm^3 functional residual capacity, at a breathing rate of 15 breaths per minute and a ventilation rate of 1.3 m^3/h. The resulting deposited surface area concentration in the gas-exchange region is hence a dose-related metric for the particle surface area. This metric has been reported to be of higher health relevance than the particle number or mass concentration [20]; however, in most toxicological studies the surface area is based on BET analysis of the powder and not of the aerosol. Initially, instruments developed for estimating the LDSA concentrations used an electrostatic ion trap with adjustable voltage so that the size distribution could be modified to reflect the deposited dose in either the alveolar or the tracheobronchial region of the human lung [21,22]. It was later shown that a single ion trap setting is sufficient to determine those two fractions as well as the fraction of the particle surface area concentration that would deposit in the entire human respiratory tract [23]. An instrument that was developed more recently therefore only provides a single value as the lung deposited surface area concentration [14]. Depending on the design of the diffusion charger, the effective charge level of the particles may be different and not proportional to the lung deposited surface area concentration. In that case the resulting metric is commonly referred to as active of Fuchs surface area, i.e., the fraction of the particle surface area that is available for ion attachment [24,25].

5.2.1.2 *Principles for Measuring Size-Resolved Concentrations*

The most commonly used measurement principle for submicron and nanoparticle size-resolved particle concentrations is based on electrical mobility analysis. This principle involves an inertial pre-separator to remove particles that are too large, followed by a particle charger, an electrical mobility classifier and a particle concentration measurement device. The most commonly used device is the scanning mobility particle sizer (SMPS), which is commercially available from several manufacturers. An SMPS uses a bipolar charger in order to establish a known equilibrium charge distribution [24,26]. Such bipolar chargers are also called neutralizers, since the overall eventual charge level of all particles is nearly zero. Most common neutralizers use radioactive materials (Kr-85, Am-241 or Po-210), although more recently, soft X-ray bipolar chargers have been introduced to the market [27] in order to overcome legislative constraints on the use of radioactive materials. The charged particles are introduced into a differential mobility analyzer, where the particles are classified according to their electrical mobility [28]. Different electrical mobilities can be classified in a DMA by applying different voltages. Various DMA types are available, which can cover a size range from approximately 3 nm to 1 μm. The mobility-classified particles are counted downstream of the DMA, usually with a CPC. The complete mobility range of interest is covered by ramping up the DMA voltage either continuously or stepwise. The primary output of an SMPS is the electrical mobility distribution of the airborne particles. The electrical mobility is a function of the particle size and particle charge. The known charge distribution of the particles can be recalculated into the number size distribution of the particles [29,30]. Since the different electrical mobilities are measured sequentially in an SMPS, the measurement of a size distribution requires between approximately two and six minutes, depending on the SMPS model, its settings and required accuracy. It has been shown in several publications that the sizing accuracy of SMPS devices is very good and on the order of ±5%, whereas the accuracy of the number concentration measurement is around ±20% [17,31−33]. ISO and CEN ISO standards have been published on SMPS and their use in workplace measurements ISO 15900 [34] and EN ISO 28439 [35], respectively.

Since the data evaluation routine makes use of the concentrations measured in different electrical mobility channels in order to correct for multiply charged particles, it requires the size distribution to be stable over the measurement cycle. An alternative to the SMPS is a system developed at the Tartu University in Estonia [36] that has been commercialized as the Fast Mobility Particle Sizer (FMPS). The FMPS is also an electrical mobility analyzer but differs from the SMPS mainly with respect to particle charging and detection principles. In an FMPS, particles are charged in a unipolar charger before being introduced into a coaxial, radial electrical mobility classifier near the inner electrode. This classifier operates with a fixed electric field distribution, i.e., it does not ramp up the voltage. Particles are deflected towards the outer electrode. The distance particles travel axially before being deposited on the outer wall is a function of the electrical mobility of the particles. The outer cylinder of the FMPS classifier consists of an array of electrometers to measure the particle-induced currents in real time. The measured currents, each one representing a certain electrical mobility range, are then deconvoluted by using the known charge distributions to obtain the particle number size distribution in a size range from 5.6 nm to 560 nm with a time resolution of 1 s. In contrast to the SMPS, the FMPS can therefore also be used to monitor transient size distributions. However, the downside of the fast measurement principle is a lower size resolution and accuracy. While Jeong and Evans [37] reported deviations from SMPS data during atmospheric measurements, Asbach et al. [32] and Kaminski et al. [17] carried out thorough laboratory investigations on the comparability of SMPS and FMPS. Both studies reported a fairly good comparability for compact sub-100 nm NaCl particles but increasing discrepancies for larger particles as well as for agglomerated particles. These discrepancies can reach up to 50% for both size and concentration measurements.

5.2.2 Principles of Aerosol Sampling

Aerosol samples are collected when one requires a more specific knowledge/understanding of the exposure than what could be obtained with particle monitors for various particle metrics. Samples can be collected for subsequent analysis of the whole sample or individual particles. The choice of analytical method, together with the specific characteristics of the occupational exposure, will determine the optimal sampling method. The whole sample may either be analyzed chemically, for example using ICP—MS, or the individual particles may be analyzed physically, morphologically or chemically in a scanning or transmission electron microscope (SEM or TEM, respectively) depending on the size resolution required by the sampled particles and the analytical method. At workplaces, nanoparticles seldom occur as individual primary particles but as aggregates and agglomerates either consisting of primary particles of the same composition and origin or mixed with other particles (either also of nano-size or much larger) occurring at the workplace. The sampling situation is relatively simple if the nanoparticles of concern consist of a very uncommon inorganic or organic compound for which the limit of quantification is very low, the source of the nanoparticles is the only possible source for this compound and only the overall mass concentration of the compound existing as nanoparticles is of interest. In such a case, any traditional sampling method with a suitable sampling medium can be used. However, this situation is very rare and usually the situation is more complex. For example, it may be necessary to determine the size distribution of an airborne compound, which consists of the nanoparticles (from primary particles up to large agglomerates — from a few nanometers to a few micrometers) and possibly also other materials. It may also be necessary to identify the engineered particles in an electron microscope.

In the following sections we will first present the different mechanisms employed in sampling of airborne nanoparticles, followed by a brief description of the currently commercially available nanoparticle samplers/devices.

5.2.2.1 *Mechanisms of Separation of a Particle from an Air Stream*

The mechanisms used for separating airborne nanoparticles from the air (streaming or quiescent) onto a collection substrate are the same as in traditional aerosol sampling. For most of the separation methods, the force acting on an airborne particle depend on its size. In Section 5.2.2.2 the different particles sizes are discussed in more detail. These separation methods are named after the mechanisms employed: diffusion, electric field, thermophoretic, impaction and interception. Most collection- or size-classifying devices use one or more of these mechanisms.

SEPARATION BY DIFFUSION

When a small particle is constantly hit by the random movement of air molecules from all directions it will absorb the momentum of the molecules and recoil. This process will force upon the particle a staggering random movement in all directions (both regarding direction and distance moved). A common name for this phenomenon is 'Brownian motion' after the scientist who first saw it in a microscope. The distance a particle can move due to this bombardment decreases with increasing particle size.

SEPARATION OF CHARGED PARTICLES BY AN ELECTRIC FIELD

Particles that have become electrically charged will retain their charge until the distribution of electrical charges on the airborne particles have been reduced to Boltzmann's equilibrium by adsorption of ions. At workplaces this occurs very slowly, and for measurement purposes the particles can therefore be considered as permanently charged. Negatively and positively charged particles that enter an electric field will move along the electric field and deposit on the surface with a high and low voltage, respectively.

THERMOPHORETIC SEPARATION

Airborne particles passing through an area where the thermal gradient is high, for example, between two plates with a temperature difference of $10-50°C$ and separated by a distance of 1 mm or less, will receive more momentum from air molecules with higher temperature/kinetic energy on the side of the particle facing the higher temperature than the side facing the lower temperature. As a result, the particles will move in the direction of the cooler surface.

SEPARATION BY IMPACTION

An air stream with (high) speed that encounters a static obstacle with a surface perpendicular to the direction of the air stream will accelerate and bend around the obstacle. Airborne particles in this stream will try to follow the bent streamlines, but with increasing inertia particles will increasingly deviate more from their original streamlines. Particles with a sufficiently high inertia will impact on the surface of the obstacle. The inertia of a particle increases with increasing size. The obstacle to the air stream does not need be in the center of the air stream but could as well be a sudden reduction in the cross-section of the air stream. Impaction is a force based on aerodynamics. In a cyclone the air stream carrying the particles is forced into a swirling, circular motion. The separation due to the centripetal acceleration in a cyclone is also based on aerodynamics.

SEPARATION BY INTERCEPTION

When an airborne particle hits a surface (either another particle or the enclosure of the system), the particle will generally stick to the surface. Only if its momentum is sufficiently high could it bounce back and remain airborne. A particle that happens to come within less than one half of its diameter away from such a surface has deposited. This is mainly of interest for particles with a high aspect ratio, that is, fibers, nanofibers, nanothreads, nanotubes, etc. Although being of nano-size (<100 nm) in diameter, they can be very long and thus easily attach to surfaces.

5.2.2.2 *Measures of Particle Size*

For all these separation mechanisms, the separation efficiency depends on the particle size. For spherical, compact, nonporous particles of a specified density, the mass is directly linked to the diameter of the particle. However, for nonspherical particles, for example, aggregates and agglomerates, the concept of particle size is not that simple. For airborne particles, the size of the particles always refers to the size of the entity that moves as a single object. The primary particles are for these purposes only the constituent of the airborne particle, and their size (distribution), their number and the external and internal shape they extend over determine the size of the airborne particle. In all cases the size is given as, for example, a separation efficiency (e.g., diffusive) equivalent diameter or volume-equivalent diameter. It is determined as the diameter of a compact, nonporous sphere (if necessary of a specified density) for which the separation efficiency or volume is equivalent to that of the nonspherical particle. The equivalent diameters for the separation mechanisms presented above are termed diffusive (or thermodynamic) equivalent diameter, (electric) mobility equivalent diameter for separation by an electric field, thermophoretic equivalent diameter, aerodynamic equivalent diameter for separation by impaction, gravity and in cyclones (for spheres of density 1 g/cm^3) and interception equivalent diameter for separation by interception. A measurement instrument is usually calibrated using spherical particles of known size and density. As a consequence, nonspherical particles can have many equivalent diameters, some with different numerical values, and it is the situation of interest that dictates which equivalent diameter to apply.

Most separation mechanisms used in aerosol sampling also occur inside the human respiratory tract, apart from thermophoretic separation. Separation by an electric field is usually low, as the electric field inside the respiratory tract is generated by image forces, i.e., the electrostatic force on a charge in the neighborhood of a conductor, which may be thought of as the attraction to the charge's electric image. The resulting electrical field is not even for very highly charged particles as strong as that applied in aerosol sampling. The particle deposition efficiency in various regions of the human respiratory tract is described by the ICRP model [19]. For specific breathing parameters the deposition efficiency has been calculated using this model and expressed as simple mathematical functions [6,38] that are specified as sampling conventions for the deposition efficiency in a European and international standard, EN ISO 13138 [39]. See also the latter part of Chapter 4, which presents the sampling conventions for the deposition efficiency.

5.2.2.3 *Measurement Devices*

To serve any of the measurement objectives, a suite of devices to measure nanoparticles has been introduced to the market, and new devices are likely to become available in the foreseeable future. In general, devices used to assess exposure to nanomaterial or nano-size aerosols can be subdivided into (near) real-time monitors that provide particle number concentration, particle mass concentration or particle surface area concentration (near)-real time and devices that are used to collect time-aggregated aerosols samples. Aerosol samplers, including the used sampling media, can be divided into categories according to i) whether they are intended for personal or static sampling; ii) collection mechanism (e.g., by diffusion/electric field/thermophoretics or by aerodynamics); iii) according to requirements for off-line characterization of the sample, i.e., the analysis of the whole sample or of individual particles in electron microscopes; and iv) the analysis of the whole sample by chemical analysis.

Devices that provide estimates of mass, number or surface area concentration have been summarized by ISO TC 229 ISO/PDTS 12901-1 [40] and are shown in Table 5.1.

Real-time handheld devices are available that provide estimates of total particle concentration, e.g., condensation particle counters, or size-specific particle concentration, e.g., optical particle counters and diffusion charger-based devices, as well as those that provide mass-based estimates (light-scattering laser photometers). Recently, a handheld

TABLE 5.1 Devices That Provide Estimates of Mass, Number or Surface Area Concentration

Metric	Devices	Remarks
Number directly	Condensation particle counter (CPC)	CPCs provide real-time number concentration measurements between their particle diameter detection limits. They operate by condensing vapor onto sampled particles and detecting/counting the droplets formed. Typically used with 1000 nm size selective inlet and able to detect down to around 10 nm
	Differential mobility particle sizer DMPS	Real-time size-selective (mobility diameter) detection of number concentration, giving number-based size distribution
	Electron microscopy: SEM, TEM	Off-line analysis of electron microscope samples can provide information on size-specific aerosol number concentration
Mass directly	Size-selective static sampler	Assessment of the mass of nano-objects can be achieved using a size-selective personal sampler with a cutoff point of approximately 100 nm and the sample analyzed by gravimetric weighing or by chemical analysis. Although there are no commercial devices of this type currently available, some cascade impactors (Berner-type low-pressure impactors or Micro-orifice impactors) have selection points around 100 nm and can be used in this way
	Tapered element oscillating microbalance TEOM	Selective real-time monitors, such as the TEOM, can be used to measure nano-aerosol mass concentration online with a suitable size-selective inlet
Surface area directly	Diffusion charger	Real-time measurement of aerosol active surface area. Note that active surface does not scale directly with geometric surface above 100 nm. Not all commercially available diffusion chargers have a response that scales with particle active surface area below 100 nm. Diffusion chargers are only specific for nano-objects if used with an appropriate inlet-pre-separator
	Electrostatic low-pressure impactor ELPI	Real-time size-selective (aerodynamic diameter) detection of active surface area concentration. Note that active surface area does not scale directly with geometric surface area above 100 nm
	Electron microscopy: SEM, TEM	Off-line analysis of electron microscope samples can provide information on particle surface area with respect to size. SEM analysis provides direct information on the projected area of collected particles, TEM analysis provides direct information on the projected area of collected particles, which could be related to geometric area for some particle shapes

device (DiSCmini monitor) that provides estimates of the surface area of the aerosol fraction deposited in the gas-exchange region [14] has become commercially available.

Stationary monitors providing estimates of size-selective particle number concentrations include DMA—CPC combinations (e.g., Scanning or Fast Mobility Particle Sizers (S/FMPS, or spectrometers), aerodynamic particles sizer (APS) and the (Electrical) Low Pressure Impactor (ELPI)). A stationary diffusion charger device provides estimates of the surface area (Nanoparticle Surface Area Monitor), while the tapered element oscillating microbalance (TEOM) measures the mass of aerosols. In general, the static samplers are rather bulky and not suitable to measure personal exposure.

The most accurate size-resolved devices using a DMA for measuring nano-scale particles usually need a radioactive source as a bipolar charger, whereas less accurate size-resolved instruments use a nonradioactive unipolar charger. An example of the latter is the Naneum Nano-ID NPS500, which uses a proprietary Corona charger system. Recently, an advanced neutralizer that features bipolar diffusion charging using soft X-ray has been introduced into the market as an alternative for traditional radioactive sources, e.g., TSI model 3087.

5.2.2.4 Samplers and Substrates for Nanoparticle Sampling and Analysis

SAMPLERS

The samplers described here collect the aerosol fraction that deposits in only one or all compartments of the human respiratory tract. The samplers try to emulate one or more of the sampling conventions from EN ISO 13138 [39]. They have several stages, each with different separation mechanisms. In the first stage(s), large particles are aerodynamically separated from the air stream. At the second stage, the remaining fraction(s) in the aerodynamic size range down to 0.3—0.5 μm are collected by aerodynamic separation. In the third stage, the particles are deposited by diffusion and collected on nylon mesh nets. The particles with an aerodynamic diameter less than 0.3—0.5 μm and that do not deposit in the mesh nets (and are exhaled) are not collected.

Cena et al. [41] have developed an add-on device to the SKC personal aluminium cyclone for the respirable fraction. This sampler is called the Nanoparticle Respiratory Deposition (NRD) sampler and consists of two stages. The sampler is inserted between the cyclone outlet and usual 37-mm filter cassette. Its first stage is an impactor with a cut-size of 0.3 μm, and the total pressure drop is low enough to allow a normal personal sampling pump. The final stage emulates the total deposition by diffusion in all compartments of the respiratory tract.

SPECTROMETERS Deposition by diffusion in nylon nets is used in two spectrometers (one static and one personal) that give the mass-weighted particle size distribution for the element(s) analyzed. By integration of the measured mass size distribution multiplied with the sampling convention for the deposition efficiency of the fraction interest [39], the deposited concentration can be calculated. In each spectrometer, four and five sets of nylon mesh nets with different characteristics are used to obtain the particle size distribution as a function of particle diffusive size. The Nano-ID Select Wide-Range Aerosol Sampler (WRAS) is a combined (static) aerodynamic and diffusive spectrometer for the size range from 10 µm down to 5 nm [42]. This spectrometer uses a cascade impactor for the aerodynamic size range.

AERODYNAMIC SEPARATION OF PARTICLES

By decreasing the pressure inside a sampler, particles nominally in the diffusive size range can be separated by aerodynamics. In cascade impactors this occurs as nozzles with decreasing diameters are needed in order to separate smaller particles. A low operating pressure can also be achieved by mounting a critical orifice upstream of the aerodynamic separation stage. Operating an aerodynamic separator at a much higher flow rate reduces the cut-size down to a few 0.1 µm. Most of these methods require very powerful (and heavy) pumps and are therefore only appropriate for static sampling. However, it must be remembered that the aerodynamic particle size is not necessarily relevant for particle deposition in the human respiratory tract, as deposition of submicron/nanoparticles is determined by their diffusive equivalent instead of the aerodynamic equivalent diameter.

CASCADE IMPACTORS The traditional static cascade impactors, i.e., MOUDI, ELPI and Berner, all have their smallest cut-size at approximately 60 nm, although modifications of all these impactors exist for obtaining the mass-weighted size distribution for airborne particles aerodynamically down to 10 nm or even lower. One personal cascade impactor is available that has a cut-size below 0.4 µm; the Sioutas Personal Cascade Impactor has a cut-size of 0.25 µm. However, this is not suitable for measuring nanoparticles.

RESPIRABLE CYCLONE PLUS AN IMPACTOR Tsai et al. [43] have developed an active PErsonal Nanoparticle Sampler (PENS) that consists of an impactor that is mounted downstream of an existing cyclone for the respirable fraction. The impactor has an aerodynamic cut-size of 100 nm and is followed by back-up filter.

HIGH-FLOW METAL MESH NET SAMPLERS (COLLECTION BY IMPACTION) Furuuchi, Otani and co-workers have developed a personal nanosampler based on impaction in a layered metal fiber mesh filter [44]. The cut-size is circa 0.2 μm, which is too high for nanoparticles [45].

LOW-PRESSURE CYCLONES Tsai and colleagues [46—49] have investigated how axial cyclones operate at very low pressures (downstream of a critical orifice but upstream of a vacuum pump). They have not developed any device but have presented results that show that it would be possible to determine an aerodynamic cut-size for a cyclone in the range 20—100 nm. As a powerful pump is required, such device would only be of use for stationary measurements.

SAMPLING MEDIA AND ANALYSIS

Gravimetric analysis generally has insufficient sensitivity for determining the mass of nanoparticles. Inductively Coupled Plasma Atomic Emission Spectrometry (ICP-AES) or Inductively Coupled Plasma Mass Spectrometry (ICP-MS) are well suited for highly sensitive kinds of analyses, although other alternatives, such as proton-induced X-ray emission (PIXE), are also available. Four international standards [50—52] give guidance on the sample treatment and sample analysis using ICP-AES and ICP-MS, respectively. Collection substrates are generally mixed cellulose ester (MCE) filters for post-separation collection (coated) Mylar films (for separation by impaction) and nylon mesh nets (for separation by diffusion).

DIFFUSIVE COLLECTION OF AIRBORNE PARTICLES Collection substrates for diffusively separated particles to be analyzed chemically as an ensemble are often nylon mesh nets, but metal mesh nets also exist. These exist in various sizes (fiber diameter and porosity), and the nylon mesh nets are easily dissolved by the sample preparation techniques for ICP-MS analyses.

For analyses with electron microscope techniques it is very important that the collection substrate does not become overloaded with collected particles. Knowledge of the particle number concentration and the specific requirements of the electron microscope are needed in order to estimate an optimal sampling period. Depending on the electric conductivity of the collected particles, the electron microscope and the kind of EM analysis, the samples might need to be covered by a conducting layer.

SAMPLING FOR ANALYSIS WITH A SCANNING ELECTRON MICROSCOPE (SEM)

With this analytical technique the reflected picture of the particles is analyzed. In order to make a good distinction between the particles and their background, the particles must be collected on a nearly perfectly flat surface. Two kinds of collection substrates are in wide use, polycarbonate filters and silicon wafers.

POLYCARBONATE FILTERS These filters have a flat surface with nearly uniformly sized holes for the air stream. Filters are available with different hole diameters and can be used with any flow rate of interest. The hole diameter and air speed across the filter surface determine which particle size will be collected by the lowest efficiency. In order to distinguish nanoparticles from the background the filter is metal-coated. Liu et al. [53] give information on the collection efficiency of various filter materials, though the data are by now more than 30 years old so they can only be seen as informative. Cyrs et al. [54] have recently investigated the collection efficiency of polycarbonate filters for nanoparticles.

These filters are available in many sizes and can be used in any filter holder, though a material that is electrically conducting is preferable. Particles (both free primary particles and aggregates/agglomerates) deposit mainly on the surface of the filter but also inside the holes. The smallest particles deposit by diffusion and the largest by aerodynamics and interception.

SILICON WAFERS Silicon wafers have an extremely flat surface. The particle deposition is perpendicular to the air stream flowing over the wafer. The particles are distributed almost randomly over the collection surface. The flow rate generally is very low. Presently no sampler based on these concepts is commercially available.

SAMPLING DIRECTLY ONTO A GRID FOR ANALYSIS WITH A TRANSMISSION ELECTRON MICROSCOPE (TEM) With this analytical technique the transmitted electron rays are analyzed. In order to be able to analyze the particles, they must be attached to the lateral side of a horizontally mounted grid (assuming a vertical electron beam). The grids used are either regular metallic quadratic grids or irregular plastic grids, so-called Lacey or Holey carbon films. The particles generally are deposited by diffusion or electrostatics. Diffusive deposition requires much lower flow rates than electrostatic deposition.

SAMPLING DIRECTLY THROUGH A TEM GRID The particles are deposited by diffusion directly from the air stream passing through the grid openings. There are many versions available of this collection principle, also as personal samplers. Currently at least two simple versions exist. The Aspiration Electron Microscopy Sampler designed by VTT Technical Research Center of Finland (from where it is commercially available) [55] and the Mini Particle Sampler (MPS) developed by INERIS and distributed by EcoMesure [56].

DEPOSITION ONTO A TEM GRID ATTACHED TO A FILTER Simultaneous samples can be collected for analyses by SEM and TEM by mounting a TEM grid onto a polycarbonate filter or a Holey film on top of part of a mixed cellulose ester (MCE) filter. The TEM grid may either be mounted so that the air stream passes through it or so that the air stream only passes above the grid (by closing the pores of the filter underneath). Variations of these methods are described by Dye et al. [57], Tsai et al. [58] and Bard et al. [59]. However, they are not commercially available, and users have to attach the TEM grids to the filters themselves. The particles deposit by diffusion into the TEM grid. The collection efficiency of a TEM grid is much lower than for deposition by an electric field (see below).

DEPOSITION ONTO A TEM GRID BY AN ELECTRIC FIELD OR THERMO-PHORESIS The particles are deposited by field forces directly on a TEM grid, which is mounted with its surface facing perpendicular to the air stream. In the electrostatic precipitator, the particles are first electrically charged by a radioactive source or a corona discharger. Downstream of being charged, the particles enter a strong electric field between two parallel plates (electrodes) generated by a high-voltage supply (5–10 kV). On one of the electrodes a TEM grid is mounted. In the thermophoretic sampler, a strong temperature gradient exists between a heated plate and a temperature sink on which the TEM grid is mounted. Due to the high-voltage supply/stability of the temperature gradient needed, these kinds of equipment only exist as static samplers. An electrostatic precipitator for TEM grids can be manufactured from the information given by Dixkens and Fissan [60]. A commercial version is the TSI Nano Aerosol Sampler 3089 (TSI, Shoreview (MN), USA). Miller et al. [61] describe a handheld version. Thomassen et al. [62] describe a thermophoretic sampler developed at the Fraunhofer Institute − ITEM in Hannover. Wen and Wexler [63] and Thayer et al. [64] describe other personal TEM grid samplers based on thermophoretic deposition. Li et al. [65] and Bau et al. [66] have determined the collection efficiency of the NAS 3089 and the FI-ITEM thermophoretic sampler.

With the UNC Passive Aerosol Sampler airborne particles are directly collected onto a copper TEM grid mounted onto a SEM stud with a few drops of liquid carbon black. These are static samplers intended for passive use (i.e., without any active suction sampling), though the authors propose without any testing that they could also be used for long-term personal sampling. Particles are deposited by diffusion from the air passing above the stud [67]. This sampling method has only been evaluated for spherical particles, for which the diffusive equivalent diameter could be estimated. For the kind of aggregates/agglomerates expected at workplaces, it would be much more difficult to estimate the diffusive equivalent diameter.

5.3 MEASUREMENT STRATEGY

5.3.1 General

In addition to the selection of the appropriate measurement device, the measurement strategy should also address which workers should be monitored, and when (and for how long). From the perspective of compliance measurements or epidemiology workers are generally grouped into so-called similarly exposure groups (SEGs) or exposure zones on the basis of an *a priori* observation and understanding of the processes and tasks and the likelihood of exposure [3,68]. However, it is acknowledged that the professional judgement of the exposure assessor may not be sufficient for the relevant exposure grouping which may even be worse for exposure to NOAA. Job-task (activity)-based sampling [69] may partly overcome this deficiency, however, this depends on the objectives of the sampling. To assess long-term exposure in view of chronic health endpoints, activity-based sampling may be less relevant and instead shift, e.g. 8 h time weighted average (TWA), or even longer-term monitoring or sampling may be more appropriate. If the actual duration of exposure is short, which is currently still the case for many exposure scenarios involving nanomaterials, then activity-based sampling will often be the preferred strategy. If activity-based personal sampling or monitoring is not feasible, concentration mapping could be used to illustrate spatial and temporal variability of the aerosol concentration distribution in a workplace as function of the work process, and to identify contaminant sources [3].

5.3.2 Sampling Objectives

The major factor that determines the measurement strategy, including the selection of the measurement devices, is the objective of the measurement. In this section possible sampling objectives will be presented briefly, followed by a more detailed description of some measurement strategies related to the measurement objects.

5.3.2.1 Exposure Analysis

This somewhat broad term may include i) exploration of whether exposure events or concentration increases are likely to occur, ii) determination of emission/release rates, iii) measurements focused on the evolvement of concentrations or size distributions over time, for exampling, for modeling purposes; and iv) surveillance of workplace air. For example, if the aim of the measurement is surveillance of workplace air, continuous monitoring by a static device with a visual or acoustic alarm set at an action level (concentration) may be sufficient. In case of monitoring of workplace air for process releases, or determination of the efficacy of a control measure, a single metric and non-size-resolved device may be sufficient.

Exposure measurements for *risk assessment* purposes may include i) measurements to show compliance with exposure limits (= safety assessment); ii) measurements used for the assessment of the lung-deposited dose; or iii) measurement for the assessment of a biological or health effect relevant exposure. Measurement for compliance purposes should meet the requirements aligned with the limit value with respect to measurement or sample device, analytical method, duration of the measurement period, number of measurements, etc. For example, the NIOSH 8 h TWA recommended exposure limit (REL) for carbon nanotubes [70] requires the measurement of the respirable fraction (e.g., by a cyclone) followed by off-line determination of the amount of EC in the sample (NIOSH Analytical Method 5040, [71]).

Assessment of lung-deposited dose and the requirements with respect to the sampling devices have been discussed previously. Assessment of a biological or health effect relevant exposure would require first of all knowledge of which (panel of) biological effects are relevant and how they can be measured. An example of such an approach is the measurement of particle-associated reactive oxygen species (ROS) (e.g., [72,73]).

5.3.2.2 *Epidemiological or Health Surveillance*

Exposure assessment has to be tailored to the design of the health surveillance or epidemiological study, e.g., panel, cross-sectional, cohort studies, and the required resolution with respect to exposure. Important considerations in epidemiological studies are contrasts in exposure and the precision of the exposure estimates. The first is determined by the design of the study, i.e., by selection of study populations, and the second is determined by the measurement effort, i.e., more measurements result in higher precision of the exposure estimates. A panel study, which often is explorative by nature and focused on short-term or reversible changes in health within (an) individuals, will benefit from a risk-assessment-designed exposure assessment. It may require intensive measurement of (health and) exposure on a small population where relatively simple devices or cheap combinations of sampling and analysis may provide a sufficiently accurate estimate of exposure variation over time. Cohort studies usually focus on long-term (health effect) differences between groups. Hence measurements aimed at giving estimates for cumulative exposure may be sufficient for cohort studies. However, as one of the aims of the cohort study may be to determine occupational exposure limits, more accurate exposure assessments will be required and consequently precise measurement devices.

A comprehensive guide for testing compliance with occupational exposure limits for airborne substances has been drafted by British Occupational Hygiene Society (BOHS) and the Nederlandse Vereniging voor Arbeidshygiëne (NVvA) [68]. Occupational exposure limits have not yet been derived for any NOAA, neither by the European SCOEL (the Scientific Committee on Occupational Exposure Limits) nor by any national OEL-setting authority. However, as was addressed above, NIOSH [70,74] has proposed mass-based recommended, time-weighted exposure limits for two mNOAA: nanoCNTs and nanoTiO$_2$.

A tiered approach for compliance measurements has been proposed, as outlined below (see also Figure 5.1). In the first tier (screening test) the workforce is divided into similarly exposed groups (SEGs), and three representative personal exposure measurements are taken from random workers in the SEG. If all three exposures are $<0.1 \times$ OEL, it can be assumed that the OEL is complied with. If at this stage or any later one any result is $>$OEL, the OEL is not complied with and you will enter the next tier (group compliance test). At least seven more samples have to be taken from the SEG, at least two per worker from workers picked at random. Use all nine (or more) samples to apply a test that establishes, with 70% confidence, that there is $<5\%$ probability of any random exposure in the SEG being $>$OEL [6]. Perform an analysis of variance on the nine (or more) results to establish whether the between-worker variance is $>0.2 \times$ total variance. If it is, the next tier (individual compliance test) must be applied by analysing the nine (or more) results to do an individual compliance test. There should be $<20\%$ probability that any individual in the SEG has $>5\%$ of exposures $>$OEL.

Commonly, a tiered approach is also used for exposure assessment in view of a risk assessment framework. Strategies are pragmatic decision schemes based on tiered-type approaches in order to avoid comprehensive exposure measurements − in the case of NOAA reflecting the lack of analytical capabilities and the costs of employing scientific instruments in the field. In these approaches, information is collected in each successive tier at a more detailed level in order to reduce the uncertainty in the exposure estimates. After each tier, a decision can be taken either to stop further collection or to continue. The use of a tiered approach will likely be different for researchers and occupational hygiene (OH) practitioners. In most tiered approaches for exposure assessment, the first tier consists of an information collection and evaluation phase to decide on the likelihood of release or exposure, so strictly speaking, this tier is not a part of a measurement strategy.

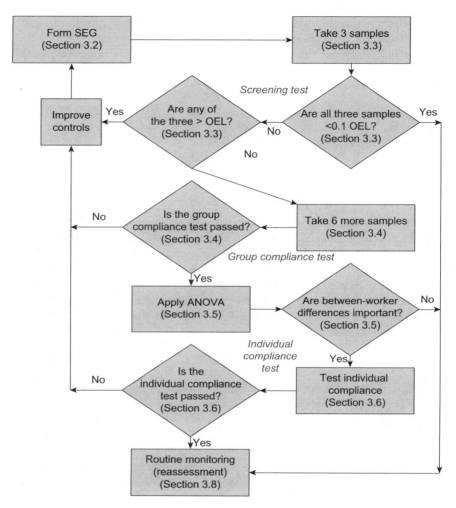

FIGURE 5.1   **Flowchart of the tiered approach for compliance testing as proposed by BOHS/NVVA.** *Courtesy of BOHS/NVVA [63].*

Presently, there are several proposed measurement strategies for (m)NOAA including the nanoparticle emission assessment technique (NEAT) [75], the approach proposed by the British Standard Institution (BSI) [76], the approach proposed by a number of institutions in Germany (BAuA, BG RCI, IFA, IUTA, TUD, VCI) [77], on which the German NanoGem project [16] based its closely related proposal, the French approach [78] and the tiered approach proposed by McGarry and coauthors [79]. All five suggest a tiered approach, starting with a relatively simple and limited set of measurements and/or gathering basic information on processes and jobs and materials in a first tier followed by extended assessment in subsequent tiers. In general, a tiered approach is consistent with modern industrial hygiene practice. The decision criteria to enter the next tier are key factors for such approaches. Each tier will use a different type of information or devices and will generate a different degree of detail and specificity of data. Evaluation of the data may lead to revisiting a certain percentage of the workplaces for the next tier assessment. A generic outline of a tiered approach based on the NanoGem tiered approach is given below.

Tier 1: Data gathering and evaluation. The task in tier 1 is to clarify, e.g., through systematic collection and evaluation of information and/or by on-site inspection, whether nanomaterials are used in the respective workplace and if they can be released from the corresponding processes. If a release cannot be excluded, the potential exposure has to be determined in tier 2. In the present tiered approaches structured information and evaluation methods as Control Banding tools [80] are not included; however, such methods can be used for a decision to proceed or finish an assessment.

Tier 2: Basic assessment. As long as no real-time devices exist that differentiate (m)NOAA from other, natural or process-generated nano aerosols, results from real-time measurements need to be compared with the background concentration levels to determine a realistic value for the exposure to (m)NOAA. The background level can either be measured or assumed from past experience. If tier 2 measurements reveal a concentration level that is significantly increased over the background, then it is concluded that a potential exposure exists and this has to be assessed further in tier 3.

A slightly different approach has been proposed by BSI [76] and adopted by van Broekhuizen [81], respectively, that makes use of benchmark levels or nano reference values. If the exposure exceeds an 'action level', actions are required, e.g., more detailed measurements or control measures. The non-substance-specific reference values are expressed as particle size integrated total particle number concentration and could also be used in the framework of a tiered approach.

The simplified exposure potential measurements in tier 2 are conducted using easy-to-use devices, measuring particle size integrating total particle number concentration. Tier 2 measurements can be conducted in two different ways. On the one hand, permanently installed particle measurement devices can monitor the particle concentration in a workplace and deliver information on potentially increased concentrations. If permanent monitoring reports a significantly increased concentration, the reasons have to be assessed in a tier 3 measurement. A second possibility of simplified exposure assessment in tier 2 is by screening of a workplace using handheld monitors. Particle concentrations are measured near potential release locations or in the breathing zone of workers and compared with the background concentration. If these measurements reveal a concentration level significantly higher than the background concentration, the reasons for the increase have to be checked in tier 3.

Tier 3: Comprehensive exposure assessment. Tier 3 measurements are intended to provide clear evidence for the presence or absence of the nanomaterials in the breathing air of a worker and, if possible, quantify exposure with respect to an estimated lung-deposited dose. Tier 3 measurements require an extended set of measurement and sampling equipment for evaluating a possible occupational exposure to m(NOAA). Measurements in tier 3 always include the determination of the particle background in the workplace, either through simultaneous measurements at a representative background location(s) or through consecutive measurements prior to and after the process under investigation or by consecutive measurements of the identical process both with and without the use of the nanomaterial.

Clearly, each tier needs decision criteria and dedicated (measurement) strategies, which enable a reliable and unambiguous decision on entering a next stage. As has been addressed in the introduction of this chapter, a challenge in the comprehensive exposure assessment is to estimate the deposited dose of a specific nanomaterial. Ideally, this would require multi-metric size-resolved personal measurement devices that provide estimates of particle deposition of a specific nanomaterial in the respiratory tract. Presently such devices exist for surface area. Diffusion charger-based portable/static devices like NSAM and Aerotrak provide estimates of the surface area of nanoparticle aerosols that would deposit, corresponding to the ICRP lung deposition curves for the tracheobronchial (TB) and alveolar (A) regions of the human respiratory tract, whereas the personal DiSCmini estimates the A-region deposition. Currently, no such devices exist for use in workplace measurements that provide lung deposition for mass and particle number metrics. Therefore, a panel of measurements and off-line analyses are used to provide data that could be used to estimate dose (either directly or using models for exposure and deposition).

5.4 NEW DEVICES (PRE-PROTOTYPES) DEVELOPED BY NANODEVICE AND IMPLICATIONS FOR MEASUREMENT STRATEGIES

5.4.1 Nanodevice Devices

A panel of novel (pre-prototype) devices, samplers, pre-separators or measuring concepts have recently been developed within the

NANODEVICE project. Pre-prototype means that real devices have been developed and benchmarked with existing devices, but features important for final marketing like sophisticated user interfaces, etc., are not available yet. These pre-prototypes are summarized in Table 5.2.

In the category of devices 'total or size specific N-S-M concentration in real-time', the low-cost total (active) surface area monitor (A3) charges

TABLE 5.2   Pre-Prototype Monitors, Samplers and Detection Systems as Developed by the NANODEVICE Project

Category	Description	Device	
A	Total or size specific N-S-M concentration in real-time	A1	Wide-range real-time classifier (10 nm–27 µm) (mobility + optical detection)
		A2	Nano aerosol classifier and short/long-term sampler system (20–450 nm)
		A3	Low-cost total (active) surface area monitor
		A4	Micro-string-based particle mass sensor (1 ng) (5–300 nm)
B	Material specific monitors in (near-) real-time	B1	Real-time CNT monitor (chemical + physical selection + counting)
		B2	Catalytic activity aerosol monitor (CAAM) quasi real-time
		B3	Aerosol fluorescence-labeled surface species monitor (300 nm–>20 µm)
C	Personal samplers for off-line particle analysis	C1	Wide-range size resolving personal sampler (2 nm–5 µm) up to 8 size fractions
		C2	Sampler* for aerosol fraction deposited in the gas-exchange region (20 nm–5 µm)
		C3	Sampler* for aerosol fraction deposited by diffusion in the anterior nasal region 5–400 nm
		C4	Electrostatic nanoparticle sampler 20–400 nm
		C5	Thermal precipitation sampler 20–400 nm
D	Personal substance specific samplers for off-line analysis	D1	Cyto-thermal precipitator
		D2	CNT specific sampler

Can also be used as pre-separator.

aerosol particles in a unipolar diffusion charger, followed by electrical detection of the total amount of charged particles using escaping charge technology that has been miniaturized. Particle size range of operation is estimated to be $0.01-3$ μm. A battery-operated version is available, which affords to operate as a personal device. Currently, the measurement results, i.e., total active surface area (approximately proportional to lung-deposited surface area), are displayed by three colored LEDs indicating a user-definable limit value.

The nano aerosol classifier and short/long-term sampler system 'NanoGuard' (A2) is a modular system that enables size-resolved particle number counting ($20-450$ nm) and lung-deposited surface area estimates, together with the option for sampling either for high concentrations/short duration, using the 'electrostatic precipitator' (C4), or for low concentrations/long duration, using the 'thermal precipitator' (C5). The first precipitator is an easy-to-use 'plug & play' substrate holder for 50 mm wafers operating at a flow rate of 0.1 LPM. The second sampler based on thermophoresis affords close to homogeneous deposition of particles (up to 300 nm) on a substrate. The sampler is well suited for long-term (8 h) sampling and enables, depending on the type of substrate, either EM analysis of TEM grids or specific *in vitro* toxicological analysis of the cell-line substrate in wells (see D1).

The size range (10 nm-27 μm) covered by 'wide-range real-time classifier' (A1) using electrical and optical principles, with currents and single particle counting (light scattering), would be extremely useful with respect to ease of use, since currently this large size range can only by covered by combination of devices.

The 'micro-string-based particle mass sensor' (A4), a miniaturized mechanical and electromechanical elements (MEMS)-based technology, would be a useful contribution to the suite of personal devices for direct monitoring of particle mass by its extreme sensitivity; however, currently there is only the proof of concept.

In the category of 'material specific monitors in (near-) real-time' (B), (pre-prototype) devices to measure 'CNT' monitors have been developed (B1), whereas two new approaches with respect to 'biological effect' have also been explored (B2 and B3).

The CNT monitor provides a number concentration; CNT particles in the air are preselected using certain physical/chemical characteristics of CNT. Individual CNT objects are then counted using optical techniques.

In general, material-specific monitors will be extremely useful to solve the issue of the 'background' aerosol and more specifically the group of CNTs of great concern from a risk perspective. However, no exact details on the specificity of the monitor are currently available.

Two concepts of biological effect-relevant measurement have been addressed within the NANODEVICE project, i.e., the 'catalytic activity aerosol monitor' (CAAM), based on specific model chemical reaction for each material and IR absorption (B2), and the 'aerosol fluorescence labeled surface species monitor', by simultaneous detection of fluorescence and light-scattering signal (B3). Both concepts have the potential to contribute to the development of new exposure 'metrics'; however, the relevance of the biological effect should be investigated in more depth.

Within the category of 'personal samplers for off-line particle analysis' (C), the precipitators (C4 and C5) form a subgroup and have been discussed as part of the modular system A2.

The 'Personal Nano Sampler (PNS)' (C1) is a wide-range size-resolving personal sampler that determines the size distribution in the diffusive range. Particles in the aerodynamic size range are separated from the air stream by up to three cyclones with a final cutoff of 300 nm. Particles with aerodynamic diameters <300 nm are collected on a series of specially designed diffusion stages with a variety of substrates.

Samplers C2 and C3 represent another subgroup of samplers that weight the deposition according to the ICRP criteria on lung deposition, which are very relevant for a more accurate estimation of the dose with respect to risk assessment. Both samplers can also be used as pre-separators in combination with real-time monitors to set the size fraction that will be monitored. Sampler C2 (GE sampler) deposits airborne particles by aerodynamics and diffusion in the gas-exchange region, whereas sampler C3 deposits airborne particles by diffusion in the anterior part of the nose (ET1 sampler). The GE sampler has three cascaded impactors. The collection substrate of the second impactor collects the fraction deposited by aerodynamics and the third impactor collects the aerodynamic particles that do not deposit in the gas-exchange region. The final stage emulates the total deposition by diffusion. In the ET sampler, two initial cyclones clear the air stream from particles depositing by aerodynamics and in the final stage emulates the total deposition by diffusion in the anterior part of the nose.

Two samplers have been categorized as 'substance-specific samplers', but the 'cyto-thermal precipitator' (D1) that collects nanoparticles directly onto a cell culture enables a more specific effect analysis than collecting a specific substance. The CNT specific sampler (D2) is based on magnetic deposition, i.e., charged carbon nanotube fibers are deposited on a surface by a magnetic field. The whole sample is analyzed with off-line Raman scattering.

5.4.2 Implications for Measurement Strategies

Recent developments related to measurement devices, including those within the NANODEVICE project, have extended the suite of devices that can be selected for various measurement objects with respect to nanoparticles. In general, potential benefits of the newest developments are:

i) reduction of size and weight of existing device concepts that afford portability (and could be used in a position closer to the receptor)
ii) increase of size ranges for both the size-resolved real-time monitors and size-resolved samplers, which increase the ease of use, since combination of devices can be avoided
iii) match with ICRP/ISO criteria for lung-deposited fraction, which enables interpretation of the results with respect to the dose
iv) material or substance specificity, which is an advantage with respect to background aerosols
v) exploration of concepts, which could improve the relevance of exposure measurement in view of biological or health effect.

The actual benefits of the samples will also be determined by other factors, such as their sensitivity, user-friendliness/user-interface, robustness, performance reliability and price. Furthermore, final market relevance of the developed instruments will be determined by regulations, e.g., agreed benchmark levels or exposure limits.

It should also be recognized that the suite of devices that are already on the market are useful and already fit into measurements strategies [1].

With respect to the tiered approach, devices of categories A of Table 5.2 that are in a stage of developing near market, i.e., A1−A3, could be useful for tier 2, since the focus in this tier is on real-time monitors providing 'total concentrations', which enable instant decisions. However, if tier 2 would also be interpreted as baseline measurements of processes (and/ or background concentrations), samplers including C1−C3, and the previously mentioned samplers MPS, PENS and NRD, can also be applied here. Undoubtedly, substance-specific monitors, e.g., the CNT monitor (B1), will have substantial benefits compared to the other monitors, since distinction from background should not be an issue for this device, which simplifies the criterion to continue the decision tree.

Tier 3 focuses on a comprehensive exposure assessment in view of risk assessment. Therefore monitors and (personal) samplers are especially very relevant that provide size-resolved concentrations or samples, respectively. Hence monitors A1, A2 (combined with C4) and B1, and (personal) samplers C1−C3 and D2 are useful. In the near future monitors B2 and B3, which provided other information, and the cyto-thermal precipitator (D1) may also have added value with respect to a comprehensive exposure assessment in view of risk.

Another perspective to evaluate the relevance of the various newly developed devices is to allocate application domains with respect to different measurement objectives, e.g., exposure analysis and risk assessment. Within the measurement objectives, the various subcategories

TABLE 5.3 Potential Application Domains of Recently Developed (Pre-Prototype) Devices

Objective	Strategy	Device Monitor	Device Sampler	Device Other	Remarks
Exposure analysis					
	Exploration	Event/activity	A1, A3, B1	C1, C4	NRD PENS	
	Emission/release	Event/activity	A1, A2, B1	C1, C4	NRD PENS	
	Modeling	Event/activity	A2	C1	NRD PENS	
	Surveillance	Process/continuous monitoring	A3		MPS	
Risk assessment					
	Compliance	Shift/TWA	A2, (A4)	C1, D2*		*CNT
	Threshold (benchmark/ reference value)	Fixed period/ activity	A3, B1			
	Lung-deposited fraction	Activity	A1, A2	C1, C2, C3, C5	NRD	
	Biological (effect) relevance	Activity	B2**	C1	ROS potential	**Catalytic activity
				D1***		***In vitro (cell lines)
	Substance specific/ characterization		B2, B3 B1			Catalytic activity Fluorescent surface species CNT

are distinguished, as discussed in Section 5.3.2. The measurement duration part of the strategy is categorized as event/task/activity-based or shift-based, and the devices are subdivided into (Nanodevice) monitors, samplers and previously mentioned other new samplers. This evaluation is summarized in Table 5.3.

5.5 CONCLUSIONS

In general, exposure assessment for NOAA is more complicated than for conventional chemicals due to uncertainty in relevant dose and exposure metrics and particular issues of measuring small particles with distinction of background aerosols.

Sampling strategies are determined by the aims of the study. For conventional substances this has been driven by regulatory issues. As currently no Occupational Exposure Limits are set for NOAA, the initial measurement strategies have been proposed, aiming to address the following issues: pragmatic, measuring fore- and background, using suite of instruments and predominantly stationary or handheld (not personal).

The newly developed devices by NANODEVICE address some of the issues further, and this will provide more opportunities to design optimal sampling strategies. Especially, the risk assessment can be improved by more accurate estimates of the dose or the assessment of (biological) effect exposure. However, the actual benefits can only be evaluated thoroughly for those devices that will actually be introduced into the market.

References

[1] Brouwer D, Berges M, Virji MA, et al. Harmonization of measurement strategies for exposure to manufactured nano-objects; report of a workshop. Ann Occup Hyg 2012a;56(1):1—9.
[2] Kuhlbusch TAJ, Asbach C, Fissan H, et al. Nanoparticle exposure at nanotechnology workplaces: A review. Particle Fibre Toxicol 2011;8, art. no. 22.
[3] Ramachandran G, Ostraat M, Evans DE, et al. A Strategy for Assessing Workplace Exposures to Nanomaterials. J Occup Environ Hyg 2011;8(11):673—85.
[4] Brouwer D. Exposure to manufactured nanoparticles in different workplaces. Toxicology 2010;269(2-3):120—7.
[5] Löndahl J, Pagels J, Swietlicki E, et al. J. A set-up for field studies of respiratory tract deposition of fine and ultrafine particles in humans. Aerosol Sci 2006;37:1152—63.
[6] Bartley DL, Vincent JH. Sampling conventions for estimating ultrafine and fine aerosol particle deposition in the human respiratory tract. Ann Occup Hyg 2011;55: 696—709.
[7] Lin MI, Groves WA, Freivalds A, et al. Laboratory evaluation of a physiologic sampling pump (PSP). J Environ Monit 2010;12(7):1415—21.
[8] Stolzenburg MR, McMurry PH. An ultrafine aerosol condensation nucleus counter. Aerosol Sci Technol 1991;14:48—65.

[9] Hering SV, Stolzenburg MR, Quant FR, et al. A laminar-flow, water-based condensation particle counter (WCPC). Aerosol Sci Technol 2005;39:659—72.

[10] Aitken J. On improvements in the apparatus for counting the dust particles in the atmosphere. Proc R Soc Edinburgh 1889;16(129):134—72.

[11] Aitken J. On the number of dust particles in the atmosphere. Trans R Soc Edinburgh 1888;35(1):1—19.

[12] Asbach C, Kaminski H, Von Barany D, et al. Comparability of portable nanoparticle exposure monitors. Ann Occup Hyg 2012a;56:606—21.

[13] McMurry P. The history of condensation nucleus counters. Aerosol Sci Technol 2000;33:297—322.

[14] Fierz M, Houle C, Steigmeier P, Burtscher H. Design, calibration, and field performance of a miniature diffusion size classifier. Aerosol Sci Technol 2011;45(1):1—10.

[15] Marra J, Voetz M, Kiesling HJ. Monitor for detecting and assessing exposure to airborne nanoparticles. J Nanopart Res 2010;12:21—37.

[16] Asbach C, Kuhlbusch TAJ, Kaminski H, et al. Standard Operation Procedures For assessing exposure to nanomaterials, following a tiered approach. NanoGem 2012.

[17] Kaminski H, Kuhlbusch TAJ, Rath S, et al. Comparability of Mobility Particle Sizers and Diffusion Chargers. J Aerosol Sci 2013;57:156—78.

[18] Mills JB, Park JH, Peters TM. Comparison of the DiSCmini Aerosol Monitor to a Handheld Condensation Particle Counter and a Scanning Mobility Particle Sizer for Submicrometer Sodium Chloride and Metal Aerosols. J Occup Environ Hygiene 2013. (http://dx.doi.org/10.1080/15459624.2013.769077) in press.

[19] ICRP. Human Respiratory Tract Model for Radiological Protection. ICRP Publication 66, 1994. Ann ICRP 24(1—3).

[20] Oberdörster G. Toxicology of ultrafine particles: in vivo studies. Philos Trans R Soc A 2000;358:2719—40.

[21] Fissan H, Neumann S, Trampe A, et al. Rationale and principle of an instrument measuring lung deposited nanoparticle surface area. J Nanopart Res 2007;9:53—9.

[22] Shin WG, Pui DYH, Fissan H, et al. Calibration and numerical simulation of nanoparticle surface area monitor (TSI model 3550 NSAM). J Nanopart Res 2007;9:61—9.

[23] Asbach C, Fissan H, Stahlmecke B, et al. Conceptual Limitations and Extensions of Lung Deposited Nanoparticle Surface Area Monitor (NSAM). J Nanopart Res 2009;11:101—9.

[24] Fuchs NA. On the stationary charge distribution on aerosol particles in bipolar ionic atmosphere. Geofis Pura Appl 1963;56:185—93.

[25] Jung H, Kittelson DB. Characterization of Aerosol Surface Instruments in Transition Regime. Aerosol Sci Technol 2005;39:902—11.

[26] Wiedensohler A, Martinsson BG, Hansson H-C. A new unipolar charger for submicron particles. J Aerosol Sci 1990;21(Suppl. 1):S571—4.

[27] Lee HM, Kim CS, Shimada M, Okuyama K. Bipolar diffusion charging for aerosol nanoparticle measurement using a soft X-ray charger. J Aerosol Sci 2005;36:813—29.

[28] Pui DYH, Liu BYH. A submicron aerosol standard and the primary absolute calibration of the condensation nuclei counter. J Colloid Interface Sci 1974;47:155—71.

[29] Hoppel WA. Determination of the aerosol size distribution from the mobility distribution of the charged fraction of aerosols. J Aerosol Sci 1978;9:41—54.

[30] Fissan H, Helsper C, Thielen HJ. Determination of particle size distribution by means of an electrostatic classifier. J Aerosol Sci 1983;14:354—7.

[31] Helsper C, Horn HG, Schneider F, et al. Intercomparison of five mobility particle size spectrometers for measuring atmospheric submicrometer aerosol particles. Gefahrstoffe - Reinhalt Luft 2008;68:475—81.

[32] Asbach C, Kaminski H, Fissan H, et al. Comparison of four mobility particle sizers with different time resolution for stationary exposure measurements. J Nanopart Res 2009b;11:1593—609.

[33] Leskinen J, Joutsensaari J, Lyyränen J, et al. Comparison of nanoparticle measurement instruments for occupational health applications. J Nanopart Res 2012;14:718.
[34] International Organization for Standardization. Determination of particle size distribution - Differential electrical mobility analysis for aerosol particles. Geneva, Switzerland: International Organization for Standardization; 2009. ISO 15900: 2009—05 (E).
[35] International Organization for Standardization. Workplace atmospheres - Characterization of ultrafine aerosols/ nanoaerosols - Determination of the size distribution and number concentration using differential electrical mobility analysing systems. Geneva, Switzerland: International Organization for Standardization; 2011. ISO 28439.
[36] Tammet H, Mirme A, Tamm E. Electrical aerosol spectrometer of Tartu University. Atmos Res 2002;62:315—24.
[37] Jeong CH, Evans GJ. Inter-comparison of a Fast mobility Particle Sizer and a Scanning Mobility Particle Sizer incorporating an ultrafine water-based condensation particle counter. Aerosol Sci Technol 2009;43:364—73.
[38] Bartley DL. Do We Really Need SEVEN New Aerosol Particle Sampling Conventions? Ann Occup Hyg 2011;55(7):692—5.
[39] Comité Européen de Normalisation. Air Quality — Sampling conventions for airborne particle deposition in the human respiratory system. Brussels, Belgium: Comité Européen de Normalisation; 2012. EN ISO13138:2012.
[40] International Organization for Standardization. Nanotechnologies — Guidelines for occupational risk management applied to engineered nanomaterials — Part 1: Principles and approaches. Geneva, Switzerland: International Organization for Standardization; 2011. ISO TC 229.
[41] Cena LG, Anthony TR, Peters TM. A Personal Nanoparticle Respiratory Deposition (NRD) Sampler. Environ Sci Technol 2011;45(15):6483—90.
[42] Gorbunov B, Priest ND, Muir RB, et al. A Novel Size-Selective Airborne Particle Size Fractionating Instrument for Health Risk Evaluation. Ann Occup Hyg 2009;53(3):225—37.
[43] Tsai CJ, Liu CN, Hung SM, et al. Novel Active Personal Nanoparticle Sampler for the Exposure Assessment of Nanoparticles in Workplaces. Environ Sci Technol 2012;46(8):4546—52. dx.doi.org/10.1021/es204580f.
[44] Furuuchi M, Choosong T, Hata M, et al. Development of a Personal Sampler for Evaluating Exposure to Ultrafine Particles. Aerosol Air Qual Res 2010;10(1):30—7.
[45] Hata M, Thongyen T, Bao L, et al. Development of a high-volume air sampler for nanoparticles. Environ Sci: Processes Impacts 2013;15(2):454—62.
[46] Tsai C-J, Chen D-R, Chein H, et al. Theoretical and Experimental Study of an Axial Flow Cyclone for Fine Particle Removal in Vacuum Conditions. J Aerosol Sci 2004;35(9):1105—18.
[47] Hsu Y-D, Chein HM, Chen TM, Tsai C-J. Axial Flow Cyclone for Segregation and Collection of Ultrafine Particles: Theoretical and Experimental Study. Environ Sci Technol 2005;39(5):1299—308.
[48] Tsai C-J, Chen S-C, Przekop R, Moskal A. Study of an Axial Flow Cyclone to Remove Nanoparticles in Vacuum. Environ Sci Technol 2007;41(5):1689—95.
[49] Chen S-C, Tsai C-J. An axial flow cyclone to remove nanoparticles at low pressure conditions. J Nanopart Res 2007;9(1):71—83.
[50] International Organization for Standardization. Workplace air — Determination of metals and metalloids in airborne particulate matter by inductively coupled plasma atomic emission spectrometry — Part 2: Sample preparation. Geneva, Switzerland: International Organization for Standardization; 2012. ISO 15202—2.
[51] International Organization for Standardization. Workplace air — Determination of metals and metalloids in airborne particulate matter by inductively coupled plasma

atomic emission spectrometry — Part 3: Analysis. Geneva, Switzerland: International Organization for Standardization; 2008. ISO 15202—3.

[52] International Organization for Standardization. Workplace air — Determination of metals and metalloids in airborne particulate matter by inductively coupled plasma mass spectrometry. Geneva, Switzerland: International Organization for Standardization; 2010. ISO 30011.

[53] Liu BYH, Pui DYH, Rubow KL. Characteristics of Air Sampling Filter Media. In: Marple VA, Liu BYH, editors. AEROSOLS in the Mining and Industrial Work Environments. USA, 1983: Ann Arbor Science Publishers, Ann Arbor (MI); 1983. p. 989—1038.

[54] Cyrs WD, Boysen DA, Casuccio G, et al. Nanoparticle collection efficiency of capillary pore membrane filters. J Aerosol Sci 2010;41(7):655—64.

[55] Lyyränen J, Backman U, Tapper U, et al. A size selective nanoparticle collection device based on diffusion and thermophoresis. J Phys 2009. Conference Series 170:012011.

[56] R'mili B, Le Bihan OLC, Dutouquet C, et al. Particle Sampling by TEM Grid Filtration. Aerosol Sci Technol 2013;47(7):767—75.

[57] Dye AL, Rhead MM, Trier CJ. A Porous Carbon Film for the Collection of Atmospheric Aerosol for Transmission Electron Microscopy. J Microsc 1997;187(2):134—8.

[58] Tsai S-J, Ada E, Isaacs JA, Ellenbecker MJ. Airborne nanoparticle exposures associated with the manual handling of nanoalumina and nanosilver in fume hoods. J Nanopart Res 2009;11:147—61.

[59] Bard D, Thorpe A, Wake D, et al. Investigations of methods for the sampling of airborne nanoparticles by electron microscopy. In: Inhaled Particles X. Sheffield, UK; 2008.

[60] Dixkens J, Fissan H. Development of an Electrostatic Precipitator for Off-Line Particle Analysis. Aerosol Sci Technol 1990;30(5):438—53.

[61] Miller A, Frey G, King G, Sunderman C. A Handheld Electrostatic Precipitator for Sampling Airborne Particles and Nanoparticles. Aerosol Sci Technol 2010;44(6):417—27.

[62] Thomassen Y, Koch W, Dunkhorst W, et al. Ultrafine particles at workplaces of a primary aluminium smelter. J Environ Monit 2006;8(1):127—33.

[63] Wen J, Wexler AS. Thermophoretic Sampler and its Application in Ultrafine Particle Collection. Aerosol Sci Technol 2007;41(6):624—9.

[64] Thayer D, Koehler KA, Marchese A, Volckens J. A Personal, Thermophoretic Sampler for Airborne Nanoparticles. Aerosol Sci Technol 2011;45(6):744—50.

[65] Li C, Liu S, Zhu Y. Determining Ultrafine Particle Collection Efficiency in a Nanometer Aerosol Sampler. Aerosol Sci Technol 2010;44(11):1027—41.

[66] Bau S, Ouf FX, Miquel S, et al. Experimental measurement of the collection efficiency of nanoparticle samplers based on electrostatic and thermophoretic precipitation. In: International Aerosol Conference 2010. Helsinki, Finland; 2010.

[67] Nash DG, Leith D. Ultrafine Particle Sampling with the UNC Passive Aerosol Sampler. Aerosol Sci Technol 2010;44(12):1059—64.

[68] BOHS/NAVVA. Testing Compliance with Occupational Exposure Limits for Airborne Substances. Derby/Eindhoven; 2011. http://www.bohs.org/library/technical-publications.

[69] Peters TM, Elzey S, Johnson R, et al. Airborne Monitoring to Distinguish Engineered Nanomaterials from Incidental Particles for Environmental Health and Safety. J Occup Environ Hyg 2009;6:73—81.

[70] NIOSH. Draft current intelligence bulletin: occupational exposure to carbon nanotubes and nanofibers. Cincinnati, OH: US Department of Health and Human Services, Centers for Disease Control, National Institute for Occupational safety and Health; 2013. DHHS (NIOSH), NIOSH Docket Number: NIOSH 161-A; Available at http://www.cdc.gov/niosh/docket/review/docket161A/.

[71] NIOSH. NIOSH manual of analytical methods (NMAM®), Diesel particulate matter (supplement issued 3/15/03). 4th ed. In: Schlecht PC, O'Conner PF, editors. Cincinnati,

Ohio: U.S. Department of Health and Human Services, Centers for Disease Control and Prevention, National Institute for Occupational Safety and Health; 1994. DHHS (NIOSH) Publication No. 94-113 [www.cdc.gov/niosh/nmam/].

[72] Crilley LR, Knibbs LD, Miljevic B, et al. Concentration and oxidative potential of on-road particle emissions and their relationship with traffic composition: Relevance to exposure assessment. Atmos Environ 2012;59:533−9.

[73] Smita S, Gupta SK, Bartonova A, et al. Nanoparticles in the environment: Assessment using the causal diagram approach. Environ Health 2012;11(Suppl. 1):13.

[74] NIOSH. Current intelligence bulletin 63: occupational exposure to titanium dioxide. Cincinnati, OH: Department of Health and Human Services, Public health Service, Center for Disease Control and Prevention; 2011. Publication No. 2011−160.

[75] Methner M, Hodson L, Geraci C. Nanoparticle emission assessment technique (NEAT) for the identification and measurement of potential inhalation exposure to engineered nanomaterials—part A. J Occup Environ Hyg 2010;7:127−32.

[76] BSI. Nanotechnologies—part 3: guide to assessing airborne exposure in occupational settings relevant to nanomaterials. London, UK: British Standards Institution; 2010 (BSI PD 6699-3:2010).

[77] VCI. Tiered Approach to an Exposure Measurement and Assessment of Nanoscale Aerosols Released from Engineered Nanomaterials in Workplace Operations. Presented by: IUTA, BAuA, BG RCI, VCI, IFA, TUD, https://www.vci.de/Downloads/Nanomaterials%20in%20Workplace%20Operations.pdf; 2011.

[78] Witschger O, Le Bihan O, Reynier M, et al. Recommendations for characterizing potential emissions and exposure to aerosols released from nanomaterials in workplace operations. Hyg Secur Trav - 1er trimestre 2012;226:41−55.

[79] McGarry P, Morawska L, Morris H, et al. Measurements of particle emissions from nanotechnology processes, with assessment of measuring techniques and workplace controls. Safe Work Australia; 2012.

[80] Brouwer DH. Control Banding Approaches for Nanomaterials. Ann Occup Hyg 2012b;56:506−14. http://dx.doi.org/10.1093/anhyg/MES039.

[81] Van Broekhuizen P, Van Broekhuizen F, Cornelissen R, Reijnders L. Workplace exposure to nanoparticles and the application of provisional nanoreference values in times of uncertain risks. J Nanoparticle Res 2012;14(4), art. no. 770.

6

Quality Control of Measurement Devices — What Can Be Done to Guarantee High-Quality Measurements?

Hans-Georg Horn [1], Dirk Dahmann [2], Christof Asbach [3]

[1] TSI GmbH, Aachen, Germany

[2] IGF Institut für Gefahrstoff-Forschung, Institut an der Ruhr-Universität Bochum, Bochum, Germany

[3] Institute of Energy and Environmental Technology (IUTA), Air Quality & Sustainable Nanotechnology, Duisburg, Germany

6.1 RESPONSIBILITIES OF THE MANUFACTURERS (HORN)

6.1.1 Introduction

Simply said, the measurement device manufacturer's one and only responsibility with respect to device quality is to ensure that each product handed over to a user fulfils all its specifications. This relates to technical specifications as well as specifications on device reliability. Technical specifications are typically verified by applying standards or guidelines. As simple as this sounds, there are several obstacles to overcome when measurement devices use principles or measure quantities for which no standards exist. In this chapter, we outline a strategy that should be used in such cases.

6.1.2 Technical Specifications

Technical specifications are the base of any measurement device description; they are crucial to define requirements for device verification, validation, calibration and evaluation of device quality. The final goal of a set of technical specifications is to describe or define:

- the measured quantities (measurands, ranges and uncertainties);
- the range of ambient and sampling conditions (e.g., temperature, pressure, relative humidity); and
- physical device properties like size, weight and power consumption.

Since ENP measurement devices are typically a combination of several subsystems like a flow system, detector(s), electronics and firmware/ software algorithms for data acquisition and inversion, secondary specifications like the inlet flow accuracy become necessary, especially to define requirements for device end control and for routine device performance checks by the user (Figure 6.1).

While device design normally starts from a set of critical requirements derived from the needs of an application — like ENP workplace monitoring — technical specifications of a device can only be determined after an extensive validation and verification program, where the influence of,

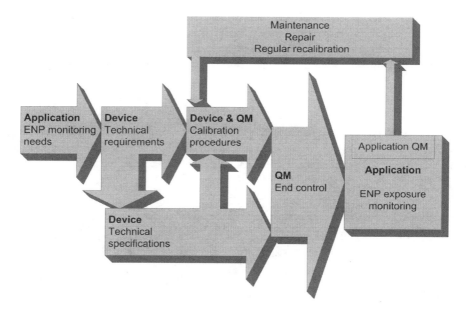

FIGURE 6.1   From application-specific ENP monitoring needs to the application of a monitoring device.

e.g., ambient conditions on the measurement range and on the measurement uncertainty, are derived by means of statistical analysis.

The Condensation Particle Counter (CPC) is a well-established instrument that measures particle number concentration in the size range from a few nanometers to several micrometers. We use this instrument as an example to explain technical specifications. The measurand is particle number concentration, i.e., the number of particles suspended in a unit volume of carrier gas. A working fluid vapor condenses on the particles, which then grow to micrometer-sized working fluid droplets. These droplets can easily be detected and counted. Particle number concentration can be calculated from count rate and measurement flow rate. The particle detection efficiency of a CPC is close to 100% over a wide particle size range; however, it drops to zero for particles too small to act as a nucleus for condensation or too large to be transported to the detection optics. The particle concentration range is limited by coincidence losses (the higher the concentration the higher the probability that more than one particle passes the detection optics at the same time, which results in undercounting). A good example of technical requirements for a CPC is found in UN-ECE Regulation 83 [1], which standardizes the use of a CPC for the particle number emission measurements for the certification of light-duty vehicles. Figure 6.2 shows the most important technical requirements derived from the application needs together with the set of technical specifications of a CPC resulting from these requirements. As can be seen, the requirements must be converted into a set of technical specifications that guarantee that the requirements are met. The technical requirements define, e.g., the tolerable range of the inlet flow rate, internal component temperatures, etc. and finally the technical requirements themselves. Technical specifications then form the base for device end control and for production quality management, as described in Section 6.1.4.

Generally spoken, application-related quality management for any measurement device will only be of use for an application if application-specific needs, related technical requirements as well as the resulting technical specifications of the measurement device are known.

6.1.3 Device Calibration

When calibrating a measurement device, the deviation of measured and 'true' values must be determined. 'True' values are normally based on traceable standards provided by national metrology institutes (NMIs). For ENP measurement devices this holds, for example, true for a measurand like particle size, where comparison of the particle size reported by a device with the NMI-certified size of the particles used for calibration is common practice (see, for example, [2] or [3]).

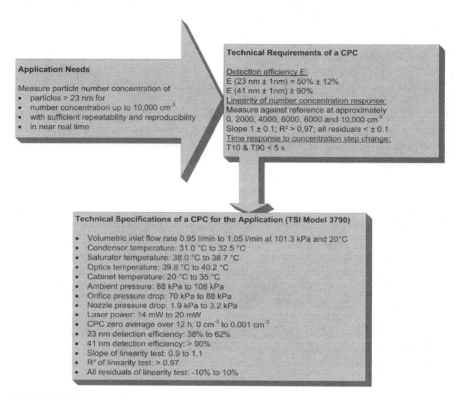

Application Needs

Measure particle number concentration of
- particles > 23 nm for
- number concentration up to 10,000 cm⁻³
- with sufficient repeatability and reproducibility
- in near real time

Technical Requirements of a CPC

Detection efficiency E:
E (23 nm ± 1nm) = 50% ± 12%
E (41 nm ± 1nm) ≥ 90%
Linearity of number concentration response:
Measure against reference at approximately
0, 2000, 4000, 6000, 8000 and 10,000 cm⁻³
Slope 1 ± 0.1; R² > 0,97; all residuals < ± 0.1
Time response to concentration step change:
T10 & T90 < 5 s

Technical Specifications of a CPC for the Application (TSI Model 3790)

- Volumetric inlet flow rate 0.95 l/min to 1.05 l/min at 101.3 kPa and 20°C
- Condensor temperature: 31.0 °C to 32.5 °C
- Saturator temperature: 38.0 °C to 38.7 °C
- Optics temperature: 39.8 °C to 40.2 °C
- Cabinet temperature: 20 °C to 35 °C
- Ambient pressure: 88 kPa to 108 kPa
- Orifice pressure drop: 70 kPa to 88 kPa
- Nozzle pressure drop: 1.9 kPa to 3.2 kPa
- Laser power: 14 mW to 20 mW
- CPC zero average over 12 h: 0 cm⁻³ to 0.001 cm⁻³
- 23 nm detection efficiency: 38% to 62%
- 41 nm detection efficiency: > 90%
- Slope of linearity test: 0.9 to 1.1
- R² of linearity test: > 0.97
- All residuals of linearity test: -10% to 10%

FIGURE 6.2   Example of technical specifications derived from the application needs
and technical requirements for nanoparticle emission measurements from combustion
engines. *UN/ECE, 2012.*

For many measurands, traceable calibration standards are not available, either because the measurand is not yet established or because a traceable standard cannot be produced. The latter is, for example, the case for the measurand 'particle number concentration of an ENP aerosol'. As described above, measurement of the number concentration of aerosol particles with the CPC is common practice. However, due to unavoidable processes like diffusion loss or coagulation, there is no way to preserve aerosol particle number concentration and use 'aerosol particles in a bottle' as traceable calibration material. One way to resolve this problem is to define a reference method for the determination of the desired measurand [4—6]. NMIs can provide calibration of devices traceable to the reference method, and these devices can then themselves serve as traceable transfer standards for further calibrations.

For the measurement of particle number concentration of aerosols with CPCs, the European Association of National Metrology Institutes

TABLE 6.1   Applicable Device Calibration Methods

Situation	Calibration Method	Example
Traceable standard material for measurand available	Calibrate device with standard material	Calibrate particle size measurement with NMI-traceable PSL particles
Traceable alternative measurement method available	Calibrate device by correlation with alternative method	Calibrate particle number concentration measurement by correlation with aerosol electrometer using singly charged particles
Neither standard material nor alternative measurement method are available for the measurand, but sufficient (at least 10) devices exist	Use statistical criteria on multiple devices of the same kind to determine average and measurement uncertainty. Use average for calibration of the devices	Device manufacturer determines average measurement values based on a batch of devices. A reference device is calibrated to measure this average value. This reference device is then used for further calibrations
Neither standard material nor alternative measurement method are available for the measurand and only a non-representative number of devices exists	Full calibration impossible, only subsystems of the device can be calibrated	Flow control system, electrical circuits and optical system of a prototype instrument are calibrated separately, but the prototype itself cannot be calibrated in this situation

(EURAMET) works towards establishing a traceable reference [7]. Generalizing this example, there are basically three ways to do a device calibration (see also Table 6.1).

If linearity of response for measured quantities is not inherent in a method, calibration becomes a two-dimensional problem. Both the correct determination of the measurand (e.g., particle size) and of a related quantity (e.g., particle number concentration at this size) become part of the calibration procedure.

For most particle measurement calibration methods, the result is only valid if the device under calibration afterwards also measures the reference material. This leads to the concept of equivalent properties. If, for example, the particle size measurement of an optical particle spectrometer was calibrated with spherical polystyrene latex (PSL) particles, the result of a measurement of other aerosol particles must be expressed as 'optical PSL equivalent size'. This means that the 'true' size of a measured particle not consisting of PSL and/or not being spherical in shape remains unknown, even with a calibrated optical particle spectrometer. It can only be stated that the measured particle scatters the same amount of light in the optical particle spectrometer as a spherical PSL particle of the reported size would do.

6.1.4 End Control and Product Quality Management

End control in a production process for particle measurement devices is typically a combination of device calibration and checking a set of critical specifications against their allowed tolerances. Which comes first depends on the nature of the calibration process. If operating parameters of a device are tuned to bring the device's measurement results as close as possible to the calibration reference, the check of the critical specifications comes last. On the other hand, if the calibration only reports the deviation from the reference or if correction algorithms are applied for this correction, the check of the critical specifications may come first. In the latter case, tolerances for allowed corrections are helpful to determine whether or not a device passes the end control. Figure 6.3 shows this process.

Unless required — and typically paid for — by an end user, the end control process is of general nature and not directly related to a specific application. Methods used for factory calibration and end control are most often chosen with respect to good repeatability and reproducibility, but costs of calibration and end control per device also play a role.

Maintaining high production standards is one of the important goals of production quality management, therefore the statistical analysis of the quality of incoming goods for production as well as the results of the end control are very important control tools [8]. While the fraction of devices passing the end control itself is simply a cost factor, a decrease of this fraction or strong variations over time should trigger analysis to find and

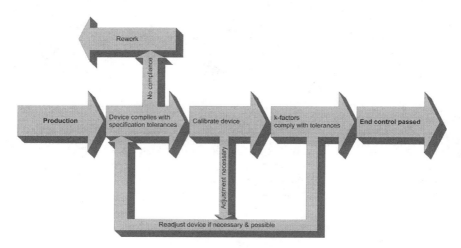

FIGURE 6.3   A typical end control process scheme.

correct the root cause. The standard deviation of the results of each parameter checked in end control (or a multiple of it, e.g., six sigma) compared to its allowed tolerances can be used as feedback into production process optimization [9].

6.2 RESPONSIBILITIES OF THE USERS (*DAHMANN*)

6.2.1 Introduction

Devices for measuring airborne concentrations of nanoparticles can be classified according to their intended use, their mode of operation (direct reading or not; 'portable'; person-mountable, which may be different from the former; battery-operated or not, etc.) and their requirements with respect to the qualification of operating personnel.

A good description of the possible ways measurement devices can be used in workplaces can be found in [10]. For example, representative measurements, worst-case measurements or measurements in order to identify and quantify source emissions should be mentioned here. The question of when to apply which measurement strategy is discussed elsewhere in the *Handbook* (see Chapter 5). If quantification of exposure is the aim of the application of measurement devices (which may not always be the case, as source identification could only require *relative* signal differences to be recorded), then EN 482 [11] defines quality requirements. The most important characteristic for a given measurement procedure is its so-called extended uncertainty as a measure for the expected combination of random and nonrandom uncertainties for sampling step and analytical step. In the case of direct reading instruments, these steps are normally integrated, whereas for procedures where a sampling method is applied to retrieve 'samples' (e.g., filter samples) for a subsequent analytical step (e.g., SEM), they are to be regarded separately. The standard EN 482 defines minimum requirements for procedures intended to be used for compliance measurements (i.e., comparison to threshold limits). Formally in the absence of almost all kinds of nationally regulated threshold limits for ENPs these requirements are not directly applicable, but it is strongly suggested to try to comply with them anyway, as it can be assumed as very likely that current quantitative exposure measurements will be the base of future threshold limits. Therefore they should follow the same quality rules as future compliance measurements will be expected to.

A current approach to a suitable and economically reasonable way to perform measurements of airborne nanoparticle concentrations is to do it in a stepped way or in so-called 'tiers' (e.g., [12,13]). When this approach is used, a first assessment step (called tier 2) would be to perform orientation or screening measurements to find out in which *potential* workplaces with nanomaterial handling (where the identification of these workplaces would have taken place depending on written information in a so-called tier 1 step) these can be localized as being present above background. In these measurements 'simple', fast and direct-reading devices would be used. It is also at this stage not yet really important to select a high-quality (and possibly expensive in equipment and manpower) method or to have exceptional regards as to which metric be used. If nano-objects were found 'above background' or if the simple tier 2 methods were not adequate, an expert assessment or tier 3 method would have to be applied. While the essence of this approach is to manage effort and in consequence costs, it should be noted that when requirements for quality control are discussed, tier 2 measurements are not to be regarded as 'measurement light', where the respective investigations could be performed by unskilled labor. Quite to the contrary, tier 2 measurements also require an adequate level of quality control and must not be treated as of minor importance. Similar approaches of 'stepped' depths of measurements have been proposed by various agencies (e.g., INRS, F; TNO, NL; NIOSH, USA).

The following text will therefore not differentiate between these two types of measurements.

6.2.2 Primary Quality Control of Devices and Procedures

There are some fundamental limiting factors if calibration of measurement devices for nano-objects in workplace air is intended. Most important is the fact that no primary standards exist. Currently it is not possible for users to generate aerosols of sufficiently well-known concentration and number distribution to calibrate their systems. Recently Koch and coworkers [14] have published a procedure to generate aerosols by Brownian coagulation with liquid droplet particles (and possibly for spark-generated solid particles as well) in number concentrations, which may be calculated from simple environmental data. Also, Yli-Ojanpera et al. have suggested a procedure leading to a traceable particle number concentration standard [15]. These procedures are currently not commercially available and are possibly also better suited for manufacturers because of the considerable effort it would create for users to implement them.

Another procedure principally if not practically suited for users to calibrate their systems is the so-called CAST (i.e., Combustion Aerosol Standard) device, which produces combustion aerosols under well-controlled conditions [16]. The system is commercially available but considered too much effort for the typical users' needs. Additionally, currently, at least for CPC number concentration calibration, CAST is accepted only as a source of primary calibration particles. DMA classification and a reference detector (CPC or electrometer) are necessary to do calibration measurements.

Another possibility is to make use of certified polystyrene latex (PSL) suspensions, which are available down to 20 nm; however, with an atomizer, only particles >100 nm can be meaningfully aerosolized (e.g., [17]). These may then be used to reproducibly produce mono-disperse aerosols with known properties. It does of course require the availability of suitable aerosol generation equipment and is basically also only a method to calibrate the sizing of instruments and not concentration.

While the practical possibilities for users to calibrate their instruments are limited, the requirements for maintenance are getting more impor-tant as a consequence. It is very important to follow the instructions for use by the manufacturer on one hand, and for the latter to consequently follow the processes described in Section 6.1. An example may be the careful monitoring of the necessary liquid (e.g., water or alcohols) in CPC devices. The user should set up a routine for performance and documentation of these maintenance procedures according to the in-structions for use.

An all-important part of the calibration process of those measurement devices that apply directed airflows to measure the aerosols (i.e., pumps) is to regularly calibrate these airflows. Experience has shown that failing air flow controls can lead to problems in measurement of nano-objects in air [18]. Readily commercially available devices for airflow calibrations include mass flow meters, bubble meters and rotameters. These devices are not further specified in this handbook.

6.2.3 Secondary Quality Control of Procedures

While the possibilities of direct calibration of measurement devices are limited (see above), all possible care should be taken to apply them in workplace measurements in a way that all data collected with them are traceable afterwards. First and most important, they should always be used following a written standard operation procedure (SOP) for the intended use. These SOPs possibly need to differentiate the various measurement purposes (see above) and the actual use of the instruments in the field and their maintenance in the lab (see above). Examples for this approach can be downloaded from the German NanoGEM project website [13], which displays general procedures for background distinc-tion screening measurements, expert assessment 'scientific' measure-ments of concentration and size distribution as well as several instrument-related individual SOPs. The overall purpose of the approach to follow written instructions (SOPs) is that all individual re-sults can be reproducibly traced back to the conditions present in the workplace for which they have been obtained. This involves a detailed description of the conditions under which the equipment has been operated (e.g., flow rates, scan times, scan directions, etc.), the location and time regime of the measurements, the detailed description of the respective workplace conditions, especially all details regarding cross sensitivities from ultrafine particles or the background in general, and all exposure determinants of the workplaces in question.

In general it should be the aim of these measures to allow a third person trying to follow the exposure data collected to completely understand under which circumstances and for which conditions they are valid. This sounds trivial, but experience shows that it is far from being universally applied.

6.3 EXTERNAL INSTRUMENT COMPARISON (ASBACH)

As mentioned in Sections 6.1 and 6.2, there is currently no standard for the calibration of aerosol instruments measuring particle number or surface area concentrations as well as particle size distributions. Despite several efforts towards the development of reference aerosol generators [14,15], the calibration of aerosol instruments is commonly only done by the instrument manufacturers using their own protocols (see Section 6.1). Due to the lack of a readily available 'reference aerosol' for calibration, the only feasible way for end users to check the plausibility of particle concentrations (number, surface area, etc.) measured by aerosol instruments is by simultaneous measurements with instruments of the same or similar type and eventual comparison of the results. For most measurement tasks this is absolutely sufficient, because the comparability of aerosol instruments is often more important than the absolute accuracy. For example, when assessing workplace exposure to airborne nanomaterials by simultaneously measuring the background aerosol in a representative background site and the workplace aerosol in or near the breathing zone of a worker, it needs to be made sure that potential differences between the two measurements are indeed caused by the aerosol and not by the employed instruments. Hence the instruments need to be checked for their comparability (and not necessarily accuracy) prior to a measurement campaign.

A simple and quick intercomparison of instruments measuring particle size-integrated concentrations can be done by operating the instruments side-by-side while sampling room air and comparing the measured concentrations. This is certainly sufficient for a quick-and-easy check of the overall performance of the instruments and to identify any obvious instrument failures. The main downside, however, is that the aerosol as well as the sampling is undefined. Even though the instruments may be located very close to each other and sampling from the same air space, the aerosol can show spatial variations of the concentration, causing potential differences of the instruments' readings. Furthermore the aerosol characteristics, especially the size distribution, are unknown. When comparing instruments that cover different particle size ranges, this can be a major problem if the particle size distribution shows contributions in the nonoverlapping size range. The use of defined, lab-produced aerosols and well-laid-out sampling techniques that assure that all instruments sample the same aerosol is hence essential for a quantitatively meaningful intercomparison of instruments. Details about such intercomparison studies can be found in [18–23]. Since different particle and aerosol properties can affect the measurement accuracy and comparability of instruments, Asbach et al. [18,19] and Kaminski et al. [20] chose a variety of different aerosols. Their studies were carried out at the nanoTestCenter at the Institute for the Research on Hazardous Substances (IGF) in Dortmund, Germany [24], which has been used as a tool for quality control in the NANO-DEVICE project. Spherical or at least very compact NaCl particles or droplets of a liquid with high surface tension (DEHS) were produced by dispersing an NaCl solution or the pure liquid, respectively, with an atomizer. The size of NaCl particles can be adjusted between roughly 30 and 150 nm modal diameter via the NaCl concentration in the dispersed solution. Atomized DEHS commonly shows size distributions with modal diameters around 200–300 nm. Agglomerated particles were freshly produced either by a commercial spark discharge generator (Palas GFG1000) or a home-built diffusion burner [25]. The size distribution and degree of agglomeration could be adjusted by means of the spark discharge frequency and dilution flow rate (spark discharge generator) and the precursor material, feed rate and the dilution flow rate (diffusion burner), respectively. Additionally, the exhaust from a diesel engine could be used as test aerosol but was only employed in two of the aforementioned studies [19,22].

The aerosols were electrically neutralized and — in the case of wet-dispersed NaCl — dried in a silica gel diffusion dryer before they were fed into a 20 m long wind tunnel with a diameter of 0.7 m, where they were mixed with a defined and controllable amount of dilution air. The amount of dilution air defines the eventual particle concentration and in case of high initial concentrations also limits the growth of particles by coagulation. The wind tunnel feeds into a 20 m^3 measurement chamber. It has been shown that the aerosol inside the chamber is homogeneously distributed such that all instruments placed inside the chamber sampled the same aerosol. In addition, instruments that require frequent attention, e.g., SMPSs, can be placed outside the measurement chamber and sampled from the chamber through a T-shaped sampling train. In their latest measurements, Asbach et al. [18] and Kaminski et al [20] produced NaCl particle size distributions with a modal diameter around 35—40 nm, DEHS droplets with modal diameter of 250 and 200 nm, respectively and spark-generated, agglomerated soot particles with modal diameters between 30 and 100 nm. These materials were chosen to cover a wide range of particle sizes and morphologies because both may affect the particle measurement. DEHS droplets were spherical, NaCl particles cubic (and hence close to spherical particles concerning their airborne behavior) and soot particles chain-like agglomerates of very small primary particles. Especially all measurements that include particle charging can be biased, because the charging characteristics of particles are affected by the particle morphology [26,27]. All aerosols were produced in at least two different particle number concentrations, usually one around $20,000/cm^3$, mimicking a common workplace concentration, and another one around $100,000/cm^3$, which is the upper limit of several CPC models. The concentrations were kept constant for at least one hour in order to obtain sufficient data sets also for rather slow-reading instruments like mobility particle sizers with time resolution of several minutes. In order to check the dynamic response of especially fast-reading instruments like CPCs or diffusion chargers, the concentrations were quickly ramped before and after each one-hour run.

6.3.1 Data Evaluation

For evaluating the data taken during times of constant aerosols, timelines of data from all instruments measuring the same aerosol quantity (e.g., number or surface area concentration) can be plotted in one graph. Such graphs help identify outliers, i.e., single instruments that showed strongly different results than the others or single data points that are out of range and obviously caused by short-lived instrument failures. Such single outliers should be removed from the data evaluation and deleted. If they occur frequently, this should be noted. Quantitative analysis of timelines is not easily possible. Instead it was shown [18] that regression analysis in which results from one instrument are plotted against the results from another instrument for equal times is a better choice. An example of such an analysis is shown in Figure 6.4 for number concentration data of an NaCl aerosol, measured with a Miniature Diffusion Size Classifier (miniDiSC, [28]) and a handheld version of a Condensation Particle Counter (handheld CPC, TSI model 3007). The data were plotted against each other and the linear fit equation determined by a least-square fit. The resulting fit equation is given in the graph. It can be seen that the slope of the fit is very close to one and the y-intersect rather low, indicating a good agreement between the two instruments.

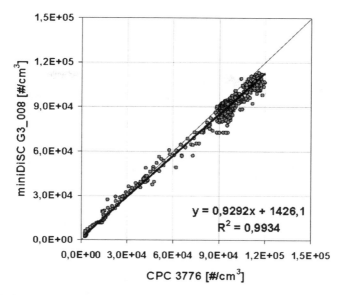

FIGURE 6.4 Regression graph showing number concentrations measured with a miniDiSC against number concentrations measured with a handheld CPC. *Data from Kaminski et al. [20].*

A regression coefficient of >0.99 furthermore indicates that there was only a very low scatter of the measured concentrations. It is hence clear that the three parameters from the linear fit, i.e., slope, y-intersect and regression coefficient, can very well describe the comparability of two instruments. The slope shows the absolute mean agreement between the two instruments. A large y-intersect indicates a permanent offset of one instrument. The R^2 value describes the quality of the fit. If all data points fall onto the fit curve, then $R^2 = 1$, and lower R^2 values indicate that the data points are more scattered. In the aforementioned studies, the regression coefficients of the analyses were usually close to one, indicating a high correlation between the instruments.

For instruments measuring particle-size-resolved concentrations, firstly the single size distributions of each instrument should be plotted in one graph per instrument. The left graph in Figure 6.5 shows examples for size distributions of a sodium chloride aerosol, measured with eight SMPSs from two manufacturers A and B. The graph in Figure 6.5 allows for the identification of single outliers. Unless they are systematic, these should again be deleted and omitted from further evaluation. In Figure 6.5 instrument SMPS-B3 consistently showed higher concentrations, whereas SMPS-B4 size distributions were continuously below the other distributions. Those differences were systematic and could hence

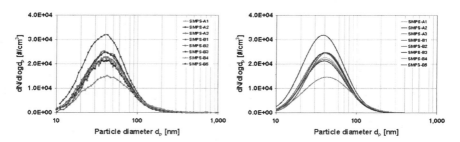

FIGURE 6.5   Averaged size distributions of a sodium chloride aerosol measured with different SMPSs from two manufacturers A and B (left) and fitted lognormal distributions (right).

not be neglected. In general, it was shown in different studies [19,20] that total number concentrations measured with different SMPSs usually agree to within approximately ±10%, if the SMPSs are properly used. To obtain the mean size distribution measured by an instrument, the mean concentration and standard deviation should be calculated for each particle size bin separately. A graphical comparison of the averaged size distributions is possible by plotting them in the same graph. Such comparison is, however, qualitative rather than quantitative. Asbach et al. [19] and Kaminski et al. [20] instead fitted the averaged size distribution to lognormal distributions:

$$n(d_p) = \frac{N_\infty}{\sqrt{2\pi}.d_p.ln(\sigma_g)}.exp\left[-\frac{\left(ln(d_p) - ln(d_{p.g})\right)^2}{2.ln^2(\sigma_g)} \right]$$

where N_∞ is the total particle concentration, d_p is the particle diameter, σ_g is the geometric standard deviation and $d_{p.g}$ is the geometric mean diameter. Since all test aerosols were freshly produced, the assumption of lognormal distributions was justified, as was expressed in the regression coefficients R^2 of the fits that were all very close to one. The fitted size distributions are shown in the right graph in Figure 6.5.

Fitting of the data allows for easy and direct comparison of the three parameters that determine a lognormal distribution, i.e., the geometric mean diameter (alternatively also the modal diameter), the geometric standard deviation and the total particle number concentration. While comparison of the geometric mean diameter provides insight into the sizing accuracy of the instruments, the reported total number concentration expresses the comparability of the CPC readings as well as the data deconvolution algorithms. In addition, the reported total number concentration may in itself be directly compared to results of 'integrating' instruments like handheld CPCs or diffusion charging instruments (see above) if their respective measurement ranges are covering the same particle size ranges.

References

[1] UNECE. Uniform provisions concerning the approval of vehicles with regard to the emission of pollutants according to engine fuel requirements. Regulation No. 83 of the Economic Commission for Europe of the United Nations (UNECE). —Official Journal of The European Union; 2012. 15.02.2012, L42/1 ff.

[2] International Organization for Standardization. Determination of particle size distribution — Single particle light interaction methods — Part 1: Light scattering aerosol spectrometer. Geneva, Switzerland: International Organization for Standardization; 2009. ISO 21501—1.

[3] International Organization for Standardization. Determination of particle size distribution — Single particle light interaction methods — Part 4: Light scattering airborne particle counter for clean spaces. Geneva, Switzerland: International Organization for Standardization; 2007. ISO 21501—4.

[4] Liu W, Osmondson B, Bischof O, Sem G. Calibration of Condensation Particle Counters; 2005. Technical Paper 2005-01-0189. http://dx.doi.org/10.4271/2005-01-0189 SAE.

[5] Liu BYH, Pui DYH. A Submicron Aerosol Standard and the Primary Absolute Calibration of the Condensation Nucleus Counter. J Colloid Interface Sci 1974;47:155—71.

[6] International Organization for Standardization. Aerosol particle number concentration — Calibration of condensation particle counters. Geneva, Switzerland: International Organization for Standardization; 2012. ISO/CD 27891.

[7] Buhr E, et al. Calibration service for mobility measurements based on reference material for monodisperse spherical particles in aerosols. EMRP ENV02 PartEmission Report: Deliverable 1.1.1, 2012-07-23. http://www.ptb.de/emrp/partemission-publications.html; 2012.

[8] Pydzek T, Keller PA. Quality Engineering Handbook. CRC Press; 2003. 2nd revised edition.

[9] Pydzek T, Keller PA. The Six Sigma Handbook. 3rd ed. McGraw-Hill Professional; 2009.

[10] British-Adopted European Standard. Workplace Atmospheres — Guidance for the assessment of exposure by inhalation to chemical agents for comparison with limit values and measurement strategy. British-Adopted European Standard; 1995. EN 689.

[11] British-Adopted European Standard. Workplace Exposure — General requirements for the performance of procedures for the measurement of chemical agents. British-Adopted European Standard; EN 482; 2012.

[12] VCI. Tiered Approach to an Exposure Measurement and Assessment of Nanoscale Aerosols Released from Engineered Nanomaterials in Workplace Operations. http://www.vci.de/Downloads/Tiered-Approach.pdf; 2011.

[13] Asbach C, Kuhlbusch TAJ, Kaminski H, et al. Standard Operation Procedures For assessing exposure to nanomaterials, following a tiered approach. NanoGem. http://www.nanogem.de/cms/nanogem/upload/Veroeffentlichungen/nanoGEM_SOPs_Tiered_Approach.pdf; 2012a.

[14] Koch W, Lödding H, Pohlmann G. A reference aerosol generator based on Brownian coagulation in a continuously fed well stirred tank reactor. J Aerosol Sci 2012;49:1—8.

[15] Yli-Ojanpera J, Mäkelä JM, Marjamäki M, et al. Towards traceable particle number concentration standard: Single charged aerosol reference (SCAR). J Aerosol Sci 2010; 41:719—28.

[16] Jing L. Standard Combustion Aerosol Generator (SCAG) for Calibration Purposes. 3rd ETH Workshop "Nanoparticle Measurement", ETH Hönggerberg, Zürich; 1999.

[17] Yook SJ, Fissan H, Engelke T, et al. Classification of highly monodisperse nanoparticles of NIST-traceable sizes by TDMA and control of deposition spot size on a surface by electrophoresis. J Aerosol Sci 2008;39:537—48.

[18] Asbach C, Kaminski H, von Barany D, et al. comparability of Portable Nanoparticle Exposure Monitors. Ann Occup Hyg 2012b;56:606—21.

[19] Asbach C, Kaminski H, Fissan H, et al. Comparison of four mobility particle sizers with different time resolution for stationary exposure measurements. J Nanopart Res 2009;11:1593—609.

[20] Kaminski H, Kuhlbusch TAJ, Rath S, et al. Comparability of mobility particle sizers and diffusion chargers. J Aerosol Sci 2013;57:156—78.

[21] Helsper C, Horn HG, Schneider F, et al. Intercomparison of five mobility particle size spectrometers for measuring atmospheric submicrometer aerosol particles. Gefahrstoffe—Reinhalt Luft 2008;68:475—81.

[22] Dahmann D, Riediger G, Schletter J, et al. Intercomparison of mobility particle sizers (MPS). Gefahrstoffe—Reinhalt Luft 2001;61:423—7.

[23] Leskinen J, Joutsensaari J, Lyyränen J, et al. Comparison of nanoparticle measurement instruments for occupational health applications. J Nanopart Res 2012;14:718.

[24] Dahmann D. A novel test stand for the generation of diesel particulate matter. Ann Occup Hyg 1997;41(Suppl. 1):43—8.

[25] Monsé C, Dahmann D, Asbach C, et al. Development and evaluation of a nanoparticle generator for human inhalation studies. Int J Occup Environ Health 2012 (in preparation).

[26] Wen HY, Reischl GP, Kasper G. Bipolar diffusion charging of fibrous aerosol particles—I. Charging theory. J Aerosol Sci 1984;15:89—101.

[27] Shin WG, Wang J, Mertler M, et al. The effect of particle morphology on unipolar diffusion charging of nanoparticle agglomerates in the transition regime. J Aerosol Sci 2010;41:975—86.

[28] Fierz M, Houle C, Steigmeier P, Burtscher H. Design, calibration, and field performance of a Miniature Diffusion Size Classifier. Aerosol Sci Technol 2010;45:1—10.

Examples and Case Studies

Christof Asbach[1], Olivier Aguerre[3],
Christophe Bressot[3], Derk H. Brouwer[4],
Udo Gommel[5], Boris Gorbunov[6],
Olivier Le Bihan[3], Keld Alstrup Jensen[7],
Heinz Kaminski[1], Markus Keller[5],
Ismo Kalevi Koponen[7], Thomas A.J. Kuhlbusch[1,2],
Andre Lecloux[8], Martin Morgeneyer[9],
Robert Muir[6], Neeraj Shandilya[3,9],
Burkhard Stahlmecke[1], Ana María Todea[1]

[1] Institute of Energy and Environmental Technology (IUTA),
Air Quality & Sustainable Nanotechnology, Duisburg, Germany
[2] Center for NanoIntegration Duisburg-Essen (CENIDE),
Duisburg, Germany
[3] Institut National de l'Environnement industriel et des RISques (INERIS),
Verneuil-en-Halatte, France
[4] TNO, Quality of Life, Research & Development, Zeist, The Netherlands
[5] Fraunhofer Institute for Manufacturing Engineering and Automation IPA,
Department Ultraclean Technology and Micromanufacturing,
Stuttgart, Germany
[6] Naneum Ltd., Canterbury Innovation Center, Canterbury, United Kingdom
[7] Danish Nanosafety Centre, National Research Centre for the Working
Environment, Copenhagen, Denmark
[8] Belgium Nanocyl s.a., Sambreville, Belgium
[9] Université de Technologie de Compiègne (UTC), Compiègne, France

Handbook of Nanosafety
http://dx.doi.org/10.1016/B978-0-12-416604-2.00007-X

223

7.1 INTRODUCTION

Nanomaterials and nanoparticles are used in a wide range of applications and have become part of everyday life [1]. With the increased use of nanomaterials, concerns have been raised about their potential negative impacts on human health, the environment and − due to contamination − on product quality. In view of a sustainable development of nanotechnologies, possible risks have to be investigated and communicated. A risk posed by a nanomaterial may only arise if exposure and a hazard potential coexist [2]. Hazard potential may be expressed as the material's human or eco-toxicity or as its potential to harm the (clean) production of products, e.g., semiconductor chips. These effects, particularly with respect to human health, are discussed elsewhere in this book, whereas the focus of this chapter is on the release and consequently the potential exposure to nanomaterials. The starting point for exposure is a release of nanomaterials during any step in their life-cycle, e.g., during their synthesis, further processing, handling, downstream use or treatment at the end of their life-cycle. Due to the vast number of combinations of nanomaterials and potential processes which may lead to their release, currently only case by case studies can be conducted. It will be desirable at some point in the future to group similar combinations such that a limited number of studies will be sufficient to obtain the required information. However, thus far too little is known about the different release processes such that only very specific case studies can be conducted. This chapter focuses on nanomaterial release to the airborne state by presenting five case studies. One major problem of release and exposure related measurements is the differentiation of released particles from ubiquitous background particles. Commonly used equipment for aerosol measurement only physically characterizes airborne particles and can hence not distinguish between background and nanomaterial. Several attempts have been developed in the past to decide based on a tiered approach whether a particle concentration measured in a workplace is statistically significantly increased over the background concentration (e.g., [3,4]). All these approaches have to deal with the deficiencies of the available on-line measurement equipment that only allows for physical characterization. Therefore they complement their measurements with sampling of particles for subsequent chemical and/or morphological analysis for obtaining a definitive proof for the presence or absence of the nanomaterial under investigation.

The first four case studies in this chapter have to deal with the background problem and found different ways to solve it. The first case study gives recommendations for systematic studies on possible particle release under different stress situations. The authors suggest the use of emission chambers with very low background noise and summarize the findings of different studies in the literature that have used such chambers to study the release of nanomaterials from products under various stress conditions, e.g., abrasion, UV irradiation, etc. The second study in this chapter looks at the possible release of carbon nanotubes (CNTs) in a CNT production facility. CNTs have been reported to be carcinogenic under certain circumstances [5] and hence, following the precautionary principle, require increased attention. The authors of this study used a combination of physical on-line characterization of airborne particles and size-resolved particle sampling for subsequent morphological TEM analyses and chemical analyses by ICP-MS. In the chemical analyses, they evaluated the cobalt (Co) content as a surrogate for CNTs, because Co is used in the facility as a catalyst for CNT growth in a CVD process. On the other hand it is an element rarely found in the environment and hence a good indicator for differentiating CNTs from background. By using this technique, they found that only very minute amounts of CNTs were released into the workplace and that CNT concentrations were significantly below the background level. The third case study also looks at release of CNT, but from CNT—polymer composites during treatment steps at the end of their life-cycle. These steps included shredding as well as incineration of the composites and were experimentally simulated using emission chambers in a laboratory. The authors of this study also used on-line physical characterization of the particles and particle collection for morphological analyses using SEM and chemical analyses of the Co content of the samples by total reflection X-ray fluorescence (TXRF). They found that during shredding, the CNTs remained embedded in the polymer matrix and no free CNTs were released. During incineration experiments, they found that following the minimum requirements for German incineration plants, the CNTs were usually fully disintegrated and no free CNTs were found in the collected airborne particles or deposited fly ash. Only in one case were CNTs still found in the slag residues. Those were, however, still embedded and no free CNTs.

The authors of the fourth case study investigated the release of and consequent exposure to nanomaterials at a paint-manufacturing workplace, where various powders, some of them including nanomaterials, were used. The authors used aerosol instrumentation with high time resolution and point out the necessity for such high time resolution, because the aerosol in most workplaces can be quickly changing, making the results of some instruments with lower time concentration meaningless. Only by using highly time-resolved measurements were the authors able to detect short-lived particle release events and differentiate them from ubiquitous background particle fluctuations.

The problem of background distinction can only be avoided if the workplace under investigation is in a very clean environment such as a clean room. The last case study in this chapter presents work done on nano-scale particle emissions from clean room equipment which may potentially harm workers in the clean room. However, the main aim of this work was to investigate how the clean room quality is affected by these emissions, because the cleanliness in a clean room also defines the quality of the products that can be produced.

The five case studies selected for this chapter are supposed to give an overview on how studies on particle release can be conducted. From the different approaches taken in the studies for the different materials and release scenarios, it is obvious that many more of such case studies will have to be conducted in the future for a better understanding of the release processes and to eventually group the different process and standardize the test procedures.

7.2 EMISSION CHAMBERS, A METHOD FOR NANOSAFETY (*LE BIHAN, MORGENEYER, SHANDILYA, AGUERRE, BRESSOT*)

7.2.1 Introduction

Determining the risk of exposure to nanomaterial inhalation at a workstation is a complex process. Many questions beg for answers, in particular concerning the source of the particles. What precisely is the source? Are there multiple sources? How do you distinguish the sources and measure background noise? How does the source work, and in particular, when and for what reasons does this source start emitting particles? What are the physical and chemical properties of the particles? What kinds of quantities are emitted?

In view of answering such questions, industrial hygiene researchers and engineers require further information and data. Measures may be taken on site, and studies carried out in the lab. For the latter, the emission chamber is a high-performance tool that may be used under specific conditions.

7.2.2 Presentation of the Concept

International standards [6,7] define emission chambers as a test method dedicated to measuring the discharge of volatile organic compounds (VOC) in indoor air. In the absence of a specific standard pertaining to the release of (nano-) particles, it becomes possible to use such documents as a starting point.

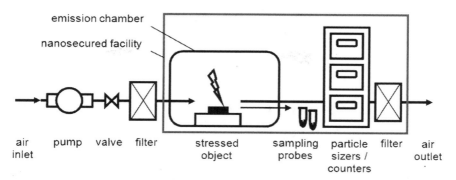

FIGURE 7.1   Schematic diagram of an emission chamber.

7.2.2.1 Description and Operation

The operating principle of an emission chamber is very simple (see Figure 7.1). Air from a clean source is conducted through a conduit to the chamber, which is sealed. The sample brought to this enclosed space is then stressed in various ways (using mechanical abrasion, etc.). The air, potentially loaded with particles, is then evacuated through a conduit leading on the one hand to on-line particle counters (for the detection of particles during the test) and on the other hand to particle extraction systems enabling a series of off-line analyses of particle composition using transmission or scanning electronic microscopy (SEM, TEM). The complete installation may be placed under a hood to guarantee confinement (nanosafety).

The goal is to create testing conditions that are similar to real conditions and to document them. Standards in relation to VOCs [6,7] establish specifications in terms of temperature and relative air humidity, as well as airflow and air velocity rates (0.1 m/s to 0.3 m/s). System leaks should not exceed 0.5% of the chamber volume per minute when over-pressurized at 1 kPa.

7.2.2.2 Definitions and Physical Principles

The main definitions supplied for the case of VOCs [6,7] are summarized below and are applicable to the study of airborne particles. *Background noise* is defined as the quantity of particles present in the chamber independent of the source under study. It is linked to the quality of the clean air injected into the chamber, and to additional sources. The lower the background noise, the lower the detection threshold for measurements.

The emission phenomenon may be observed through time depending on the *clean air exchange rate* of the chamber (flow rate of clean air to the chamber divided by its air volume (1/h)). *Specific air flow* (e.g., per sample surface under study $((m^3/h)/m^2))$ is a fundamental unit which enables comparison of *emission rates* across various studies, that is, the mass of particles emitted per unit time (mg/h). The *specific emission rate q* (also referred to as *emission factor*) may be defined relative to the dictates of the study — per unit q_U (e.g., the number of objects introduced into the chamber, (mg/h/1)), length q_L (e.g., long objects (mg/h/m)), surface q_S (default scenario, if release of particles from surfaces is under study $(mg/h/m^2))$, volume q_V (mg/h/m^3) and mass q_M (mg/h/kg).

A simple time-dependent mass balance may be defined assuming a 'uniform' flow rate: $V \frac{dC(t)}{dt} = R(t) - QC(t) - S(t)$, where V is the volume of air present in the chamber (m^3), $C(t)$ is the mass of particles evacuated with the loaded air (mg/m^3), $R(t)$ is the rate of release from the object to the chamber (mg/h), Q is the clean air flow rate feeding into the emission chamber (m^3/h), $S(t)$ are stray particles on the walls of the chamber, conduits or the object itself (mg/h), and t is time measured in hours, for example, starting when the object is introduced into the emission chamber.

Stray particles S may be limited with an appropriate layout of the air inlet and of air sampling in the chamber, for example, directly at the level of an abrasion site.

7.2.3 Case of Intentional and Non-Intentional Particles

Studies dedicated to ultrafine particles (non-intentional particles with a diameter less than 100 nm) in indoor air [8] used a large volume emission chamber (e.g., 70 m^3 [57], 32 m^3 [9]) or a smaller volume (e.g., 2.5 m^3 [10]). Except for Koponen et al. (20.6 m^3 [11]), studies dedicated to nano-materials all used chambers with smaller volumes. Vorbau et al. [12] worked with a 1.5 L volume, and INERIS used 80 L to 500 L chambers [13,14]. All of the studies were conducted with a filtered air inlet, with varying self-defined requirements concerning background noise (e.g., < 20 p/cm^3 [15]).

A minority of authors precisely qualify experimental conditions (air exchange rate, etc.), evaluating and factoring in stray particles on the walls and sampling lines (e.g., [10]). There are to date only a few researchers who express their findings in terms of 'emission factors' and who supply repeatability data (i.e., consolidation of their findings with repeated testing under same conditions), although this is one of the advantages of emission chambers (e.g., Figure 7.2d).

In the majority of cases, the authors sought to characterize and quantify the particles emitted from the materials-stress couple under consideration. The point is also to determine whether the nanoloads used with materials are yielding individual free nanoparticles, or heterogeneous matrix-nanomaterials (e.g., polymers).

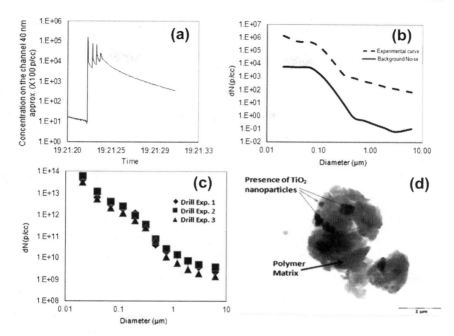

FIGURE 7.2   (a) Time-based monitoring during four successive grinding tests; (b) medium-size distribution during pumicing in comparison to background noise; (c) number of particles emitted during drilling — three tests; (d) example of a heterogeneous particle consisting of the (polymer) matrix and TiO_2 nano-charge.

The stress tests under study were varied [13]: pumicing [13,14,16], UV radiation and friction combined [15], abrasion [12,17], sanding [11], piercing [14].

In the majority of cases, real-time measurement tools are used: particle counters (CPC) measuring total number concentration for sizes ranging from a few nanometers to a micron; spectrometers for measuring particle number size distributions from a few nanometers to a micron (SMPS, FMPS, DMS) and beyond (ELPI); optical counters (OPC) measuring particle sizes from around 0.3 μm up to 15—20 μm.

This real-time measure makes it possible to determine background noise in the emission chamber and to observe the impact of the events triggered by the scientist.

It should be noted that mechanical units used for the application of stresses to a material in an emission chamber can also considerably influence and hence contaminate the particle concentration in the chamber. They themselves act as a potential source of particle emission, for example, motor brush, rotors, etc. The ideal solution to avoid such a situation is to place these units outside of the chamber (e.g., in [13]).

In the example presented in Figure 7.2a–c, the scientist performed a series of four pumicing tests in succession on a nanostructured object. Impact within the emission chamber is clearly observable (Figure 7.2a), since as a result of these operations, the total concentration of the chamber changes from $10^3/cm^3$ (background noise) to 10^6–$10^7/cm^3$. Particles emitted, in the present case, concern the totality of the size spectrum (Figure 7.2b). Knowing the initial background noise makes it possible to calculate the impact of the source.

Sampling for a microscopic analysis is used in more than half of the studies, and is conducted mainly via electrostatic precipitation.

This approach will likely become systematic. It is facilitated by an easy and new technique, the filtration by porous TEM grid (e.g., by MPS® [18]). Indeed, it enables the study of nanocharge, and in particular whether they remain within the matrix as heterogeneous particles (see Figure 7.2d), or whether they are released as free particles.

7.2.4 Discussion

Because of the controlled atmosphere that is afforded, an emission chamber makes it possible to focus on a single source, i.e., to ignore background noise in the atmosphere and from other sources. A very large range of measuring materials for aerosols may be used as this is laboratory testing. In particular this makes it possible to use devices that are otherwise difficult to use in the field (e.g., SMPS). Consequently, the physical and chemical characterization of the aerosol is very comprehensive.

An emission chamber is a small device that is relatively easy to use, which facilitates the carrying out of tests. The installation can easily be changed and modified.

However, this approach may only be adopted in a situation that can be modelled on an experimental basis. This actually sets the limit, for example, for complex and/or voluminous processes.

7.2.5 Conclusion and Future Prospects

Despite a limited number of references, it is already possible to conclude positively as regards the usefulness of an emission chamber for the study of NM emissions.

To date, use of this device remains new and 'exploratory'. The device's potential is still not fully utilized. This holds true for the determination of 'emission rates', which enables the modelling of particle dispersion and the study of exposure scenarios.

Harmonization of practices via standardization would facilitate the comparison of studies and promote the development of tests recognized by all members of the professional community, or even at a regulatory level.

Acknowledgements

The authors would like to thank the European Commission (Project 'Nanodevice'), the ANSES (project 'Nano-EMIS'), the French ministry of the environment (Programme 190, DRC 33) as well as the Nanoledge company and the Picardie Region.

7.3 EXPOSURE OF WORKERS TO CARBON NANOTUBES IN A COMMERCIAL PRODUCTION FACILITY; PRELIMINARY RESULTS IN THE FRAME OF RISK ASSESSMENT AND RISK MANAGEMENT (*LECLOUX, GORBUNOV, BROUWER, MUIR*)

7.3.1 Introduction

Carbon nanotubes (CNTs) define a family of nanomaterials made up almost entirely of carbon. The most common industrial-scale manufacturing processes are laser ablation, chemical vapor deposition (CVD), and arc discharge [19]. In the family of CNTs, multi-walled carbon nanotubes (MWCNTs) are of special interest to industry. Structurally, MWCNTs consist of multiple layers of graphene superimposed and rolled in on themselves to form tubular shapes. Such cylindrical graphitic polymeric structures have novel or improved properties that make them potentially useful in a wide variety of applications in electronics, optics and other fields of materials science.

Three properties of MWCNTs that are especially interesting for industry are: the electrical conductivity (as conductive as copper), their mechanical strength (five times lighter than steel and 20–100 times stronger, depending on the occurrence of defects in the structure) and their thermal conductivity (same as that of diamond and more than twice that of copper). In order to exploit these properties it is essential for industry to ensure sustainable development of these materials, particularly with respect to Health, Safety and Environment (HSE).

For appropriate measures to be put in place to ensure the safe production and uses of CNT, an understanding of the hazards and exposure risks during manufacture is essential. Potential hazardous properties of CNT are matters of ongoing research but there are indications that certain types of CNT may be hazardous if workers or users are exposed through inhalation pathways [5].

CNT are chemical substances, and as such their production and use is covered by existing regulations on chemicals, and workers' protection practices. Current good practice for the handling of CNTs is to minimize potential exposure of workers, particularly via inhalation and dermal exposure routes. To date, measurements of concentration levels of airborne CNT are limited to research-scale production, e.g., Maynard et al.,

Han et al., Bello et al., Tsai et al. [20—23], or during release by aerosol generation, e.g., Myojo et al. [24]. There is very limited information in the literature directly relevant to exposure levels and properties of airborne CNT at production facilities.

To understand and manage risk of exposure in manufacturing and downstream use of carbon nanotubes in an effective fashion, it would be a significant advantage to have knowledge of the levels of exposure of workers in a typical industrial environment and to identify the processes which represent the greatest exposure risk. This sub-chapter describes first measurements of airborne nanoparticles taken at a number of locations at a typical CNT plant operated by Nanocyl SA at their production site in Sambreville, Belgium. The aim was to measure levels of CNT in the air, for the purpose of risk assessment and risk management related to the exposure of workers to CNT.

7.3.2 Methods and Instrumentation

7.3.2.1 Instrumentation

Nanocyl uses a CVD-based process for industrial-scale production. As there are no published good practice guides for choice of metrics or instruments [25] appropriate for exposure measurements at a CNT production and processing site, a range of nanoparticle characterization instruments were deployed at a number of locations throughout the production site. Samples were taken close to the manufacturing unit, in research and quality control laboratories, during bagging and handling of CNT, in administration offices and close to a downstream extruder producing CNT/polymer compounds (Figure 7.3).

A 12-stage electrical low-pressure impactor, ELPI model 97 2E (Dekati, Finland), was used to collect size-resolved particle concentration data.

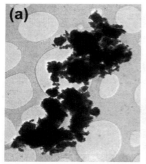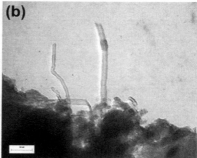

FIGURE 7.3 Personal air sample from the reactor at the product site. (a) Low magnification (5.6—6.3K), showing mostly metal catalyst residue; (b) higher magnification, indicating possible nanotubes.

Two condensation particle counters (CPCs), a UCPC model 3786 size range 2.5 to 3000 nm (TSI, USA) and a Portacount model 8020 (TSI, USA) with particle size range from 20 to 1000 nm were employed to measure workplace particle total number concentrations. In addition, near-real-time surface area concentration was measured by two different types of diffusion chargers, i.e., LQ1 (Matter Engineering, Switzerland) and a Nanoparticle Surface Area Monitor (model 3550, TSI, USA). The first device measures the active surface area concentration, whereas the latter one mimics the geometric surface area of lung-deposited particles [26]. Additionally, particle number concentrations close to the workers' breathing zones were measured using CPC (TSI, model 3025) located next to the workstation with instrument inlets (tubing) positioned near to the workers' breathing zones.

For characterization of morphology, samples on TEM grids were collected by an electrostatic precipitator (Nanometer Aerosol Sampler model 3089, TSI, USA). In addition to the static sampler, a specially developed 25-mm gold-coated polycarbonate filter/TEM grid (3 mm Holey Carbon film 400 meshes Ni) combination was placed in an open-face filter holder and used for personal sampling (PAS) at a flow rate of 2 L/min. The filter/TEM grid combination is very similar to the filter/TEM grid combination described by Tsai et al. [27]. The personal TEM-grid sampling was an activity-based process [28], i.e., focused on workers' activities associated with handling of CNTs, whereas the static sampling conducted included periods of activity and non-activity. TEM analysis included X-ray mapping, analysis of transmitted or scanned images for size and agglomeration of collected particles. Elemental analysis was performed using Energy Dispersive X-ray spectroscopy on individual particles.

In addition, a size selective sampler (Nano-ID Select, Particle Measuring Systems, USA) was used to collect aerosol samples in the wide size range from 1 nm up to 30 μm in 12 size fractions [29] for off-line analysis. The sampler is based on a unique technology that mimics the deposition in the human respiratory tract combining inertial deposition for particles above 250 nm with diffusion deposition below 250 nm. Samples were collected at various locations over extended periods. The aim was to detect levels of CNTs at workplaces against comparatively high levels of background aerosol concentrations and to evaluate exposure to CNTs.

Total mass concentrations of aerosols were measured using a mass reader accessory to Nano-ID (MR250, PMS, USA).

7.3.2.2 *Measuring Sites and Production Activities*

During various activities measurements have been carried out at four different places at Nanocyl facilities: during quality control sampling at the reactor, at the bagging and packaging unit, at the compounding station and in the research laboratory (R&D). The background concentration was determined by performing one background full-scale sampling in an office and by quantifying particles in ambient background conditions (reference background).

7.3.3 Results

7.3.3.1 Particle Concentration Measurements

BACKGROUND MEASUREMENTS

Nanoparticle number concentrations are known to be subject to considerable spatial and temporal variations [30]. Measurements of the background concentration at Nanocyl confirm this and show substantial within-day (factor 2.5) and between-days (factor 5) variations (Table 7.1). Such levels of variations have been reported earlier, e.g., Yeganeh et al. [31]. Unintended additional external sources as well as natural variations in background aerosol characteristics can have major impacts on these variations [25].

MEASUREMENTS DIRECTLY RELATED TO THE PRODUCTION OF CNT

There is some evidence that specific activities, e.g., taking quality control samples, changing bags, etc., results in an increase in total particle number concentration as measured using CPCs, which suggests there is an (additional) emission of particles as a result of these activities (Table 7.1). However, the observed, statistically significant, increases of concentrations were small, i.e., differences below 10%.

Examination of TEM samples show that CNTs are only observed on TEM grids used during sampling and during bagging (Figure 7.4). These low levels can be explained by the fact that the production is carried out in a closed process and that the bag filling station is state of-the-art technology, designed to minimize the potential for particle releases.

MEASUREMENTS DURING COMPOUNDING OF CNT

It was not possible to distinguish changes from baseline particle concentrations and the particle concentrations associated with the various activities carried out during compounding. The particle concentration—time plots showed many 'peaks' but these could not be directly associated with handling CNT. It has been demonstrated clearly how the release of (polymer) fumes increases total number concentrations of particles during compounding [27]. It is a reasonable assumption that the extruder heating phase is responsible for elevations of total number concentrations.

TABLE 7.1 Number Concentration Measurements at Various Locations Obtained with Different Instruments

Measurement	ELPI ≤100 nm (dN#cm³)	ELPI ≥100 nm (dN#cm³)	Portacount (#/cc)	Reference Portacount (#/cc)	CPC (#/cc)
Office (background) day 1	14,642	1424	3914	na	na
Office (background) day 2	137,940	5722	23,485	20,862	34,593
Production (sampling)	No difference (7524/6971)	Increase 3.4% (3020/2922)	Increase 5.6% (7209/6824)	No difference (2723/2702)	Increase 7.2% (9632/8988)
Producution (bagging)	No difference (9378/9241)	No difference (3335/3347)	No difference (8560/8552)	No difference (6680/6771)	No difference (11,877/11,940)
Compounding (bag emptying)	10,312	3754	18,036	11,660	34,919
Compounding (calibration)	9628	3718	13,164	13,354	21,582
Compounding (close to extruder)	45,343	7925	30,499	24,082	38,168
R&D Research reactor [1]	26,841	4362	16,051	12,160	28,302
R&D Research reactor [2]	17,535	3680	10,839	10,010	19,754
R&D Research reactor [3]	22,492	5058	12,679	10,620	27,985

na: not applicable.

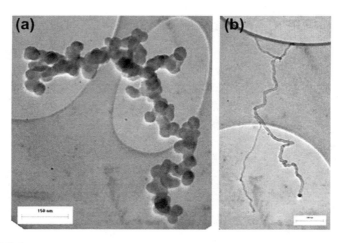

FIGURE 7.4   TEM pictures of a personal air sample during bagging of CNTs at the packaging station in the production site (a); (b) at higher magnification indicating possible nanotubes.

On the TEM grids (as part of air sampling filter), most of the particles detected proved to be complex structures, mainly agglomerates and aggregates, some large particles and very few small distinct particles. Based on observed differences between the 'reference background' location and the workstation, it can be concluded that compounding activities result in a small elevation of the nano-sized particle concentration levels. However, there is no evidence that particles causing this increase are CNTs.

SIZE-RESOLVED MASS CONCENTRATION OF CNT MEASURED WITH NANO-ID

Carbon and carbonaceous material are persistent in the atmosphere. Since it is difficult to analyze directly trace quantities of CNT carbon against carbon in background aerosols, chemical analysis of other associated elements was used to quantify concentrations of CNT in the air. CNTs usually contain traces of other chemicals including catalysts, e.g., Fe, Al, and Co. In the Nanocyl process, a significant component of the catalyst used is Cobalt. Background levels of Co are low; therefore, Co concentrations measured using, e.g., ICP trace chemical analysis can be used as a method for quantification of CNT concentrations. The ratio of the catalyst residue contained in bulk CNT samples was measured by the manufacturer (Nanocyl) and used as the baseline for aerosol CNT concentration calculations. Therefore, a mean Co/CNT mass ratio was used (undisclosed figure). The background aerosol mass concentration was quantified using MR250 (PMS, USA). This accessory to

Nano-ID is quasi-on-line and provides total mass concentration in a few minutes.

Samples were collected using the Nano-ID instrument and analyzed for Cobalt carried out using ICP. Samples were collected over a number of hours at sites within the Nanocyl facility. Sampling times varied between 8 hours and 200 hours according to the time needed to collect sufficient material to be able to detect catalyst residues using ICP analysis techniques. Size-resolved mass concentration size distributions of CNT were derived. These were used to calculate total exposure and respirable mass concentrations of CNT in the workplace atmosphere.

The respirable concentration of CNT measured at the packaging facility was 1.4 μg CNT/m^3. The sampling time was 72h sampling. The size distribution of CNT particles at this site is similar to distributions observed at other sites. The majority of CNT particles are in the micron size range; see Figure 7.5. However, a relatively small mass of particles was detected in the sub-micron and nano-range.

The respirable concentration of CNT measured near the extruder of the R&D compounding unit was relatively small, 1.0 μg CNT/m^3. A continuous sampling of 144 hours was necessary. The majority of CNT particles were detected in the micron size range. The quantity collected in the nano-range was below the detection limit of the chemical analysis technique. The size distribution of CNT particles was similar to that shown in Figure 7.5.

The respirable concentration of CNT measured in the office was 0.25 μg CNT/m^3. All CNT particles detected were found in the micron size range. The quantity collected in the nano-range was below the detection limit even after 200 hours of continuous sampling. The size distribution of CNT particles was similar to that shown in Figure 7.5.

In addition, the average background mass concentration in the office was measured using Nano-ID and a mass conversion accessory, Mass Reader MR250 (PMS, USA), and found to be approximately 40 μg/m^3. This confirms that typical mass concentrations of CNT are much lower than concentrations of background aerosols. Therefore, chemical identification of CNT is crucial for health risk evaluation at workplaces.

Chemically selective CNT measurements confirm low levels of CNT particles at all measuring sites including compounding, laboratory, and production sites and in the office area. The measured concentration of a few μg CNT/m^3 was considerably lower than the average total background aerosol particle concentration. This explains why conventional methods of measuring number concentration of aerosol particles detect very little or no changes in comparison to the background.

Data obtained with Nano-ID indicate that CNT at workplaces can be successfully detected and quantified at low levels against background aerosol concentrations.

TABLE 7.2   Surface Area Concentrations

Locations	NSAM		LQ-1
	A Fraction	TB Fraction	
Production reactor room	23.1	6.7	61
Production packaging	10.1	4.8	68
Outside near the production facilities	30.6	6.9	na
Outside near the reception (background)	19.0	5.0	na
R&D extrusion room	46.5	19.5	69–137
Lab reactor room	87.5	20.7	41–55
Corridor near to the lab	76.7	16.5	na
R&D offices	24.0	5.3	29.8

The Second and Third Columns Show the NSAM Mean Surface Area Concentrations ($\mu m^2/cm^3$) of the Alveolar (A) and Tracheo-Bronchial (TB) Fraction, Respectively. The Last Column Shows the Results of the LQ-1 Diffusion Charger. The Measurements with the Two Devices were Conducted on Different Days

ACTIVE SURFACE AREA CONCENTRATIONS

The results for the active surface area concentrations and the surface area of the lung deposited particles are listed in Table 7.2. Both devices were used on very different days, so comparison of the data is inappropriate. The data for the surface area of the lung deposited particles (NSAM-data) show that the surface area concentrations at the production site were similar to those in the offices or at the reception. This observation was not confirmed by the LQ-1 data, where the office showed the lowest surface area concentration.

EVALUATION OF VARIOUS INSTRUMENTS

On-line instrumentation (CPCs, ELPI and surface area instruments) failed to detect CNT. There was an indication in CPC data that there might be an increase in total nanoparticle concentrations associated with specific activities associated with handling CNT, but the increases were small and it was not possible to state whether they were due to CNT emissions or incidental emissions of other nanoparticles. This level of difference could also be attributed to natural variations in background aerosol concentrations. The ELPI was not able to distinguish between background aerosol concentrations and the target nano-objects. Instruments measuring surface area gave conflicting information.

In general the on-line instrumentation used proved to be of limited use for CNT detection; see Table 7.3. Ideal on-line instrumentation would need to be specific to CNT and sensitive enough to detect very low concentrations against high ambient background concentrations.

TABLE 7.3   Evaluation of Various Instruments for CNT Detection and Health Risk Evaluation at Workplaces

Instrument	On-line Measurement	Selectivity Against Background	Sufficient Sensitivity to CNT	Can the Instrument Be Used to Quantify Exposure to CNT?
ELPI	Yes	No[a]	No[a]	No
Portacount	Yes	No	No	No
CPC	Yes	No	No	No
NSAM	Yes	No	No	No
LQ-1	Yes	No	No	No
Nano-ID	No	Yes	Yes[b]	Yes

[a]Samples from each stage may later be analyzed for CNT and/or Co by SEM/TEM and EDX.
[b]Co is detected as a surrogate for CNT.

Off-line techniques, namely Nano-ID sampling plus chemical analysis for catalyst residues and TEM, gave strong indications of the presence of low concentrations of CNT. Data from Nano-ID samples gave average concentrations of CNT in the air ranging from 0.25 $\mu g/m^3$ (office) to 1.4 $\mu g/m^3$ (extruder) against a general background aerosol mass concentration of approximately 40 $\mu g/m^3$. Of the instrumentation used only Nano-ID coupled to chemical analysis gave data suitable for estimating airborne CNT concentrations at workplaces.

Comparison of various instruments used (Table 7.3) reveals that Nano-ID has only one drawback. It is not an on-line instrument. The sampling time is normally from 10 min to a few hours. However, it should be appreciated that the sampling time can be reduced considerably with high sensitivity analysis such as atomic adsorption or Raman spectroscopy. For instance, the sampling time for general background aerosol mass concentration measured at the site was 10 min. In addition, sampling over a long period of time provides time-averaged exposure and could be considered to be more representative of the real situation.

CNT EXPOSURE AND HEALTH RISK ASSESSMENT

The data obtained with Nano-ID enables CNT concentration to be compared with the 'no effect level' defined on the basis of inhalation toxicological studies. In a separate inhalation study carried out on rats using commercial samples of CNTs produced in the Nanocyl production site, a low observed adverse effect concentration (LOAEC) of 0.1 mg/m^3 has been determined [60]. At this concentration, some effects are still observed but could be reversible. On this basis, by applying a conservative safety factor of 40, it is possible to estimate an internal Occupational Exposure Level (OEL) of 2.5 $\mu g/m^3$.

In the offices, the measured respirable CNT concentration in air of $0.25 \ \mu g \ CNT/m^3$ is then ten times lower than the OEL. At the site near the extruder and in the production area the respirable CNT concentration was $1.0 \ \mu g \ CNT/m^3$ and $1.4 \ \mu g \ CNT/m^3$ respectively. This level is also lower than the OEL. If we consider that operators do not stay near the packaging unit or in the extrusion room more than a quarter or a half of their working time, the actual exposure dose is 3 to 6 times lower than the estimated OEL even without personal protective equipment. Therefore, according to the measurements made, the exposure to CNT at the plant is below the defined safety limit.

It is also important to take into account the actual deposited fractions in the lungs; see Table 7.4 and Figure 7.5. It is considerably lower than the total airborne concentration.

TABLE 7.4 CNT Particle Concentration in the Air and Deposition Fractions Calculated from Size-Resolved Data in Figure 7.5

Aerosol Particle Mass Concentration, ng/m³				
Total in the Air, m_{ta}	In the Respiratory tract	Alveolar and TB Regions	PM_{10}	Respirable Fraction
3200	2170	227	1520	1200

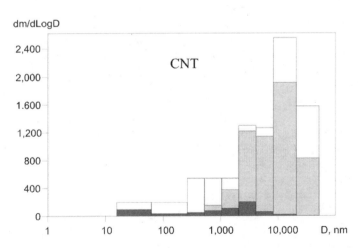

FIGURE 7.5 CNT aerosol particle mass size distributions dm/dlogD in ng/m³ sampled at a production site. Aerosol in the air size distribution shown with white bars, size distribution corresponding to particles deposited in the entire respiratory system, grey bars, and size distribution corresponding to particles deposited in the alveolar and trachea bronchial ranges, dark bars. For D > 250 nm, D represents the aerodynamic diameter. For D < 250 nm, D is the diffusion diameter.

However, as we cannot exclude the possibility of having concentration peaks from time to time, on the basis of the precautionary principle the workers are requested to use personal protective equipment at these two locations in order to ensure that the risk is negligible.

IMPORTANCE OF SIZE-RESOLVED MEASUREMENTS

In this chapter the health risk was assessed employing size-resolved CNT mass measurements. These measurements enable the respirable mass fraction to be evaluated. Along with respirable mass fraction some other metrics have been suggested as measures for health risk, for example: metrics based upon mass, number or surface concentrations, e.g., Hinds [32], Maynard and Kuempel [33], and Rodes and Thornburgs [30]. With the instrumentation deployed in this study, number- and surface-based measurements failed to detect CNT particles as shown above. As far as mass-based metrics is concerned, several conventions have been suggested: e.g., thoracic, PM_{10}, and regional deposition in various compartments of the respiratory tract (trachea, bronchial, alveolar and upper respiratory tract) [34]. The advantage of size-resolved measurements is the ability to provide a range of information relevant to exposure and dose which can provide a comprehensive data set suitable for health risk evaluation.

For example, it is known that only some fraction of airborne particles entering the respiratory system is deposited [32,35]. Various metrics have been calculated according to Hinds [32] and ICRP [36] using mass concentration size distributions obtained with Nano-ID (e.g., Figure 7.3). The difference is clearly seen where the size distribution corresponding to particles deposited in alveolar and trachea bronchial range loses most of micron size airborne particles. In the sub-micron size range between 100 and 1000 nm (1 μm) deposited particle concentrations are substantially less than total concentrations of airborne particles. The same is true for particles with sizes greater than 10 μm (10,000 nm). Very small concentration can be seen in the micron size range a mode located at 3000 nm and a very small amount in the nano-range between 10 and 100 nm. It is important to highlight that only size-resolved measurements enable all these metrics to be calculated [30].

It was found that, using measurements made at the production site, the deposited equivalent mass concentration of CNT particles in the alveolar and trachea bronchial areas of 0.227 μg/m^3 [35] was much lower than the calculated respirable mass concentration (1.2 μg/m^3) or PM_{10} (1.52 μg/m^3). These concentrations are also much lower than the estimated OEL (2.5 μg/m^3). Even smaller values were obtained at other sampling sites. This confirms the conclusion that the health risk to CNT exposure at the plant is negligible.

Data obtained with Nano-ID demonstrate the multidimensional nature of CNT particles at the workplaces. These data can be used to obtain various conventional metrics and likely to be compatible to any future legislation.

7.3.4 Conclusions

Off-line methods (Nano-ID and TEM) show that concentrations of CNT in the air at the investigated workplaces are low and much lower than background concentrations of atmospheric aerosols. The level of measured mass concentrations proved that by applying appropriate handling measures it is possible to minimize the exposure of workers in such a way that the health risk of exposure to CNTs is negligible. The health risk is assumed to be associated with mass of CNT particles deposited in the alveolar region. Measured values were found to be lowest in the office at 0.25 $\mu g/m^3$ and highest in the extruder/compounding area at 1.4 $\mu g/m^3$. Estimated exposure to CNT in the Nanocyl workplace is found to be between 3 and 10 times lower than the proposed Occupational Exposure Levels even without personal protective equipment. The most successful technique to measure concentration of CNT involved sampling with Nano-ID and subsequent chemical analysis. Currently Nano-ID is the most appropriate instrument available for health risk evaluation at working places associated with CNT exposure. On-line number concentration and sizing methods indicated some additional release of nanoparticles associated with specific manufacturing and handling activities, but it was not possible to say if these particles were CNT. On-line instruments failed to specifically detect CNTs at workplaces. In general, on-line instruments can provide useful information on short-term release of particles but for this they should be able to detect CNT selectively above background aerosols.

Acknowledgements

The Belgian Federal Agency for Workers Welfare is acknowledged for providing the NSAM data, and David Green for assistance with Nano-ID measurements.

7.4 INVESTIGATIONS ON CNT RELEASE FROM COMPOSITE MATERIALS DURING END OF LIFE (STAHLMECKE, ASBACH, TODEA, KAMINSKI, KUHLBUSCH)

7.4.1 Introduction

The use of carbon nanotubes (CNTs) in different kinds of experimental and commercial composite materials has steadily increased over the recent years due to their high electrical conductivity and mechanical strength. Such CNT composite materials are expected to be mass-produced in the near future. At the same time concerns have been raised about possible adverse health effects of inhaled CNTs [5]. Although CNTs in general are very diverse in their nature and potential health outcomes should hence not be generalized, the question on possible release of CNTs from composite materials into air was brought up with a focus on those processes that may lead to inhalation exposure. CNTs may be released from their matrix material during their entire life-cycle, i.e., during production and handling (see Section 7.3), manufacture of actual products, e.g., composites, downstream use as well as recycling [37]. The work presented here focused on the potential CNT release during the end of life of CNT composites. Shredding and incineration were investigated as a typical mechanical and thermal process step, respectively. Both processes were experimentally simulated in laboratory-scale experiments. Three different composite materials — polycarbonate (PC), polyamide (PA) and polyethylene (PE) — each with 0 wt%, 5 wt% and 7.5 wt% CNT content that were manufactured particularly for this study were investigated. The CNTs used in these composite materials were Baytubes C150P® with a typical diameter of approximately 20–30 nm and a typical length after post-production treatment below 5 μm.

7.4.2 Shredding of CNT Composites

A small-scale industrial shredder (Fritsch Kraft-Schneidmühle 'pulverisette 25') was used to study emissions of CNTs during shredding of the test composites. The shredder was located inside an enclosure in order to separate the potentially generated aerosols from the laboratory background. The enclosure was completely sealed and connected to an exhaust system, which caused continuous ventilation of the enclosure. The supply air was sucked through two HEPA filters which ensured a very low particle background level ($< 1000/cm^3$) inside the enclosure. A photograph of the enclosure with the shredder is shown in Figure 7.6. The side walls of the enclosure were made from conductive acrylic glass and grounded in order to avoid electrostatic particle losses. The particle number size distributions of the released aerosols were characterized in quasi-real time with a scanning mobility particle sizer (SMPS, TSI model 3936, [38]) to cover the particle size range from approximately 15 to 750 nm. The relative humidity and temperature inside the enclosure was additionally logged.

7.4.2.1 Experiments

The raw material was cut to pieces of approximately $10 \times 10 \times 3$ mm^3 prior to the shredding. Exemplary images of raw and shredded polycarbonate are shown in Figure 7.7. Each experimental run lasted for

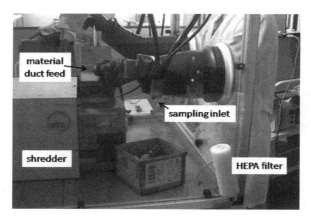

FIGURE 7.6   Enclosure with shredder.

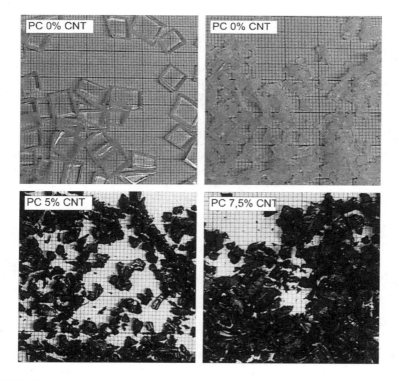

FIGURE 7.7   Pure polycarbonate (PC) prior to (top left) and after (top right) shredding; shredded PC with 5 wt% (bottom left) and 7.5 wt% CNT (bottom right).

20 minutes. Ten g of the material under investigation was added through the feed duct (see Figure 7.6) at the beginning of each minute, resulting in a total of 200 g shredded material during each run. This sequence was chosen to homogenize the temporal particle generation rate and the spatial particle distribution within the chamber.

7.4.2.2 *Results*

Figure 7.8a−d shows the number size distributions of airborne particles during shredding of the different materials, measured with an SMPS. Please note that for a better comparability of the data the scale of the vertical axis was deliberately kept equal in all graphs.

Figure 7.8a exemplarily shows the number size distributions in the enclosure during no activity (background), with idling shredder and during shredding of pure PC. A clear increase of the particle number concentration compared with the background can only be observed during shredding of material, while the idling shredder does not notably emit particles. However, since the increase of the particle number concentration is mainly in the size range <100 nm, it is very likely that these particles do stem from the shredder motor when a higher load is applied to it. Figure 7.8b−d shows the submicron number size distributions

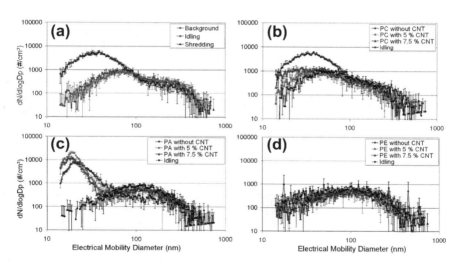

FIGURE 7.8   Number size distributions measured during shredding, each average of three runs: (a) during shredding of pure PC; background measurement was conducted prior to the experiment, idle measurement while shredder motor was running but without material in the shredder; (b) during shredding of PC with different CNT content; (c) during shredding of PA with different CNT content; (d) distribution during shredding of PE with different CNT content.

measured during shredding of the different materials in comparison with the number size distribution of particles in the chamber with idling shredder. The figures clearly show that although an increase of the particle concentration is obvious during shredding of most materials, none of the increases can be clearly linked to the presence of CNTs in the matrix materials. To the contrary, the PC material with CNT content showed lower number concentrations than the pristine PC (Figure 7.8b), while the PE material (Figure 7.8d) showed more or less no noticeable particle release on top of the particle background in the chamber with idling shredder, independent of CNT content. All PA materials (Figure 7.8c) with different CNT contents showed similarly increased particle concentrations around 30 nm.

In general, all materials containing CNTs only showed concentrations higher than background for electrical mobility diameters below approximately 50 nm. From the sizes of the CNTs used in the composite materials, it can be expected that these have electrical mobility diameters larger than 50 nm and hence the increased concentrations below 50 nm stem from different particle source and are not free CNTs. This finding is confirmed by electron microscopic analyses of particle samples that also showed no free, single CNTs. These samples were collected electrostatically by first unipolarly charging the airborne particles in a custom made corona charger, followed by deposition of the particles onto silicon substrates in an electric field in an electrostatic precipitator (TSI, model 3089 [39]). Similar results were previously also found by Wohlleben et al. [40]. Hence the measurements did not show any evidence for the release of free CNTs from these materials. This conclusion is supported by the fact that none of the particle samples taken during the shredding experiments showed single CNTs.

7.4.3 Incineration of CNT Composites

With increasing landfill taxes and fees and even the ban of landfills for combustible wastes in most European countries [41], incineration is becoming the major final treatment step of wastes. It is hence likely that at the end of their life-cycles, composite materials including those containing nanomaterials will end up in an incineration plant. The aim of the case study presented here was to investigate whether CNTs may be released from composite materials (same as described above) under the simulated conditions in an incineration plant. CNTs may be released because of incomplete combustion as part of the (fly) ash or slag.

In the present study, defined amounts of the composite material were placed inside a tube furnace. The same CNT composites as for the shredder experiments (see Section 7.4.2) were investigated here. Operating conditions were chosen in order to meet the minimum requirements for incineration plants set forth in the German legal framework (17th BImSchV) [42].

7.4.3.1 Experiments

To simulate an incinerator in the lab, two tube furnaces (Carbolite, models CTF 12/65/550 and TZF 12/65/550, each with a heated tube length of 550 mm) were used in series, connected with a quartz glass tube with 50 mm diameter. The experimental setup is schematically shown in Figure 7.9. While the first furnace was used to mimic the incineration of the composite material, the second furnace was used to simulate thermal after treatment of the exhaust gases. A ceramic crucible was used as a sample container and means of transport of the composite material in the first furnace. A primary airflow of 14 L/min was fed to the quartz tube from the entrance of the first furnace. A second airflow of 7 L/min was added to the second furnace to maintain the minimum legal requirements regarding the minimal O_2 concentration of 6% for thermal after treatment. The gas flow rates through the quartz tube were maintained with a mass flow controller (MKS Instruments Inc., Typ 647C). Clean, compressed air with 21% oxygen content was used at atmospheric pressure for both gas flows. The hot zone of the after treatment was maintained at 850°C and

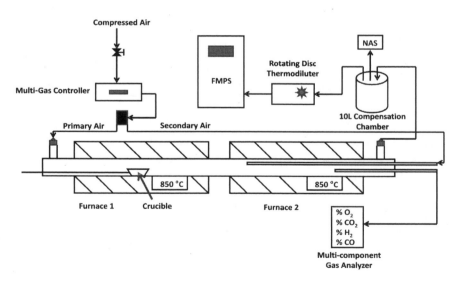

FIGURE 7.9 Schematic of the experimental setup for mimicking incineration of composite materials.

the resulting residence time of at least 2 s was kept as defined in the legal requirements.

The oxygen content was measured with 1 s time resolution at the end of the second furnace along with other gas components using a multi-component gas analyzer (NGA 2000 MLT 5). In parallel the particle number size distributions were measured in a particle size range from 5.6 to 560 nm with a Fast Mobility Particle Sizer (FMPS, TSI model 3091 [43]) with a time resolution of 1 s and 10 L/min sample flow rate. An electrostatic precipitator (TSI NAS model 3089) was used to sample particles at 0.5 L/min sample flow rate for consecutive morphological SEM analysis. To adjust for the different flow rates the total flow from the quartz tube was led through a compensation chamber from which FMPS and NAS were sampling (see Figure 7.9). Due to the very high particle number concentrations in the flue gas, the aerosol was diluted 1:1000 using a rotating disc thermo diluter (TSI model 379021). In addition samples were taken from deposited fly ash collected towards the end of the quartz glass tube and from the slag that remained in the crucible for consecutive morphological SEM and chemical TXRF analyses.

The crucible with the weighed test materials was manually placed in the center of the first furnace (PC: 500 g; PA: 480 g; PE: 290 g). The amount of test material was determined such that the oxygen content in the second furnace did not drop below 6%. After the crucible with the test material was placed in the furnace, a short time passed before the material ignited. The time until ignition as well as the burning duration was recorded. After the flame died out, the materials remained in the furnace for another 5 minutes during which particle size distribution and oxygen content were measured. A total of 10 runs were conducted for each material and two NAS samples were collected on silicon substrates for each material with a sampling time of 1 minute. The residue in the crucible was collected separately after each run, whereas the fly ash was collected after all 10 runs were terminated.

7.4.3.2 Results

After introduction of the materials into the first furnace, the material initially melts before it ignites due to the heat and presence of oxygen in the furnace atmosphere. The combustion is obviously incomplete as a lot of airborne soot could be observed. The burning of the material usually lasted for approximately 30–60 s, but the residue in the crucible continued to glow for a while after the fire died out. A typical time series of the particle number concentration during an experimental run with PC with 5% CNT content is shown in Figure 7.10.

An exemplary particle number size distribution, measured during the incineration of PC with 5% CNT content is shown in Figure 7.11. The size

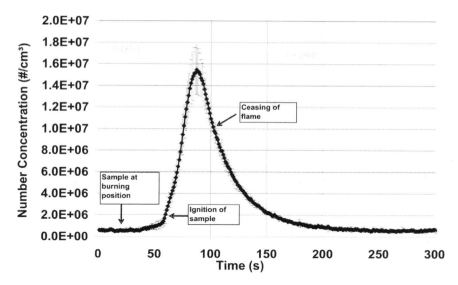

FIGURE 7.10   Typical time series of the number concentration during an incineration experiment with PC+5% CNTs.

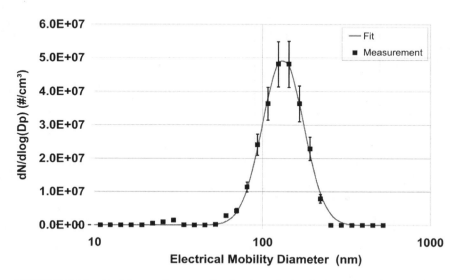

FIGURE 7.11   Particle number size distributions during incineration of PC with 5% CNT content (average of 10 runs).

TABLE 7.5   Overview of Incineration Experiments, i.e., Material Type, Amount, Time
Until Ignition and Duration of Burning (Average of 10 Runs)

Material (n = 10)	Amount mg		Time Until Ignition s		Burning Duration s	
	Mean	StDev	Mean	StDev	Mean	StDev
PC	502	5.0	43	2.1	25	1.5
PC + 5% CNT	502	4.8	43	1.2	37	2
PC + 7.5% CNT	502	5.9	41	2.2	36	1.5
PA	483	7.3	34	2.3	27	1.5
PA + 5% CNT	479	5.2	33	2	56	4.3
PA + 7.5% CNT	481	3.9	33	2.1	55	2.6
PE	289	5.9	23	3.3	29	3.4
PE + 5% CNT	295	8.7	19	2.4	53	3.9
PE + 7.5% CNT	292	6.0	22	2.2	52	2.7

distributions stemming from the other materials under investigation were
very similar.

Table 7.5 summarizes the experimental parameters as mean values of
10 runs and their standard deviations. It can be concluded from the table
that materials containing CNTs burn significantly longer than the pure
polymer matrices. The amount of slag remaining in the crucible depended
strongly on the material used. For example, pure PA and PE left no significant residues.

All results of the particle and gas measurements during incineration
experiments are summarized in Table 7.6. It can be seen that the minimum
requirements for the oxygen contents ($\geq$6%) as defined in the German
legal framework (17th BImSchV) were always met.

Particle number concentrations and count median diameters are given
as averages of 10 s periods around the detected maximum in the particle
number concentration (see Figure 7.10) and averaged over all 10 runs.
Measurements of particle size distributions showed very comparable
results with count median diameters ranging from 113 nm to 132 nm and
the total number concentration ranging from $1.1*10^7/cm^3$ to $2.1*10^7/cm^3$.
All materials showed slightly lower number concentrations and larger
particle count median diameter if the materials did contain CNTs. The
differences are, however, rather small and should not be overinterpreted,
taking into account the different amounts of material burnt and the
measurement uncertainties of the FMPS, especially for agglomerated

TABLE 7.6 Measured Average Oxygen Content as well as Total Number Concentration and Count Median Diameter (CMD) of the Particle Size Distribution Along with Standard Deviations

Material (n = 10)	O₂ Content in Flue Gas %		FMPS (#/cm³)*10⁶		nm	
	Mean	StDev	Conc.	StDev	CMD	StDev
PC	12.4	0.38	21.1	1.4	131	0.80
PC + 5% CNT	16.0	0.34	14.8	2.1	132	1.1
PC + 7.5% CNT	15.6	0.23	15.9	1.8	132	1.1
PA	7.4	0.24	18.3	1.1	113	1.5
PA + 5% CNT	15.5	0.41	10.7	1.6	127	1.7
PA + 7.5% CNT	15.1	0.27	10.7	1.9	124	2.5
PE	7.9	0.65	19.9	3.4	121	4.1
PE + 5% CNT	15.0	0.47	15.1	2.7	128	1.2
PE + 7.5% CNT	16.4	0.33	15.5	1.9	128	1.7

particles >100 nm [44,45]. In any case it can be concluded that the FMPS measurements revealed quite typical number size distributions of combustion generated particles and did not show abnormalities which may be tracked back to CNTs. However, it should also be noted that it is unclear how CNTs would be represented in size distributions measured with an FMPS. Hence the various samples taken from airborne particles, from fly ash collected inside the quartz tube and from the residue in the crucibles were analyzed by SEM and TXRF.

7.4.3.3 Analyses of Residues from Crucibles

The residues from the crucible were exemplarily analyzed for CNTs with an SEM. The samples were dispersed on a layer of conductive carbon glue on an SEM stub. While samples from incineration of PA and PE showed no evidence for the presence of non-combusted CNTs, the PC samples did show CNT-like structures as exemplarily shown in Figure 7.12. Although the images cannot give definitive proof for the presence of CNTs, the morphology of the found structures in fact do agree very well with the typical structures of the CNTs (Baytubes) incorporated in the PC material. It remains unclear why the CNTs did not completely combust in the PC material; however, it is quite likely that this was due to insufficient oxygen supply to the lower regions in the PC pile in the crucible. When the higher layers melt, they may seal the lower layers and therefore prevent oxygen from reaching these

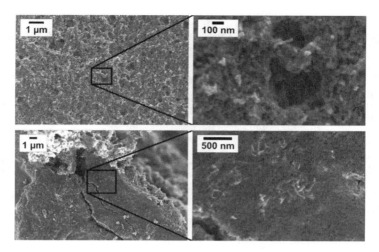

FIGURE 7.12   SEM images of residues from the crucible after incineration of PC with 5% (top) and 7.5% (bottom) CNT content; arrows indicate CNT-like structures.

regions. It should be noted that in real incineration plants, the fuel (waste) is continuously kept in motion to improve its combustion by assuring sufficient oxygen supply. The experimental conditions may hence not have perfectly matched the real conditions in an incineration plant.

7.4.3.4 Analyses of Collected Fly Ash

Samples from the fly ash deposited towards the end of the second furnace were collected for each material after completion of all 10 experimental runs. The amount of collected fly ash significantly differed depending on the material under investigation. As can be clearly seen in Figure 7.13, the highest amount of fly ash was collected for the PC materials, followed by PE and PA materials. These samples were then further analyzed for the presence of CNTs by using an SEM; however, no CNTs were found in the analyses. All analyzed particles showed typical soot-like structures, indicating that the combustion in the furnaces may have been under-stoichiometric.

7.4.3.5 TXRF Analyses

Collected fly ash and residue samples were analyzed by TXRF for the presence of cobalt, which is used as precursor material during the manufacture of CNTs. The collected powders were therefore dispersed on a thin layer of silicone paste (Korasilon M-S 2-270) on an acrylic glass substrate and analyzed with a TXRF spectrometer (Fa. Bruker, Type PicoTAX with 300 mm^2 detection area). Using TXRF it is possible to

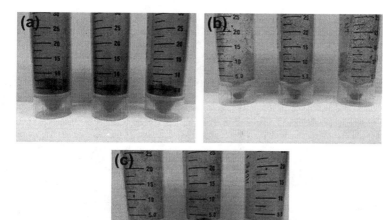

FIGURE 7.13   Fly ash collected at the end of the second furnace after 10 experimental runs; (a) PC, (b) PA, (c) PE; left to right: 0%, 5%, and 7.5% CNT content.

analyze a mass of only a few micrograms with a detection limit of sub-nanograms for heavy metals like Fe, Ni or Co.

Expectedly, the samples from the pure materials showed no cobalt content, whereas the residues from the crucible from PC material with CNT content revealed cobalt in the lower µg range. Since PE and PA disintegrated completely and left no residue in the crucible, no TXRF measurements could be conducted for these materials.

Analyses of the fly ash samples only showed cobalt contents in the lower ng range only for PC with 7.5% and PA with 5% and 7.5% CNT content, whereas all pristine materials as well as PC with 5% CNT and PE with 5% and 7.5% CNT content showed no indication of cobalt.

7.4.3.6 Analyses of Collected Airborne Particles

As mentioned above, downstream of the second furnace, airborne particles were electrostatically sampled onto substrates for consecutive analyses. While the aforementioned analyses investigated whether residues that remain in the incineration plant and will only be removed during the next revision of the plant may include CNTs, the aim of this investigation was to study whether CNTs may become airborne and consequently be released with the exhaust gas from the plant. If airborne CNTs are released they may eventually cause environmental or human exposure in case they are not captured with 100% efficiency in downstream gas cleaning steps.

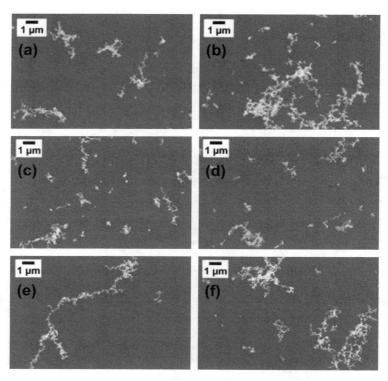

FIGURE 7.14   SEM images of particles collected during incineration of the different materials; (a) PC with 5% and (b) 7.5% CNTs; (c) PA with 5% and (d) 7.5% CNTs; (e) PE with 5% and (f) 7.5% CNTs.

Samples were taken with the NAS with a custom-made upstream unipolar charger to increase the sampling efficiency. Particles were deposited onto silicon substrates that were analyzed with an SEM for the presence or absence of CNTs. As the images in Figure 7.14 exemplarily show, none of the analyzed samples revealed CNTs or fiber-like structures. Also no CNTs embedded in other, larger particles were found and it can hence be concluded that CNTs are fully combusted under the conditions in an incineration plant.

7.4.4 Conclusions

Experiments were conducted for two simulated end-of-life-cycle processes, shredding and incineration of different polymer materials with different CNT contents. None of the experiments showed a release of free CNTs. Generally no major differences in the resulting particle quantities, sizes and morphologies were found for the same materials with different CNT contents. During shredding of the different materials, CNTs were only found embedded in the matrix material. During incineration experiments, only residue of PC materials that remained in the crucible after incineration showed CNT-like structures embedded in the slag. The latter likely was caused by an incomplete combustion of the material due to experimental deficiencies.

For the incineration experiments it can be summarized that burning of the different polymers always released very high particle number concentrations, independent of the CNT content. This is not surprising, as it has been known for a long time that any combustion releases large amounts of small particles. However, since none of the measurements revealed any evidence for airborne free CNTs, it can be concluded that under the given experimental conditions, the CNTs in the materials decomposed completely. Similar results were also found by Roes et al. [46]. Taking into account that the flue gases from an incinerator get filtered before they are released into the atmosphere, it can be concluded that a release of CNTs from incineration plants into the environment is rather unlikely. This assumption is supported by the findings of Wang et al. [53], who reported that CNTs get removed in fiber filters with a higher efficiency than spherical particles of the same mobility diameter.

Acknowledgement

The research presented here has been conducted within the CarboSafe (03X0043D) project that has received funding from the German Federal Ministry for Education and Research (BMBF) as part of the InnoCNT initiative.

7.5 THE NEED FOR SPEED: DETECTION AND CHARACTERIZATION OF PARTICLE RELEASE DURING POWDER HANDLING USING ON-LINE MONITORS (JENSEN, KOPONEN)

7.5.1 Introduction

Contributions from different particle sources combined with influence of ambient air-pollution are some of the major challenges in the interpretation of on-line monitoring data of nanoparticles and fine dusts in workplaces. In principle, process-related dust release events may vary from seconds to numerous hours. Moreover, the emission may result in negligible to highly significant particle dust concentrations in the occupational environment. An additional complication is that workers may handle several different powders and hence be exposed to emissions from various amounts of these and other particle sources during a single work-shift or even in a single work cycle.

Due to the potentially complex exposure situation in industries producing or applying manufactured nanomaterials and fine particulates, assessment of exposure to specific materials, or the potential exposure, therefore may require more in-depth analyses than traditionally applied in occupational hygiene [47,48]. In some scientific studies, workplace measurements have been conducted using several different on-line measurement instruments to enable better temporal and spatial information on the airborne particle concentrations [49,50]. Recently, it has been shown that parallel time-series data from near-field activity and far-field workroom or outdoor background positions may enable identification of particle release during powder handling from the general variations in airborne dust in the workroom (e.g., [51,52]). From these studies it appears that high temporal resolution of time-series data may be essential to enable proper identification of the activities resulting in particle release and enable characterization of the associated airborne particles. The time resolution required to resolve and characterize specific events may be in contrast to the ability of, e.g., the scanning and stepping mobility particle sizers, which are some of the most commonly applied size-distribution monitoring devices. Number concentrations may be measured at very high time resolution. For particle size-distribution measurements, longer durations of stable size distributions with fixed concentrations are in principle favorable for reaching high confidence in the data. However, such conditions may rarely be met in the workplace.

In this chapter, we give an example of the time-series evolution in particle concentrations using high-time-resolution instruments monitoring during handling of both nano- and sub-μm size particle-powders at a mixer in a paint-producing factory. We demonstrate the possibility to identify process-specific emissions from background dust using high time-resolution monitoring devices. However, although clear links can be established between dust release and activity with these devices, detailed characterization and quantification of the exposure will require a much more in-depth off-line analysis including use of imaging and chemical analysis. The work is based on data presented in Koponen and Jensen [52].

7.5.1.1 *The Workplace and Contextual Activity Information*

The study was completed at a paint manufacturer during preparation of a paint mixture with 3446.8 kg dry filler materials. The filler recipe is listed in Table 7.7 and includes the addition of 130 kg cellulose, 175 kg silica (nanosilica), 280 kg talc (nanoflakes) and 2675 kg fine TiO_2 pigment powders. The paint ingredients were added to the mixer over a period of ca. 5.5 hours, including a 1 h 47 min meeting break. The only extra activity in the work area was operation of an electrical forklift, which mainly was used to bring ingredients to the mixing station.

TABLE 7.7 Fillers Added to the Mixer During Paint Production

Materials	(kg)	Note
Cellulose Bags (2 types)	(65 + 65)	Powder
Silica Bags (Sipernat)	175	5 × 25 kg
TiO$_2$ Big-bags (RDS)	2500	5 × 500 kg
TiO$_2$ Bags (RDS)	175	5 × 25 kg
Talc Bags	280	11 × 25 + 10 kg
K 24-01 Bags (Unknown)	26.8	Powder
Propylenglycol (MPG) Bags	160	Powder
Total amount of powder	3446.8	

The volume of the workroom was 150 m^3 (20 m wide, 30 m long and 2.5 m high) and it contained a single paint mixing unit. Powder was poured manually into the mixer through a 0.8 × 0.8 m^2 feeding hatch at the top of the mixer. This allows powder to fall into the mixer tank, which was placed below floor level. Local exhaust ventilation was constructed along the two sides and the back of the feeding hatch to reduce dust release during pouring and mixing. The volume flow and efficiency of the LEV and general air-exchange rate in the workroom was unknown.

7.5.1.2 Measurements

On-line activity (Act) and far-field background (FF) measurements were performed using different particle measurement devices to document the variation in total particle number concentrations and particle size distributions. The activity measurement was placed within 1.5 m from the worker on the right-hand side of the mixer unit. The background measurement was performed at the left side of the workroom, ca. 5 m away from the activity measurement, close to two tanks where there was little influence from forklifts or any other activities. Shelves for storage of empty bags and different powders were located ca. 5 m south and ca. 5 m south-west of the FF background and Activity locations, respectively (normalized geographical directions) (Table 7.7).

Temporal activity particle number concentrations and number size distributions were measured using a TSI Fast Mobility Particle Sizer (Model 3091; 1 second time resolution, measurement range: 5.6—560 nm) and a GRIMM Dust Monitor (DM) (Model 1.109; 6 second time resolution, measurement range: 0.28—30 μm). FF background measurements were conducted using a TSI Condensation Particle Counter (CPC) (Model 3022A; 1 second time resolution) and another GRIMM Dust monitor

Model 1.109. The CPC was used instead of a second FMPS, which unfortunately had problems upon arrival at the field location.

Both stationary and personal samples of fine particulate matter smaller than 1 μm (PM_1) were sampled using Triplex cyclones (Model SC.1.106). Cellulose-acetate filters were used for electron microscopy and Teflon filters for gravimetric analysis. Finally, humidity and temperature were monitored using Gemini Data Logger (TinytagPlus) instruments. In this report, we only use the filter data to demonstrate differences in mass-concentrations between fixed stations and personal samples as well as to show the major types of particles collected on the filters by scanning electron microscopy (SEM) using a Quanta 200 FEG MKII Scanning Electron Microscope (SEM) equipped with an energy dispersive spec-trometer (EDS) (Oxford Instruments, High Wycombe Bucks, UK).

7.5.2 Results

7.5.2.1 Description of the Activity-Related Particle Release

The total concentration time-series data for both Activity and FF background are plotted in Figure 7.15. General co-variation was observed between the FF background and the baseline concentrations measured at the Activity location. However, the general variations in the FMPS-CPC measurements did not always co-vary with the measurements per-formed with the GRIMM DM's. This suggests that the particles in the FMPS-CPC and GRIMM DM sizing regions may originate from different sources and also behave differently at this work site.

The time-series data also show clear indications of numerous activity-related particle release events. It is of special interest that the episodically increased particle concentrations in the Activity FMPS measurements are rarely detected in the CPC FF background measurements. However, in the GRIMM DM measurements, the release events appear to be generally observed, but with lower concentrations at the FF background location. Since the concentration peaks in the FMPS and CPC measurements rarely exceed $1e5$ cm^{-3}, coagulation and scavenging should be of moderate importance [54]. Instead, the high mobility of the nano- and near-nano-size particles may cause rapid dilution into the workplace air and/or clearance by local ventilation before reaching the FF background, whereas the coarser particles in the GRIMM DM range have lower diffusion rates and may preferentially spread in plumes.

Considering the specific powder handling events, a high number of episodic peaks were observed in the number concentrations, especially in the FMPS and CPC measurements. Due to the high time resolution, it can be seen that several peaks may occur during some of the episodes. To better resolve the activity-related particle release, the time-series FMPS activity concentrations were subtracted by the CPC FF background

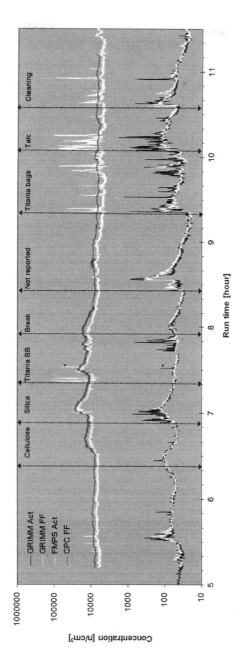

FIGURE 7.15   GRIMM DM, FMPS and CPC time-series concentration data from the Activity (Act) and Far-Field background (FF) locations. Periods with the powder handling activities start as indicated by the broken arrow lines. Titania BB denotes Titania in big-bags.

concentrations. Analyses of the CPC and FMPS data obtained before the work was initiated in the morning showed a systematic 2000 cm^{-3} higher particle concentration in the CPC than in the FMPS, so the data were adjusted for this difference. The resulting 'differential' time-series data are shown in Figure 7.16 for both the FMPS-CPC and GRIMM DM data. In addition to clear signatures from the activity peaks, the differential time-series data clearly show a few long-lasting background events. These are indicated by negative values without being preceded with a peak-event in the Activity measurement.

In the following analysis of the specific emissions related to the specific powder handling activities, we discriminate between particle release events and episodes. We define an emission episode as a confined period where particle concentrations at the Activity location generally exceed the FF background concentrations by at least 1000 cm^{-3}. The events are defined at the measurement resolution (1 sec and 6 sec for the FMPS-CPC and GRIMM DM, respectively). The number of episodes in each powder handling activity was identified by visual inspection of the time-series plots. The number of events, on the other hand, was determined automatically using the 'WHAT-IF' function in Microsoft Excel. In this work, we report the number of episodes and events for each powder handling activity where Activity measurement exceeded the FF background concentrations by 1000, 5000 and 10,000 cm^{-3} for the FMPS and 100, 500 and 1000 cm^{-3} for the GRIMM DM, respectively. The results from this analysis are listed in Tables 7.8 and 7.9.

In the first activity 2×65 kg cellulose were added to the mixer. Two corresponding release episodes were observed in the FMPS measurements and a total of 130 measurements (seconds) had concentrations above 1000 cm^{-3}. For comparison only one short release episode was observed in the GRIMM DM measurements and elevated particle concentrations exceeding 100 cm^{-3} were only observed for 12 seconds (two 6 seconds measurement). The highest GRIMM DM and FMPS peak concentrations were 147 and 2360 cm^{-3}, respectively (Tables 7.8 and 7.9).

Addition of the 175 kg Sibernate silica (7×25 kg bags) was associated with only 4, maybe 5, release episodes in the FMPS measurements, whereas 16 short release episodes could be observed in the GRIMM DM measurements. Interestingly, the particle release in the FMPS size range was only observed in connection with the second period of particle release when compared to the GRIMM DM data. Concerning the durations of the elevated particle concentrations, the FMPS particle concentrations exceeded the background by at least 1000 cm^{-3} during a total of 144 seconds. Even higher concentrations were observed resulting in 35 and 5 seconds with more than 5000 and 10,000 cm^{-3}, respectively. In the GRIMM DM size range, the particle concentration at the Activity location exceeded the FF background by 100 cm^{-3} for 216 seconds and 5000 cm^{-3} for only 6 seconds.

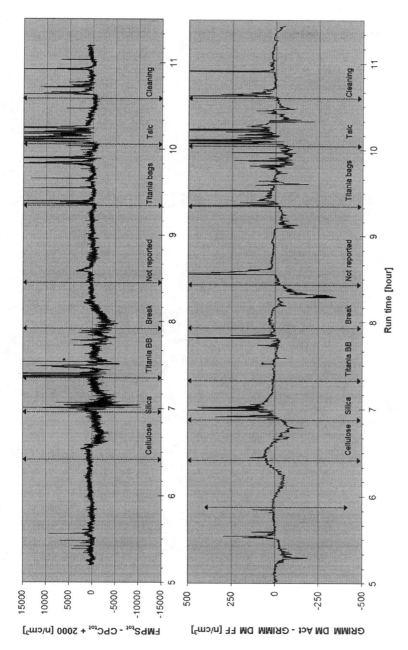

FIGURE 7.16 Time-series plots showing the difference between the Activity FMPS and FF background CPC particle concentrations and the Activity minus FF background GRIMM DM particle concentrations, respectively. Pre-activity measurements showed that the CPC concentrations were ca. 2000 cm^{-3} higher than the total from the FMPS measurements, which explains the adjustment by 2000 cm^{-3} in the top graph. *Denotes a short-term measurement at a leak observed in wall of the feeding unit.

TABLE 7.8   Summary of Activity-Related Dust Particle Concentration Data (C) Collected During Specific Powder Handling Activities Using the FMPS-CPC Time-Series Data

Powder	Duration (sec)	Maximum (cm^{-3})	Average (cm^{-3})	N (sec) C > 1e3 cm^{-3}	N (sec) C > 5e3 cm^{-3}	N (sec) C > 1e4 cm^{-3}	Release Episodes
Cellulose	716	2360	662	130	0	0	2
Silica	1749	16,700	−964	144	35	5	4 to 5
TiO$_2$ BB	1905	117,300	1115	303	168	85	12*
TiO$_2$ Bags	2571	47,620	1157	549	139	79	19
Talc	1456	150,260	6369	767	365	260	20

Including a short-term leak measurement at the side of the feeding funnel.

Addition of the five 500 kg TiO$_2$ big-bags was associated with 11 and 6 release episodes observed in the FMPS and GRIMM DM Activity measurements, respectively. Elevated FMPS Activity particle concentrations ($+1000$ cm^{-3}) were observed for 303 seconds of which the concentration exceeded the background by 5000 and 10,000 cm^{-3} for 168 and 85 seconds, respectively (Table 7.8). In the GRIMM DM measurements, the Activity concentrations exceeded 100 cm^{-3} for only 120 seconds and exceeded 500 cm^{-3} for only 1 second. The FMPS peak concentrations reached up to 117,300 cm^{-3} as compared to 598 cm^{-3} in the GRIMM DM measurements. Consequently, the results strongly indicate release of high amounts of fine particles during the TiO$_2$ powder handling. This is also indicated by the differences in the average differential particle concentrations for the FMPS and GRIMM DM measurements.

TABLE 7.9   Summary of Activity-Related Dust Particle Concentration Data (C) Collected During Specific Powder Handling Activities Using the GRIMM DM Time-Series Data

Powder	Duration (sec)	Maximum (cm^{-3})	Average (cm^{-3})	N (sec) C > 1e2 cm^{-3}	N (sec) C > 5e2 cm^{-3}	N (sec) C > 1e3 cm^{-3}	Release Episodes
Cellulose	716	147	49,5	12	0	0	1
Silica	1749	715	44,5	216	6	0	16
TiO$_2$ BB	1905	598	8,7	54	6	0	7*
TiO$_2$ Bags	2571	1299	3,8	120	24	12	11
Talc	1456	1823	62,6	282	48	12	19

Including a short-term leak measurement at the side of the feeding funnel.

Interestingly, the subsequent addition of TiO_2 powder from ordinary bags (5 × 25 kg) resulted in longer durations (549 seconds in the FMPS and 120 s in the GRIMM DM measurements) with elevated particle concentrations than addition from the big-bags. Eleven and six particle release events were observed during pouring of these bags by the FMPS and GRIMM DM measurements, respectively (Figure 7.15 and Figure 7.16). Compared to the big-bag handling, the addition of TiO_2 from ordinary bags also resulted in higher concentrations in the coarser size-fraction measured with the GRIMM DM. The particle concentration exceeded 500 and 1000 cm^{-3} for 24 and 12 seconds out of the 120 seconds elevated GRIMM DM particle concentrations, respectively. The maximum GRIMM DM peak concentration was also higher (1299 cm^{-3}) during pouring from bags than from the big-bags (598 cm^{-3}). However, this was opposite in the case of the FMPS measurements (47,620 cm^{-3} vs. 117,300 cm^{-3}).

The last powder handling activity was pouring eleven 25 kg bags with talc ('nanoflakes') into the mixer. During this activity a total of 20 and 19 episodes of particle release events were observed in the FMPS and GRIMM measurements, respectively (Figure 7.15 and Figure 7.16). Addition of talc was associated with the highest particle concentrations and longest accumulated time with elevated particle concentrations in both the FMPS and GRIMM DM measurements. The FMPS and GRIMM DM peak concentrations were 150,260 and 1823 cm^{-3} and the accumulated times of elevated (+1000 cm^{-3}) particle concentrations were 767 and 282 seconds, respectively (Tables 7.8 and 7.9). In the FMPS measurement the particle concentration exceeded the FF background concentrations by 10,000 cm^{-3} for 260 out of a total of 767 seconds.

7.5.2.2 Duration of Release Events and Needs for High Time Resolution

As indicated in Figure 7.15 and Figure 7.16 and the data derived in Tables 7.8 and 7.9, the variations in particle concentrations and activity-related release events typically varied from seconds to a few minutes. In a few cases elevated particle concentrations lasted for almost 10 minutes (see the addition of talc in Figure 7.15 and Figure 7.16). Fluctuations in the FF background concentrations showed a more gradual variation with time and were typically also of longer durations. Durations of up to 30 minutes are clearly observed.

The ability to detect a specific release event depends on the time resolution of the measurement devices applied. However, the duration or identification of a peak event may also be influenced by rapid concentration changes or instrumental scatter in the measurements. From the measurement example given in this case study, the high time resolution of 1 second for the FMPS-CPC and 6 seconds for the GRIMM DM measurements appear to be very valuable for detection of particle release associated with powder handling. However, the high time resolution is also valuable for observing the influence from background events and observing the migration of dust between the two measurement locations. Without sufficient time resolution, it may not be possible to detect a specific particle release event or to even distinguish whether particles originate from the activity in the FF background locations.

Considering the results from our time-series data in Tables 7.8 and 7.9, we on average identified between 0.08 and 0.82 activity-related general release events per minute during the powder handling activities. However, within these general release events, additional peak events were observed. For example, at least 30 concentration peaks occurred in the 19 general release events identified during handling of the talc (Figure 7.15 and Figure 7.16). In addition, we observed FF background events, which add additional variations to the particle concentrations. Consequently, the concentration changes and possibly also variations in particle size distribution may occur in very short time-scales depending on the source characteristics. Analyzing the most significant peak durations in the FMPS-CPC measurements showed that the average peak duration was on the order of 26 ± 24 seconds (48 peak events). During these events the concentrations increased and dropped from a few thousand to more than one hundred thousand particles per cm^3.

The short duration of the concentration peaks is a challenge for detailed concentration measurements and proper size-distribution analysis of the airborne dust. Averaging of data is a very common procedure to eliminate data scatter, but in this case, where we used high time-resolution instrumentation, the data reveals that this may not (at least not always) be appropriate. Averaging of data results in smoothing of the peak events and will ultimately mask individual events.

As an example, we assessed the influence of data-averaging using the results from the GRIMM DM measurements. The results are shown in Figure 7.17, which shows the time-series data of the total particle concentration at the original 6 sec resolution and averaged over 1 min, 3 min, 15 min, and 30 min, respectively. It is shown that integration times of up to 3 minutes enable identification of the major fluctuations and inferred particle release events. However, smaller short-term release events are averaged out and cannot be easily detected. Hence, considering the average peak durations in time-series data obtained by the FMPS mentioned above, most individual events would no longer be detectable.

The consequences of increased GRIMM DM integration times are clearly demonstrated in Figure 7.17b, where it is evident that an averaging time of 3 minutes is not optimal for resolving the multiple and even several-minutes-long release events observed between run-time 9 h 20 min and 11 h 00 min. The even longer averaging times of 15 and 30 minutes completely

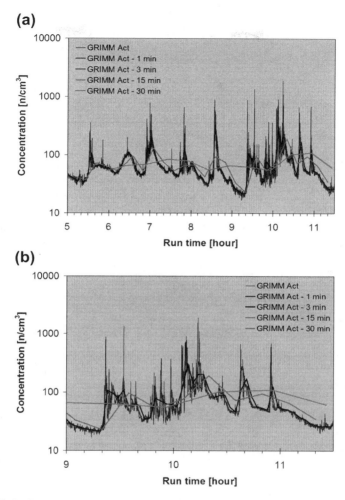

FIGURE 7.17   GRIMM DM time series plotted with the standard 6 seconds time resolution and different longer averaging times to demonstrate the influence of time resolution on the ability to detect individual release events and episodes. (a) Original and averaged time-series data for the whole work-shift. (b) Original and averaged time-series data focusing on the period where talc was handled (see Figure 7.16 for reference). Note the rapid loss of resolution with increasing averaging times.

hide the fluctuations and release events detected by the GRIMM DM shorter than at least 10 minutes. Hence, simplification of time-series data by longer averaging times may be associated with unwanted masking of actual exposure. This is further supported by the averaged differential Activity-related particle concentrations in Table 7.8 and Table 7.9, which in some cases even became negative due to background events.

The overall conclusion from this analysis is that, if a workplace measurement, in addition to concentration measurements, also aims towards characterization of particle size distributions of the emissions and workplace dust, high time resolution is mandatory.

7.5.2.3 Link Between On-line Monitoring and Personal Exposure

Finally, we discuss the PM_1 filter measurements in comparison with the on-line measurement data. The results and an SEM image of the dust collected on the personal sample are shown in Table 7.10 and Figure 7.18, respectively. It is evident that there is a very big difference between the PM_1 concentrations obtained by stationary measurements and personal sampling. The ratio between the Activity and the FF PM_1 concentrations was only 1.2, showing only minor differences in dust concentrations at the two stationary sampling points. In contrast the personal PM_1 concentration was 8.5 and 7.2 times higher than the stationary FF background and NF activity locations, respectively. From SEM analysis most of this difference may be explained by high apparent personal exposure to dust from handling TiO_2 (Figure 7.18).

For comparison with the on-line concentration measurements over the entire duration of the powder handling activities, the ratio between the average FMPS Act and CPC FF background particle concentration measurements was only 0.9, but 1.2 for the GRIMM DM. Interestingly, the ratio of the GRIMM DM measurements corresponds well to the gravimetric measurement data, whereas the nano-size particle concentration was influenced by the systematic difference in particle numbers between the CPC and the FMPS as well as the influence from various sources masking the effects of powder handling in the averaged data.

TABLE 7.10   Measured Mass Concentrations* and High, Low and Average Particle Number Concentration Values from the Duration of Powder Handling, Including the Meeting Break

Measurement Position	PM_1 (Stationary) (mg/m^3)	PM_1 (Personal) (mg/m^3) (time)	Average Number Concentration (cm^{-3})	Max (cm^{-3})	Min (cm^{-3})
Activity	0.090 (6.5h)	0.647 (6.4h)	$9.33e3^\in$ $75.7^\$$	$1.55e5^\in$ $1.89e3^\$$	$3.52e3^\in$ $16.8^\$$
FF background	0.076 (6.4 h)	—	$1.05e4^\pounds$ $64.8^\$$	$2.88e4^\pounds$ $0.39e3^\$$	$5.60e3^\pounds$ $18.0^\$$

*Data from Koponen and Jensen (submitted).
$\in$FMPS measurements.
$\pounds$CPC measurements.
$\$$GRIMM DM measurements.

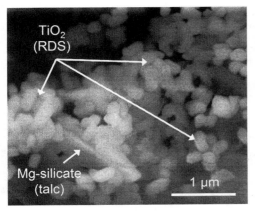

FIGURE 7.18 Scanning Electron Microscopy (SEM) image of the most abundant dust particles detected on the personal PM_1 filter sample. In addition to titanium dioxide and talc, we observed cellulose and dolomite $((Ca,Mg)(CO_3))$ from another source as the most abundant materials.

Altogether, the results from on-line particle measurements and PM_1 sampling strongly indicate that on-line measurement with high time resolution is able to detect particle release events and possibly also for size-distribution measurements. However, conclusions from average particle number concentrations should be used with care. Most importantly, our case study shows that personal sampling or monitoring in the breathing zone must not be excluded in occupational exposure assessment for either nanomaterials or fine dust. In this case, relying on stationary measurements would underscore the PM_1 exposure to about 1/6 of the measured value.

7.5.3 Conclusions

- Time-series analyses of parallel near-field (Activity) and far-field (background) measurement data of particle concentrations in the workplace are suitable for identification of particle release events and influence of background contributions if collected with high time resolution.
- The duration of activity-related particle release events may vary from seconds to hours. In the present case with powder handling at a mixer for paint production, the average peak durations were 26 ± 24 seconds (48 peak events) in measurements with a Fast Mobility Particle Sizer. During these events the concentrations increased and dropped from a few thousand to more than one hundred thousand particles per cm^3.
- Due to rapid changes in concentrations and short durations of specific handling activities, high time resolution is essential. However, averaging time of up to 1 and 3 minutes may be acceptable for detection of most important dust release events and release episodes, respectively.

- The short peak durations and rapid changes in particle concentrations make size-distribution measurement challenging using traditional size-distribution monitoring devices. In principle, complete simultaneous size-distribution measurements at a few seconds' resolution are needed to obtain reliable data.
- Stationary filter sampling and on-line measurements at the near-field activity and far-field background appear to be unable to estimate the personal exposure. Large differences were observed between the data obtained by all measurements made at the stationary locations as compared to the mass-concentration collected in the personal dust sample. Consequently, measurement in the breathing zone of the worker is necessary for proper assessment of worker exposure to nanoparticles and fine dust.

Acknowledgements

We gratefully acknowledge the financial support from the 'NANOKEM' project (grant #20060068816 from The Danish Working Environment Research Fund) and the NANO-DEVICE project (co-funded by the European Commission under the 7th Framework Programme for RTD project number FFP7 211464) that made this work possible. We also express our gratitude to the Danish Paint and Lacquer Organization for facilitating the industrial contact and the participating company for open access to their work facilities. Finally, we thank our technician S.H. Nielsen for her assistance in connection with the field measurements and managing the filter weighing.

7.6 PARTICULATE EMISSIONS FROM EQUIPMENT USED IN ULTRA CLEAN AREAS OF THE SEMICONDUCTOR INDUSTRY (GOMMEL, KELLER)

7.6.1 Introduction

The classification of clean rooms regarding airborne particulate matter is defined in ISO 14644-1. The lowest particle size that has to be measured for classification purposes are particles with an equivalent diameter of 100 nm, right at the border regarding the definition for nanoparticles according to ISO/TS 27687 [55]. Equipment used in ultra clean manufacturing processes emits particles during operation due to friction processes. These emitted particles have a major impact on the clean room class according to the systems set in ISO 14644-1 and can damage the processed component, for example, a semiconductor wafer. The requirements of cleanliness levels for the semiconductor industry increase steadily because of the continuous decreasing pitch widths processed on the wafers. Currently, half pitch widths of 22 nm or even below are discussed using extreme ultraviolet lithography [56,57].

Based on this fact, the definition of cleanliness levels according to ISO 14644-1 will be probably extended well into the nanometer range in the future.

Trends in clean room technology show that within the last 10 years a significant international interest of clean room users and equipment manufacturers came up who are in need of a classification system for equipments in clean rooms. This has been verified by a survey done by Fraunhofer IPA that 95% of clean room users require clean room suitable equipment. The certainty to choose the appropriate equipment — in regard of its cleanliness behavior — for a specific clean room class has to be enabled. This leads to the development of the German standard VDI 2083 part 9.1: Clean room technology — Compatibility with required cleanliness and surface cleanliness [58]. Within this standard, the defined equipment classification system is linked closely to the cleanliness classes within ISO-14644-1. In addition guidance is given for the design and construction of the equipment.

Currently, no international standard exists that enables an equipment cleanliness classification. An internationally accepted classification system for the clean room suitability of equipment is needed, which is directly related to the already established classification system of clean rooms. Therefore a new international standard 'Clean room Suitability of Equipments' is currently under development within the ISO 14644-family based on VDI 2083 part 9.1. This new international standard will provide the missing link between equipment and the respective cleanliness classes in the near future. This standard will provide — for the first time on an international level — a classification system for equipment and tools.

7.6.2 Experimental Design

The clean room suitability of equipment is determined by the particle emission behavior, the outgassing behavior, the chemical resistance and its cleanability. In this case study, it is only focused on particle emission. The parameters of the clean room suitability test are determined by process-specific aspects or cleanliness specifications. The examination of the compatibility with the required cleanliness aims at establishing the suitability of equipment for use in rooms where air cleanliness and other parameters are assessed on the basis of technical rules. This requires the definition of environmental conditions for testing, execution and assessment, and the documentation for determination of the compatibility with the required cleanliness. The statistical analysis of the results obtained serves as a tool for quality control. As the clean room suitability is dependent on different parameters like the equipment load, the set of parameters has to be documented in advance very well. To obtain test results that can be compared to each other, the set of load has to be identical.

7.6.2.1 Requirements

The requirements for the clean room suitability of equipment have to be fixed concerning the load states (e.g., movement cycle and movement speed) and environmental conditions (e.g., temperature, relative humidity, gaseous atmosphere). The requirements in regards to airborne particulate contamination derive from particle cleanliness specifications in accordance to ISO 14644-1.

As particulate emission from equipment is expected to be singular events with prolonged intervals without particle emission, mainly optical laser particle counters are used. The detection limit is currently around 200 nm equivalent particle diameter. Regarding the detection of possible emission of nanoscale particles, there is a need for continuously measuring particles below 100 nm for the examination of the clean room suitability of equipment especially used for critical applications in semiconductor industries.

7.6.3 Measurement Procedure

All tests have to be performed under typical clean room conditions (22 ± 1°C, 45% ± 5% relative humidity). The examinations shall be carried out in a clean room with a unidirectional flow (avoiding cross contamination). The airborne particulate contamination that is detected in a clean room without any test objects running is stated as 'background emission'. For the detection of airborne particulate contamination, several optical particle detectors are used (LasAir II 110, Particle Measuring Systems, Inc., Boulder, United States). The cleanliness class of the test environment shall be at least one class better than the desired airborne particulate cleanliness class for which the piece of equipment is classified for. It must be ensured at all times that no other sources of particles can affect any of the measurements. Excluding extraneous particles will make sure that only those particle emissions generated by the piece of equipment are detected. Equipment shall be decontaminated using suitable cleaning techniques before being brought into the test environment. The actual surface cleanliness achieved shall be tested using suitable methods. Prior to testing, all relevant parameters of operation of the equipment have to be defined and specified. A representative application has to be selected to ensure that particle emissions, which are to be expected during actual operation of the equipment, are detected. This means that a mode of operation will be chosen, which reflects the intended use of the equipment under controlled conditions.

7.6.4 Test Schedule

The procedure for the assessment of the clean room suitability of equipment can be summarized as follows [59]:

1. *Specification of parameters of operation.* The release of particles from moving elements due to tribological stress is dependent on speed, acceleration, forces that the friction element encounters, material pairing, surface quality, surrounding atmosphere and lubricant used. Because of that, the parameters of operation have to be set accordingly. If, for example, a robot system will later be used to move a load of 10 kg with a defined movement pattern, the best would be to analyze this pattern. For practical reasons, the measurements will be done with, for example, three different sets of parameters: minimal payload and slow movements, mean values and maximum payload and fastest movements possible.

2. *Localization of particle sources.* By scanning all moving elements of a robot system to be qualified, with the measuring probe of an optical particle scanner manually, all relevant measuring points with detected particle emission are localized for later classification measurement (see Figure 7.19). Each localized particle emission source is given a number and has to be classified by the classification measurements.

3. *Classification measurements.* The classification measurement is done by positioning the measuring probe with a suitable stand directly under the manually localized source of particle emission. After positioning, all personnel have to leave the clean room. The measurements will start with a defined delay in order to prevent cross contamination from particles released from the clean room personnel itself. Each measuring

FIGURE 7.19 Qualification measurements of industrial robot systems: Delta kinematic robot system (left) and six-axis clean room robot system (right). All systems are set up in an ISO 1 class clean room at Fraunhofer IPA, Stuttgart, Germany.

point will be scanned for particle release. The particle emission has to be recorded over a time period of at least 100 minutes (see Figure 7.19).

7.6.5 Classification and Statistical Analysis

All classifications are referenced to the national standard VDI 2083 part 9.1. The evaluation of clean rooms in regard of airborne particles is carried out according to cleanliness classes, as they are defined in ISO 14644-1 (see Table 7.11).

The procedure and statistical analysis is exemplarily explained for a liquid handling robot system. The measuring point with the highest emission of airborne particles will finally give the total classification number of the whole liquid handling robot system according to VDI 2083 part 9.1. With the help of preliminary localization measurements, the classification will be performed on all localized measurement points. For example, if during the manual scanning of the liquid handling robot system for the release of airborne particles, four measuring points are localized, all these spots have to be classified during the measuring campaign. An example is given in Figure 7.20 on measurement points 1–4 (MP01–MP04). MP01 and MP02 are measuring points directly under

TABLE 7.11   ISO 14644-1 [57]: Selected Airborne Particulate Cleanliness Classes for Clean Rooms and Clean Zones

ISO Classification Number (N)	Maximum Concentration Limits (Particles/m³ of Air) for Particles Equal to and Larger Than the Considered Sizes Shown Below (Concentration Limits Are Calculated in Accordance with Equation (1) in 3.2)					
	0.1 µm	0.2 µm	0.3 µm	0.5 µm	1 µm	5 µm
ISO Class 1	10	2				
ISO Class 2	100	24	10	4		
ISO Class 3	1,000	237	102	35	8	
ISO Class 4	10,000	2,370	1,020	352	83	
ISO Class 5	100,000	23,700	10,200	3,520	832	29
ISO Class 6	1,000,000	237,000	102,000	35,200	8,320	293
ISO Class 7				352,000	83,200	2,930
ISO Class 8				3,520,000	832,000	29,300
ISO Class 9				35,200,000	8,320,000	293,000

NOTE: Uncertainties related to the measurement process require that concentration data with no more than three significant figures be used in determining the classification level.

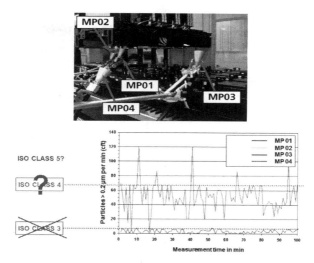

FIGURE 7.20   Time-resolved particle emission on four decisive measurement points.

moving elements of a liquid dispenser. MP03 and MP04 scan the release of airborne particles during the passage of the portal. During the classification measurements, MP01 (red line) shows the highest particle emission for particles > 0.2 μm into a defined air volume of 1 cubic foot. The air volume is based from the airflow through the measurement device of 1 cubic foot per minute. Each minute, one measuring point is displayed. The dashed lines show the limit values for particle concentration regarding all particles > 0.2 μm equivalent particle diameter for ISO class 3 and ISO class 4. The concentration of particles of predefined particle sizes will be used for the classification.

Then, the measurement results will be mathematically and statistically evaluated: For each measurement point, four size channels are considered: particles equal or larger 0.2 μm, 0.3 μm, 0.5 μm and 5.0 μm. For each particle size class, the obtained particle number concentration is correlated with the classification in ISO 14644-1; see Table 7.11. As soon as there is a certainty of 95% that the given upper class limits of the ISO 14644-1 will not be exceeded, the particle size class of this measuring point is given a classification number according ISO 14644-1. This is done for all four defined particle size classes. The worst classification number of all four particle size classes gives the classification number for this measuring point. So measuring point MP01 is classified as ISO 4, MP02 as ISO 1, MP03 as ISO 3 and MP04 as ISO 1. The worst classification number of all classified measuring points will give the final classification number for the tested equipment, here ISO 4 from measuring point MP01 (Table 7.12).

TABLE 7.12 Statistical Evaluation of the Measured Particle Emission, as Shown in Figure 7.20

Statistical Parameters		Measuring Point			
		MP1	**MP2**	**MP3**	**MP4**
Mean values for the detection size [particles/cft]	0.2 μm	51.7	0.0	3.2	0.0
	0.3 μm	20.1	0.0	1.3	0.0
	0.5 μm	3.5	0.0	0.0	0.0
	5.0 μm	0.0	0.0	0.0	0.0
Standard deviation for the detection size [particles/cft]	0.2 μm	19.3	0.2	2.3	0.2
	0.3 μm	13.7	0.2	1.4	0.1
	0.5 μm	2.9	0.2	0.0	0.0
	5.0 μm	0.0	0.0	0.0	0.0
Air cleanliness class [ISO 14644-1]		4	1	3	1
Limiting values of particles in corresponding air cleanliness class for the detection size [particles/cft]	0.2 μm	67	0.0	7	0.0
	0.3 μm	29	0.0	3	0.0
	0.5 μm	10	0.0	1	0.0
	5.0 μm	0.0	0.0	0.0	0.0
Probability of exceeding limiting values for the detection size [%]	0.2 μm	<0.1	3.0	1.7	3.0
	0.3 μm	<0.1	3.0	4.5	1.0
	0.5 μm	<0.1	3.0	<0.1	<0.1
	5.0 μm	<0.1	<0.1	<0.1	<0.1
Statistical certainty of keeping within the required limiting value for the given air cleanliness class [%]		>99.9	97.0	95.5	97.0

Finally, the equipment will be stated as 'clean room suitable for its use in air cleanliness class ISO 4 according to ISO 14644-1'.

7.6.6 Conclusion

When using the tested and classified equipment in a laminar flow clean room, the equipment will not contaminate the surrounding air worse than the given classification number. For that, it is mandatory that the used clean room is at least one class better than the given classification number. Using the described example of the tested liquid handling robot system in a clean room ISO class 3 or better, the particle concentration of the laminar airflow after passing the tested equipment will be no worse than ISO class 4 (see Table 7.11).

7.6.7 Outlook

An extension of the ISO 14644-1 classification into the nanometer range is currently in discussion. It is expected that a new ISO 14644 guideline will cover this topic in the near future. Especially for the semiconductor industry, product surface contamination due to sedimented nanoparticles is an increasing concern for later chip failure. A classification of equipment according to the described procedure needs new real-time measurement devices that can cope with the extension of the measurement range into the nanoscale.

References

[1] Woodrow Wilson Database. Available from http://www.nanotechproject.org/inventories/consumer/analysis_draft/; 2012 [accessed 21.12.2012].

[2] Krug HF, Klug P. Impact of nanotechnological developments on the environment. In: Nanotechnology. Weinheim, Germany 2: Wiley-VCH; 2008. p. 291–303.

[3] VCI - German Chemical Industry Association. Exposure Measurement and Assessment of Nanoscale Aerosols. Available from https://www.vci.de/Downloads/Tiered-Approach.pdf; 2011 [accessed 20.12.2012].

[4] NanoGEM: Standard Operation Procedures for assessing exposure to nanomaterials, following a tiered approach. Available from https://www.vci.de/Downloads/Tiered-Approach.pdf; [accessed 20.12.2012].

[5] Poland CA, Duffin R, Kinloch I, et al. Carbon nanotubes introduced into the abdominal cavity of mice show asbestos-like pathogenicity in a pilot study. Nat Nanotechnol 2008;3:423–8.

[6] ASTM International. Standard Practice for Full-scale Chamber Determination of Volatile Organic Emissions from Indoor Materials/Products; 2007. D6670–01.

[7] International Organization for Standardization. Indoor Air – Part 9: Determination of the emission of volatile organic compounds from building products and furnishing – Emission test chamber method. NF EN ISO 2006:16000–9.

[8] Ji X, Le Bihan O, Ramalho O, et al. Characterization of particles emitted by incense burning in an experimental house. Indoor Air 2010;20(2):147–58.

[9] Afshari A, Matson U, Ekberg LE. Characterization of indoor sources of fine and ultrafine particles: a study conducted in a full-scale chamber. Indoor Air 2005;15:141–50.

[10] Gehin E, Ramalho O, Kirchner S. Size distribution and emission rate measurement of fine and ultrafine particle from indoor human activities. Atmos Environ 2008;42:8341–52.

[11] Koponen IK, Jensen KA, Schneider T. Sanding dust from nanoparticle-containing paints. J Phys Conf Ser 2009:151. http://dx.doi.org/10.1088/1742-6596/151/1/012048.

[12] Vorbau M, Hilleman L, Stintz M. Method for the characterization of the abrasion induced nanoparticle release into air from surface coatings. Aerosol Science 2009;40:209–17.

[13] Gheerardyn L, Le Bihan O, Malhaire JM, et al. (2010). Nanoparticle release from nano-containing product, introduction to an energetic approach. Technical proceedings of the 2010 NSTI Nanotechnology conference 1, p. 736–9.

[14] Le Bihan O, Schierholz K. Assessment of nanoparticles emission from manufactured products: feasibility study; 2008. Proceeding, Nanosafe.

[15] Hsu LY, Chein HM. Evaluation of nanoparticle emission for TiO_2 nanopowder coating materials. J Nanopart Reş 2007;9:157–63.

[16] Zimmer A, Maynard A. Investigation of the aerosols produced by a high-speed, hand-held grinder using various substrates. Ann Occup Hyg 2002;46(8):663—72.

[17] Guiot A, Golanski L, Tardif F. Measurement of nanoparticle removal by abrasion. J Phys Conf Ser 2009;170(1). http://dx.doi.org/10.1088/1742-6596/170/1/012014.

[18] R'mili B, Le Bihan OLC, Dutouquet C, et al. (2012). "Particle sampling by TEM grid filtration". Nanosafe Conference, Books of Abstract, Grenoble.

[19] Isaacs JA, Tanwani A, Healy ML, Dahlben LJ. Economic Assessment of single-walled carbon nanotube processes. J Nanopart Res 2010;12:551—62. http://dx.doi.org/10.1007/s11051-009-9673-3.

[20] Maynard AD, Baron PA, Foley M, et al. Exposure to carbon nanotube material: aerosol release during the handling of unrefined single-walled carbon nanotube material. J Toxicol Environ Health, Part A 2004;67:87—107. http://dx.doi.org/10.1080/15287390490253688.

[21] Han JH, Lee EJ, Lee JH, et al. Monitoring multiwalled carbon nanotube exposure in carbon nanotube research facility. Inhal Toxicol 2008;20:741—9.

[22] Bello D, Hart AJ, Ahn K, et al. Particle exposure levels during CVD growth and subsequent handling of vertically-aligned carbon nanotube films. Carbon 2008;46(6):974—7.

[23] Tsai SJ, Hofmann M, Hallock M, et al. Characterization and Evaluation of Nanoparticle Release during the Synthesis of Single-Walled and Multiwalled Carbon Nanotubes by Chemical Vapour Deposition. Environ Sci Technol 2009;43:6017—23.

[24] Myojo T, Oyabu T, Nishi K, et al. Aerosol generation and measurement of multi-wall carbon nanotubes. J Nanopart Res 2009;11(1):91—9. http://dx.doi.org/10.1007/s11051-008-9450-8.

[25] Maynard AD, Aitken RJ. Assessing exposure to airborne nanomaterials: Current abilities and future requirements. Nanotoxicol 2007;1(1):26—41.

[26] Asbach C, Fissan H, Stahlmecke B, Kulbusch TAJ, Pui DYH. Conceptual limitations and extensions of lung-deposited Nanoparticle Surface Area Monitor (NSAM). J Nanopart Res 2009a;11:101—9.

[27] Tsai SJ, Ashter A, Ada E, et al. Airborne nanoparticle release associated with the compounding of nanocomposites using nanoalumina as fillers. Aerosol Air Qual Res 2008;8:160—77.

[28] Peters TM, Elzey S, Johnson R, et al. Airborne monitoring to distinguish engineered nanomaterials from incidental particles for environmental health and safety. J Occup Environ Hyg 2009;6:73—81.

[29] Gorbunov B, Priest ND, Muir RB, et al. A novel size-selective airborne particle size fractionating instrument for health risk evaluation. Ann Occup Hyg 2009;53:225—37.

[30] Rodes CE, Thornburgs JW. Breathing zone exposure assessment. In: Ruzer LS, Harley NH, editors. Aerosols handbook. Measurement, Dosimetry and Health Effects 2005. p. 61—73.

[31] Yeganeh B, Kull CM, Hull MS, Marr LC. Characterization of airborne particles during production of carbonaceous nanomaterials. Environ Sci Technol 2008;42:4600—6.

[32] Hinds WC. Aerosol technology. Properties, behaviour and measurement of airborne particles. New York, NY: John Wiley and Sons; 1999. p. 233—259.

[33] Maynard AD, Kuempel ED. Airborne nanostructured particles and occupational health. J Nanopart Res 2005;7:587—614.

[34] Price PN. Assessing uncertainties in the relationship between inhaled particle concentrations, internal deposition and health effects. In: Ruzer LS, Harley NH, editors. Aerosols handbook. Measurement, Dosimetry and Health Effects 2005. p. 157—88.

[35] Ruzer LS, Apte MG, Sextro RG. Aerosol dose. In: Ruzer LS, Harley NH, editors. Aerosols handbook. Measurement, Dosimetry and Health Effects 2005. p. 101—12.

[36] ICRP - International commission on radiological protection. "Human respiratory tract model for radiological protection". Annals of the ICRP, Publication 66. Tarrytown, NY: Elsevier Science Inc; 1994.

[37] Kuhlbusch TAJ, Asbach C, Fissan H, et al. Nanoparticle exposure at nanotechnology workplaces: a review, Part. Fibre Toxicol 2011;8:22.

[38] Wang SC, Flagan RC. Scanning electrical mobility spectrometer. Aerosol Sci Technol 1990;13:230—40.

[39] Dixkens J, Fissan H. Development of an electrostatic precipitator for off-line particle analysis. Aerosol Sci Technol 1999;30:438—53.

[40] Wohlleben W, Brill S, Meier MW, et al. On the Lifecycle of Nanocomposites: Comparing Released Fragments and their In-Vivo Hazards from Three Release Mechanisms and Four Nanocomposites. small 2011;7(16):2384—95.

[41] Confederation of European Waste-to-Energy Plants (CEWEP). Landfill taxes and bans. http://www.cewep.eu/media/www.cewep.eu/org/med_557/955_2012-04-27_cewep_-_landfill_taxes__bans_website.pdf; 2012.

[42] Siebzehnte Verordnung zur Durchführung des Bundes-Immissionsschutzgesetzes (Verordnung über die Verbrennung und die Mitverbrennung von Abfällen - 17th BImSchV), in der Fassung der Bekanntmachung vom 14.08.2003, BGBl. I. S. 1633.

[43] Tammet H, Mirme A, Tamm E. Electrical aerosol spectrometer of Tartu University. Atmos Res 2002;62:315—24.

[44] Asbach C, Kaminski H, Fissan H, et al. Comparison of four mobility particle sizers with different time resolution for stationary exposure measurements. J Nanopart Res 2009b;11:1593—609.

[45] Kaminski H, Rath S, Götz U, et al. Comparability of Mobility Particle Sizers and Diffusion Chargers. J Aerosol Sci 2013;57:156—78 doi. http://dx.doi.org/10.1016/j.jaerosci.2012.10.008

[46] Roes L, Patel MK, Worrell E, Ludwig C. Preliminary evaluation of risks related to waste incineration of polymer nano-composites. Sci Total Environ 2012;417-418:76—86.

[47] Ramachandran G, Ostraat M, Evans DE, et al. A strategy for assessing workplace exposures to nanomaterials. J Occup Environ Hyg 2011;8(11):673—85.

[48] Methner M, Beaucham C, Crawford C, et al. Field application of the Nanoparticle Emission Assessment Technique (NEAT): Task-based air monitoring during the processing of engineered nanomaterials (ENM) at four facilities. J Occup Environ Hyg 2012;9(9):543—55. http://dx.doi.org/10.1080/15459624.2012.699388.

[49] Brouwer DH, Gijsbers JHJ, Lurvink MWM. Personal Exposure to Ultrafine Particles in the Workplace: Exploring Sampling Techniques and Strategies. Ann Occup Hyg 2004;48(5):439—53. http://dx.doi.org/10.1093/annhyg/meh040.

[50] Tsai CJ, Huang CY, Chen SC, et al. Exposure assessment of nano-sized and respirable particles at different workplaces. J Nanopart Res 2011;13(9):4161—72.

[51] Zimmermann E, Derrough S, Locatelli D, et al. Results of potential exposure assessments during the maintenance and cleanout of deposition equipment. J Nanopart Res 2012;14:1209.

[52] Koponen IK, Jensen KA (submitted). Worker exposure and high time-resolution analyses of nanoparticle and pigment dust release at different mixing stations in two paint factories, J Occ Environ Hyg

[53] Wang J, Kim SC, Pui DYH. Measurement of multi-walled carbon nanotube penetration through a screen filter and single fibre analysis. J Nanopart Res 2011;13:4565—73.

[54] Schneider T, Brouwer D, Koponen IK, et al. Conceptual model for assessment of inhalation exposure to Manufactured Nanoparticles. J Exposure Sci Environ Epidemiol 2011;21:450—63.

[55] International Organization for Standardization. Nanotechnologies — Terminology and definitions for nano-objects — Nanoparticle, nanofibre and nanoplate; 2008. ISO/TS 27687.

[56] International Technology Roadmap for Semiconductors, 2011. Available at http://www.itrs.net/Links/2011ITRS/Home2011.htm.
[57] International Organization for Standardization. Cleanrooms and associated controlled environments—Part 1: Classification of air cleanliness; 1999. ISO 14644—1.
[58] VDI, The Association of German Engineers. "Cleanroom technology — Compatability with required cleanliness and surface cleanliness". VDI 2083, part 9.1. Available at http://www.vdi.eu/engineering/vdi-guidelines/vdi-guidelines-details/rili/89791/; 2006.
[59] Dennekamp M, Howarth S, Dick CA, et al. Ultrafine particles and nitrogen oxides generated by gas and electric cooking. Occup Environ Med 2001;58:511—6.
[60] Gommel Udo, Bürger Frank, Keller Markus. Reinraum- und Reinheitstauglichkeit – Begriffe, Testverfahren, Prüfungen. In: Lothar Gail, Udo Gommel, Hans-Peter Hortig. Reinraumtechnik. Berlin, Heidelberg: Springer Verlag; 2012. DOI 10.1007/978-3-642-19435-1_15.

Risk Assessment and Risk Management

Markus G.M. Berges [1], Robert J. Aitken [2,3], Sheona A.K. Read [2], Kai Savolainen [4], Marita Luotamo [5], Thomas Brock [6]

[1] Institute of Occupational Safety and Health of the German Social Accident Insurance (DGUV-IFA), Sankt Augustin, Germany
[2] Institute of Occupational Medicine (IOM), Edinburgh, United Kingdom
[3] Institute of Occupational Medicine (IOM), Singapore
[4] Nanosafety Research Centre, Finnish Institute of Occupational Health, Helsinki, Finland
[5] European Chemicals Agency (ECHA), Helsinki, Finland
[6] Berufsgenossenschaft Rohstoffe und chemische Industrie (BG RCI), Heidelberg, Germany

8.1 INTRODUCTION

Nanotechnologies are being viewed by the European Commission as one of the critical technologies of the 21st century [1] and have been identified as a KET (key enabling technology) in the Horizon 2020 strategic plan. At the same time others, considering potential and plausible occupational and environment risk, urge more restraint. For example the European Parliament has indicated the need for a stronger, more effective regulation. Others, such as ETC Group and Greenpeace, have gone further, asking for a moratorium of nanotechnologies [2–4]. These seemingly contradictory opinions may inherently be both based on the same assertion of the new properties of nanomaterials and may even agree on the possible disruptive nature of nanotechnologies. Inherently the critics do believe that by virtue of stronger regulations, perhaps after a moratorium has given time to generate more knowledge on risks of specific nanomaterials, safe handling of nanomaterials might be achievable. Safety however is like risk, 'another ambiguous word which has multiple meanings' [5], especially in democratic societies where the freedom of choice does not necessarily result in the wisest or most coherent decisions on what safety really means and what risks are acceptable and which are not. From a more scientific perspective the challenge for governments, regulatory bodies and agencies raised by nanomaterials is to make balanced, sound decisions even when relatively little nanomaterial-specific hazard and exposure data are available to inform such decisions, in particular given the evidence that many products containing nanomaterials are already on the market [6]. The urgency to act arising from the latter claim depends strongly on the definition of a nanomaterial. Applying a broad definition, the EU Commission states in a Staff Working Paper [7] that 11.5 million tons (t) of nanomaterials are marketed per annum globally. However, two very widespread commodity materials, i.e., carbon black (9.6 million t) and synthetic amorphous silica (1.5 million t), dominate the market. In contrast to these commodity nanomaterials, nanomaterials like carbon nanotubes and carbon fibers are currently marketed at quantities of several hundreds of tons and nanosilver is estimated to be marketed in quantities of about 20 tons. Both commodity nanomaterials and new nanomaterials like carbon nanotubes belong to the group of what Roco coined as 'first-generation passive nanomaterials'. Roco [8] classified engineered nanomaterials into 1) first-generation passive nanomaterials; 2) second-generation active nanomaterials; 3) third-generation systems of nanosystems; and 4) molecular nanosystems.

Since nanomaterials are already in use it is critical that effective approaches to manage the risks are known about and are used properly. In this chapter the focus is on providing information towards the risk assessment and management for the first-generation passive nanomaterials at the workplace. The information provided is drawn from the considerable numbers of already available guidance documents (both for nano and for other materials) from which the authors have attempted to draw out the key features and principles that apply. This chapter should not in itself be regarded as a 'guidance document'. In all cases, national and regional regulation and guidance should be complied with. We are not aware that second-generation active nanomaterials have entered the market, despite a rise in publications on these materials [9]. However Roco et al. state there are uncertain or unknown implications for these second- to fourth-generation materials, 'mostly because the products are not yet fabricated' [10].

8.2 BASIC APPROACHES TO CONTROL EXPOSURE TO HAZARDOUS SUBSTANCES

The risk to the health of an individual or a population arising from exposure to a chemical agent is generally considered to be a function of the intrinsic harmfulness of the chemical (its toxicity) and the dose (amount), which accumulates in the specific biological compartment (e.g., the lungs). For occupational exposures it is often very difficult to quantify the dose directly, specifically in the case of insoluble particulates. Therefore, in order to quantify and manage the risks, it is usual to assess exposure as a proxy for dose. Occupational exposure to chemicals can occur via inhalation, dermal or ingestion routes. Control of exposure (to zero) effectively removes the risks. Thus, regardless of how hazardous a substance is, without exposure there is no risk.

The OHS Framework Directive, 89/391/EEC is the cornerstone of European safety and health legislation [11], and together with additional directives like the Chemical Agents Directive [12], it specifies a well-defined approach to achieve control of exposure. This is exemplified in National Guidance in member states, for example, in the UK by the Control of Substances Hazardous to Health (COSHH) assessment [13]. The essential steps of the COSHH assessment are indicated in Figure 8.1 and may be summarized as follows:

1. Gathering information about the substances, the work and the working practices (or finding out what the problems are);
2. Evaluating the risks to health (or looking at the problems that are found);
3. Deciding on the necessary measures and implementing them;
4. Recording the assessment;
5. Reviewing the assessment.

This approach is generic to all particles and chemicals and can be applied to nanomaterials. One grouping that is of value to separate those nanomaterials that are potentially of most interest is nano-objects, their aggregates and agglomerates (NOAA) [14]. Following this approach for NOAA, it is necessary to take into account the greater uncertainty associated with the handling of NOAA at the workplace.

8.2.1 Nanomaterial Hazards

While nanomaterials have been the subject of massive financial investment worldwide, it has been recognized that they may also represent hazards to the health of workers, consumers or the environment. One of the first reports to highlight the potentially hazardous nature of nanomaterials was that of the Royal Society & Royal Academy of Engineering [15] in 2004, which has since been followed by more than 100 major reviews and position papers discussing the potential risks to health and to the environment from exposure to particulate

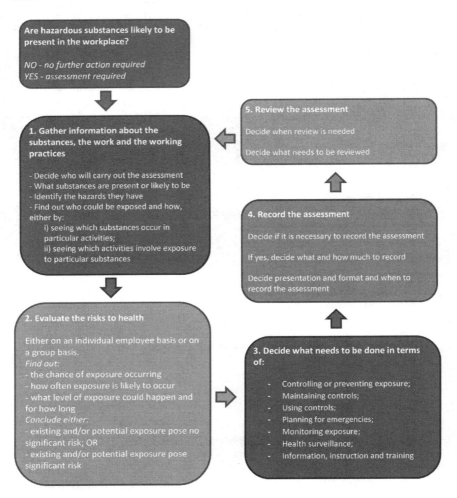

FIGURE 8.1   **Essential structure of the assessment.** *Modified from [13].*

nanomaterials. Plausible toxicology issues for nanomaterials may be summarized as follows:

- **High specific surface area.** One of the main reasons nanomaterials tend to be more reactive than their corresponding bulk counterparts is that, per unit mass, they have a much higher surface area. If surface area is a driver for toxicity this clearly implies potentially increased toxic effects.
- **Increased translocation potential.** As a result of their small size, nanoparticles and other nano-objects can reach parts of biological

systems that are not normally accessible by other larger particles. This includes the increased possibility of crossing cell boundaries, or of passing directly from the lungs into the bloodstream and so on to all of the organs in the body; or even through deposition in the nose, directly to the brain. This effect is probably more relevant for nanoparticles smaller than 30 nm [16].

- **Fiber analogy.** Based on toxicological research on asbestos and other industrial fibers, the 'fiber paradigm' states that fibers that are biopersistent in the lungs and longer than 15–20 micrometers with a diameter less than 3 micrometers are considered to be hazardous to human health. Asbestos is an example of such a biopersistent fiber. Some high aspect ratio particulate nanomaterials (HARN) (e.g., some carbon nanotubes, nanowires) have similar morphology (shape) and durability and are therefore likely to persist in the lung, if inhaled. Recently Schinwald et al. [17] have reported that nanoplatelets, such as graphene, may pose a novel hazard and structure–toxicity relationship in particle toxicology, extending the problem of morphology beyond the fiber paradigm.
- **Solubility.** It has been shown that some types of nanoparticles become more soluble as particle size decreases. This could imply increased bioavailability for particles previously considered to be insoluble.

These toxicological issues have to be kept in mind when starting with the previously mentioned approach to achieve control of exposure to NOAA at the workplace.

8.2.2 Exposure to Nanomaterials

Increased production and use of nanomaterials leads to the potential for greater exposures to workers, consumers and the environment. Human exposure to engineered nanomaterials and environmental release of these materials can occur during all life-cycle stages. The main life-cycle stages for engineered nanomaterials can be summarized as:

- manufacturing of nanoparticles;
- formulation of nanomaterials and nano-enabled products;
- industrial use of nanomaterials or products;
- professional and consumer uses of nanoproducts;
- service life of nanoproducts; and
- waste life stage of nanoproducts [18].

Throughout the life-cycle of a single material, there are multiple exposure scenarios, which may or may not occur depending on the details of manufacture, use and disposal of that material. Throughout these scenarios, the population exposed, the levels of exposure, the duration of exposure and the nature of the material to which people are exposed are all likely to be different. Specifically, in contrast to many chemical substances, the properties of nanomaterials and NOAA may be transformed throughout their life-cycle, and the approaches to control exposure at the various workplaces need to take this additional complexity into account.

8.2.3 Information Gathering

An important first step in assessing the risks is to gather information about the substances, the work and working practices. An indication of the type of information to be gathered is listed below.

8.2.3.1 What Hazardous Substances Are Present or Likely to Be Present

Identify and list:

- What substances are coming into the business and where they are used, worked on, handled or stored? Substances hazardous to health include gases, vapors, liquids, fumes, dusts and solids and can be part of a mixture of materials.
- What substances might be produced during any process as intermediates, by-products or finished products or what might be given off as wastes, residues, fumes, dusts, etc.?
- What might be transported, collected, poured, weighed, packed, discharged or disposed of?
- What substances are used in, or arise from maintenance, cleaning, repair work, research or testing laboratories, etc.? They can also arise from work on the structure of the building, e.g., removal of insulating materials or sandblasting during facade cleaning.

8.2.3.2 Identify and Evaluate the Hazards

- Identify what information you have available relating to the materials. This could include information provided by the manufacturers in the first place (e.g., labels, safety data sheets) or after your special request, information provided by national and international agencies, information provided in industry specific information services like the IMDS 'International Material Data System' in the automotive industry or the 'IEC PAS 61906' in the electrotechnical and electronic industry and public information available from third parties (material specific reports and reviews);

- Identify any information gaps and uncertainties you may have. Do you consider that the information adequately addresses the nano characteristics of the materials, i.e., does it make specific reference to the nanomaterial form?
- Seek more information if necessary. Sources of information could include any of the above or could extend to seeking external advice.

8.2.3.3 Identify Sources and Who Could Be Exposed and Any Existing Exposure Data

- Consider all possible sources in the workplace;
- Walk around and inspect;
- Identify the tasks and look at all the exposures in each case:
  - What are the processes that could lead to the release of particulate nanomaterials into the air or onto a surface?
  - What are the tasks where people are potentially exposed to particulate nanomaterials (e.g., production, cleaning, accidental releases, maintenance, transport, storage and disposal)? Exposure could be by inhalation, dermal or ingestion.
  - Identify individuals and groups who could be exposed in each task.
  - Who can potentially be exposed during each task? The individual undertaking the task, adjacent workers, visitors, contractors, managers and others might be exposed.
  - Collect any information (data) that you have regarding exposure.

8.2.4 Evaluate the Risks

8.2.4.1 Hazard Assessment: Key Considerations

Hazard assessment consists of evaluating the hazard of NOAA on the basis of a comprehensive evaluation of all available data on this material, taking into account parameters such as toxicity, *in vivo* biopersistence and factors influencing the ability of particles to reach the respiratory tract, their ability to deposit in various regions of the respiratory tract and their ability to elicit biological responses. These factors can be related to physical and chemical properties such as surface area, surface chemistry, shape, particle size, etc.

Key questions to consider would be:

- Is the NOAA solubility in water higher than 0.1 g/L?
- Has the bulk material already been classified and labeled according to national or regional legislation or GHS?
- Has the NOAA already been classified and labeled according to national or regional legislation or GHS?

- Does the NOAA contain biopersistent fibers or fiber-like structures? Is it appropriate to apply the fiber toxicity paradigm to the NOAA? Nanoplatelets like graphene may need special consideration.
- Is there specific hazard data for the NOAA available?

8.2.4.2 Exposure Assessment: Key Considerations

In the exposure assessment, it is important to assess the likelihood of exposure occurring, how often the exposure is likely to occur, the level of exposure that could happen and the duration of such exposure. It is appropriate to consider the following:

- The nature of the process for which the nanomaterial is being used or generated. Is it a high-energy process (grinding and milling) or low energy (molding)?
- What is the nature of the nanomaterial (dry powder, in solution/suspension, as a surface coating, bonded as part of matrix)?
- How much material is being used?
- Are aerosols deliberately formed in the process?
- What controls are being used?

Potential conclusions at the end of the process could be:

- Exposure is negligible;
- Exposure is possible and I have enough information to conclude that I need to improve the level of control;
- I need to collect more information through a program of measurement. If so the following approach is recommended.

8.2.4.3 Strategy for Exposure Assessment

At present, although there is no standardized way to determine nanomaterial release and exposure, there is a consensus on a tiered approach emerging as reported in Section 5.3. Therefore no further discussion of this tiered approach and other problems (i.e., background distinction, choice of metric) tied to the exposure measurement of NOAA at the workplace will be provided here.

8.2.4.4 Assessment of Risk

Consider the possible exposures and what is known about the toxicity and evaluate the risks.

Working out the risk involves combining the answers to the following questions:

- What is the potential of a substance or the combined potential of two or more substances to cause harm (i.e., the hazard)?
- What is the chance of exposure occurring?

- How often is exposure liable to occur?
- What levels are people exposed to and for how long?

In a general sense, of most concern are where the hazard and the potential for exposure are both high.

The assessment of risk will depend on the amount on information held for a given substance and could include comparison with:

- the prevailing national occupational exposure limit (OEL);
- proposed national or international benchmark levels specifically for a given or a group of NOAA;
- other self-imposed (in-house) exposure standards, considering any proposed margin of safety to take account of known or assumed differences in the toxicity of particulate nanomaterials when compared to larger (bulk) versions of the same material.

The less information you have, the more cautious the risk assessment should be. Risks should not be dismissed as negligible unless there is certain and valid evidence to do so. In all other cases, there are risks to health from exposure that must be identified and that must trigger precautions to protect people's health. If you conclude that there are residual risks, a control strategy must be developed and implemented.

8.2.5 Control Strategies

8.2.5.1 Prioritizing Control Approaches

At this stage, assessment should have reached decisions on what the problems are. The next stage is to decide what to do about them. Complete the assessment by considering the precautions needed in the light of the risks. If the risks are significant now or could foreseeably become so, then further precautions are required.

Not all problems can be solved immediately, and priorities for action will be required. Deciding priorities involves a mixture of the following:

- What are the most serious risks to health?
- What are the risks that are likely to occur soonest?
- What is the number of workers exposed?
- What are the risks that can be dealt with soonest?

The most important of these is the seriousness of the risk. If a risk is serious it should be dealt with immediately. Less important matters should not assume greater priority merely because they can be dealt with more easily or occur more quickly.

8.3 HIERARCHY OF CONTROL

The main approach to control the risk for chemicals as outlined in the Chemical Agent Directive (xii) is often called the hierarchy of control or the STOP principle. It consists of the stepwise approach of:

- Substitution (including Elimination)
- Technical (or Engineering) control measures
- Organizational (or Administrative) means
- Personal protective equipment.

In practice, an appropriate combination of these strategies will provide the best approach for a workplace to control exposures. Each of these control options are discussed in further detail below.

8.3.1 Substitution (Including Elimination)

Elimination would involve avoiding using the hazardous substance or the process that causes exposure. Consideration should be given as to whether the improved properties of the nanomaterial justify any enhanced risks associated with its use. Since the specific properties of engineered nanomaterials are usually required for manufacturing a novel product, it is unlikely that this option will often be feasible or practicable [19]. Similar statements appear in most guidance documents. However, when it comes to designing nanomaterials in the first place, ISO/TR 13121 [20] describes a process for identifying, evaluating and communicating the potential risks of developing manufactured nanomaterials in order to protect the health and safety of the public, consumers, workers and the environment, hereby advising the manufacturer about criteria for substitution before the product even reaches the market.

Substitution is usually considered to be an effective way to reduce risks to health and safety in the workplace. While the unique chemical and physical characteristics of individual nanoparticles are likely to limit possibilities for straightforward substitution of one nanomaterial for another, there might be substitution opportunities in processes involving nanomaterials. ISO/TR 12885 [21] states the following possibilities in relation to substitution or modification:

- Replacing more toxic raw materials with less toxic raw materials, including the modification of particles to reduce the hazardous properties;
- Changing the physical form of the powder or material by using dispersions, pastes, granules or composites instead of dusty powders or aerosols;

- Process changes are very effective risk control methods, for example, changing from dry processes to wet processes, and the use of water can reduce dust emissions at some dry material drop-off or transfer points;
- Substitution of spraying processes and instead use of coating or printing techniques or use of immersion baths;
- Substitution of equipment that involves the use or production of smaller quantities of toxic materials, or fewer toxic materials;
- Substitution of equipment to avoid emissions, or to reduce and better control emissions.

8.3.2 Technical Control Measures

Owing to nanoparticles' new and in some cases surprising properties, it is often assumed that the usual OSH protective measures would not be effective against nanoparticles and ultrafine particles. This is not always the case (and indeed it may be that the reverse of this is the norm). For example, where filtration measures (such as breathing masks or air collection installations) are used, the image of a sieve is frequently cited, through the large holes of which the small particles would slip. This image is not consistent with physical principles, however: particles larger than 300 nm are most effectively separated by impaction on the filter material, entrapment by the filter material, or sedimentation owing to gravity. For particles smaller than 300 nm, separation by diffusion (Brownian movement) and electrostatic forces become increasing relevant with decreasing size [22]. Evidence from the literature clearly shows that filters of the appropriate quality are suitable for the elimination of nanoparticles and ultrafine particles.

8.3.2.1 Enclosure

One way of separating the possibly exposed from desirably unexposed workers is the solid separation of working areas where nanomaterials are handled by means of walls or closed doors. Operations in which there is deliberate release of particulate nanomaterials into the air or that involve the likely release of particulate nanomaterials into the air should be performed in contained installations, or where personnel are otherwise isolated from the process (e.g., in a cabin). This includes gas phase nanomaterial production and spray drying. All other processes involving the use of dry nanomaterials should be performed in enclosed installations where possible. Evidence for the practical performance of enclosed systems is available from a number of studies. In most of the studies where an enclosed or sealed process has been used, containment was effective as long as it was maintained [23,24]. As with other materials, containment can be compromised during product removal, cleaning or maintenance unless effective processes for these activities are put in place.

8.3.2.2 *Ventilation*

All processes where there is a likelihood of dust formation should be carried out with extraction ventilation. Many types of extraction ventilation systems are available, including fume cabinets, fumehoods and dust extractors. Selection of appropriate controls will depend on the level of risk. Regular maintenance and performance testing of extraction facilities should be carried out. Extracted air should not be recirculated without exhaust air purification. General ventilation may also be appropriate. Methner et al. demonstrated recently that improperly designed, maintained or installed engineering controls may not be completely effective in controlling release of NOAA from nanocomposite materials [25].

In conclusion one can state that all the control technologies to handle dusty materials including dusty nanomaterials are in place and provide good control if implemented correctly. Even for nanomaterials with a higher hazard like carbon nanotubes, the use of dual component butterfly valves in transferring carbon nanotubes has been demonstrated [26]. The question remains on the appropriateness of the various engineering controls [27], and we will return to this when discussing the evaluation of the effectiveness of controls.

8.3.3 Organizational (or Administrative) Means

It is recommended that procedural controls should accompany the technical controls, though the risk assessment might indicate that procedural controls alone are sufficient in some circumstances [14]. Procedural controls include reducing the number of personnel exposed or the time spent by personnel on the process, limiting the process to specified areas and denying unauthorized persons access to these areas. The personnel involved should be informed of the specific hazards of free nanoparticles, the need for special measures and the potential health effects of exposure to dusts. Relevant information in the operating instructions might be included. Procedures must be in place for dealing with spills or leakages of transfer lines or valves.

Cleaning of workplaces should be carried out regularly, in line with risk control plans. Pre-cleaning of contaminated devices should be done in the work area to prevent contamination of adjacent work areas. The use of sticky doormats serves the same purpose.

The uses of medical surveillance should also be considered. Recommendations on the use of medical surveillance for workers potentially exposed to NOAA have been published by NIOSH [28]. In derogation from their 2009 recommendations, NIOSH recently (April 2013) viewed the advance in toxicological evidence on carbon nanotubes and nano-fibers to be so conclusive to make specific recommendations for medical surveillance and screening of exposed workers, including an initial evaluation with a baseline chest X-ray [29]. This advice and the other recommendations given in the document should be seriously considered. However the legal constraints on imposing certain (risky) medical surveillance measures onto the personnel may differ in various countries.

Overall the training of the personnel on the implemented procedures, the proper use of engineering controls and personnel protective equipment are of utmost importance.

8.3.4 Personal Protective Equipment (PPE)

Contrary to commonly encountered workplace practices, the use of PPE is according to the hierarchy of control the last option or a supplemental option to help support all of the other methods of exposure control.

8.3.4.1 Protection from Inhalation Exposure

Certified respirators have been shown to provide a stated levels of protection for particulate nanomaterials [30,31] and so are likely to form an important element of a control strategy where full control of emissions at source is not practical. Information on the selection and use of respirators can be found in various national guidance documents, for example the UK's Health and Safety Executive's (HSE) HSG53 [32] or in NIOSH's respirator selection logic [33]. Appropriate types of respiratory protective equipment (RPE) include disposable filtering facepieces, half and full face masks and a range of powered (air-supplied) hoods, helmets, blouses and suits. All wearers of RPE should undergo facepiece fit testing to ensure correct fitting and proper wearing [32]. The applicable restrictions upon wearing time and provisions governing occupational medical check-ups must be observed when respiratory protection is worn. PPE, especially respiratory protection, needs a significant investment in training, supervision and maintenance if it is to provide the intended level of protection. Incorrect selection or fitting or insufficient use can render it ineffective.

8.3.4.2 Protection from Dermal Exposure

Use of Skin Protective Equipment (SPE) is recommended where the possibility of dermal exposure cannot be excluded at all times. However, because of the small diameters of nanoparticles, the various kinds of SPE might have limited effectiveness [21]. Simply selecting gloves solely on the basis of glove manufacturers' published data is probably insufficient in ensuring adequate protection provided no data are given by the manufacturer for the nanomaterial of concern.

The Nanosafe 2 project confirmed the effectiveness of breathing masks, disposable protective overalls and gloves against graphite nanoparticles (20—100 nm) [34]. Nonwoven fabrics seem much more efficient against nanoparticle penetration. The use of protective clothing made with cotton fabrics should be avoided. These results do not prejudge the efficiency of gloves against colloidal dispersion liquid.

There are four basic criteria for the selection of protective gloves: i) they should be appropriate for the risk(s) and conditions where they are to be used; ii) they should be suitable for the ergonomic requirements and state of health of the intended wearer; iii) they should fit the intended wearer correctly; and iv) they should prevent exposure without increasing the overall risk. This, of course, assumes that the gloves are worn and maintained correctly. Attention must also be paid to the adequate mechanical stability of gloves and damage to the glove material in order to prevent skin contact. The overlap of the gloves with other protective clothing and the correct way in which they are put on and removed are more important for the avoidance of possible skin contact than the permeation properties of the material. Additional protection against chemicals may be necessary under certain circumstances.

The development of a glove management system, which emphasizes and reinforces the factors that need to be considered and addressed, how these interlink with each other and when they should be reviewed, should help ensure adequate protection.

8.3.4.3 Protection from Other Routes of Exposure

Eye protection is recommended where there is potential for exposure to nanomaterials. Goggles, safety glasses and full-face shields are all used in practice, though the use of a full-face shield was not always recommended for reasons related to the handling of nanomaterials specifically (e.g., it was also recommended when there is increased exposure to solvents or hot material).

It is considered that ingestion exposure in the workplace results primarily from hand-to-mouth contact (although it might also occur via the mucociliary escalator after inhalation). Strategies that tend to reduce dermal exposure to nanomaterials in the workplace will also tend to reduce exposure by ingestion [21].

8.3.4.4 Excursus: Regulation on General Dust in the German Hazardous Substances Ordinance

National laws may already regulate the handling of 'ordinary' dusty materials or the release of non-nano dust at the workplace. The German

Hazardous Substances Ordinance [35], for example, already implements in Annex I (2.3) some rules for tasks involving 'particulate hazardous substances':

1. The risk assessment for activities involving substances, preparations and articles that may release dusts shall be conducted with due consideration of their dust formation behavior.
2. In the case of activities involving exposure to inhalable dusts for which no substance-related occupational exposure has been laid down, the protective measures according to the risk assessment shall be laid down in such a way that at least the occupational exposure limit for the inhalable dust fraction and for the alveolar dust fraction is complied with.
3. Machines and devices shall be selected and operated in such a way that as little dust as possible is released. Dust-emitting installations, machines and devices shall be fitted with an effective extraction system where this is possible given the state of the art and the release of dust cannot be prevented by other means.
4. In the case of activities involving dust exposure it shall be prevented that the dust spreads to unexposed working areas where this is possible given the state of the art.
5. Dusts shall be collected and disposed of safely as far as possible at the place where they are escaping or developing. The air extracted shall be conducted in such a way that as little dust as possible passes into the workers' breathing air. The air extracted shall only be returned to the working area if it has been adequately cleaned.
6. Deposits of dusts shall be avoided. If this is not possible the dust deposits shall be removed using moist or wet processes in accordance with the state of the art or using suction processes using suitable vacuum cleaners or dust removers. Cleaning the working area by sweeping without dust-binding measures or by blowing the dust deposits with compressed air shall invariably not be permissible.
7. Equipment to separate, collect and precipitate dusts shall be in accordance with the state of the art. When these devices are first put into operation, it shall be checked that they are adequately effective. At least once a year the devices shall be inspected with respect to their proper functioning, serviced and, where relevant, repaired. The results of the inspections as recorded in accordance with sentences 2 and 3 shall be retained.
8. For dust-intensive activities, suitable organizational measures shall be taken in order to shorten the duration of exposure as far as possible. If the risk assessment reveals that the occupational exposure limits referred to in paragraph 2 cannot be complied with, the employer shall make available suitable personal protective equipment, especially

respiratory protective equipment. This must be worn by the workers. The workers must be provided with separate storage facilities for work clothing and street clothing as well as washrooms.

As can be seen this is already a comprehensive list overlapping with the recommendations given in the initial paragraph, raising the point that what has to be done practically in the field when handling a nanomaterial goes beyond the control measures already in place for the hazardous bulk form or non-nano form of the nanomaterial — where this form exists.

8.3.5 Evaluation of the Effectiveness of Control

The purpose of applying controls as part of a precautionary approach, when hazard information is unavailable or when there is limited hazard information, is to ensure that exposure of the workforce is as low as reasonably practical. Collection of exposure information associated with the implementation of controls enables demonstration, and documentation, that effective control has been achieved. Judgments considering whether effective control has been achieved could be made by comparison of measured levels with:

- the prevailing national occupational exposure limit (OEL);
- proposed national or international benchmark levels specifically for types of particulate nanomaterials;
- other self-imposed (in-house) exposure standards, considering any proposed margin of safety to take account of known or assumed differences in the toxicity of particulate nanomaterials when compared to larger versions of the same material.

Measurement methods to assess the effectiveness of control are provided in Chapter 5.

8.3.6 Documentation, Information, Instruction and Training

8.3.6.1 Record the Assessment

In recording the assessment it is important to delineate which decisions about risks and precautions have been made and the reasons behind these decisions. In particular, the requirements of hierarchy of controls are subject to the criteria of reasonable practicability. For instance, the use of personal protective equipment is only acceptable as an additional measure if prevention or adequate control of exposure cannot be achieved practicably by other more effective means. If only PPE is relied on, the assessment should make it clear why other means were not considered feasible.

8.3.6.2 *Information*

Everyone who is involved or could be affected should be provided with the information, instruction and training required to ensure their safety. It is necessary to inform and involve the employees in the risk assessment process. Without the informed and competent participation of employees, any risk management measures identified as necessary in the risk assessment are unlikely to be fully effective. It is therefore necessary that the employees know at least:

- the names of the substances to which they are liable to be exposed and the risks to health created by exposure;
- any relevant occupational exposure limit (OEL) or similar self-imposed (in-house) exposure standard that applies to the substances;
- the information on any safety data sheet that relates to the substances;
- the significant findings of the risk assessment;
- the precautions they should take to protect themselves and their fellow employees;
- the results of any monitoring of exposure, especially if these exceed any OEL; and
- the collective results of any health surveillance.

8.3.6.3 *Training*

Arrangements should be put in place to ensure that all control measures are properly and fully applied. Clear allocation of managerial responsibilities and accountabilities is particularly important in this respect. The arrangements should include training/refresher training of those individuals who have to use the control measures and procedures for ensuring measures are working as they should.

8.3.6.4 *Review the Assessment*

The assessment must in any case be reviewed at regular intervals and immediately if:

- there is any reason to suppose that the original assessment is no longer valid, e.g., evidence from the results of examining and testing engineering controls, reports from supervisors about defects in control systems;
- any of the circumstances of the work should change significantly and especially if they may have affected employees' exposure to a hazardous substance; or
- new information comes to light.

Given the emerging nature of information about nanomaterial risk it is important to keep apprised of new information as it becomes available.

New data on nanomaterials or new nanomaterials usually emerge from laboratories. Even active (second—fourth generation) nanomaterials, though probably not on the market yet, may already be handled in laboratories [9]. Laboratories are regarded and treated in most countries as a special workplace. Some safety guidelines for laboratories based on the available literature and their own expertise are discussed in the special excursus below [36—38].

8.4 EXCURSUS: SAFE HANDLING OF NANOMATERIALS IN THE LABORATORY (BY THOMAS BROCK)

8.4.1 General Considerations

A conclusive assessment of the risks in the laboratory cannot yet be performed based on the current level of knowledge. The principle of precaution therefore requires pragmatic solutions to be found for effective protective measures in lab work. The behavior of airborne nano-objects tends to be situated somewhere between fine dusts and gases or vapors. This means that the standard protective measures for such material states in the lab can also be used for nanomaterials. However, free fibrous structures or those that can potentially release fibers (nanotubes, nano-rods) should be handled with increased caution and care to keep the exposure almost to zero.

During processing, breaking up the bonds in aggregates or agglomerates can cause free nano-objects to be released in certain circumstances, for example through grinding in a (high-performance) mill. Nano-objects can also be released in certain circumstances when processing composite nanomaterials, e.g., when removing or breaking up the matrix (for example by removing a solvent or by grinding a material with embedded nano-objects). Based on the — limited — current knowledge, however, the nano-objects created in the process of breaking up matrices with embedded nano-objects seem to be formed predominantly from the decomposed material of the matrix, not from the embedded nano-objects.

Combustible nano-objects are potentially explosive when mixed with air. Indeed, their ignitability can be much higher than more coarse-grained materials (significantly lower minimum ignition energy than dusts). Dangerous, explosive atmospheres can be created with relatively low quantities. As little as 100 mg of fine material (or even less) can be sufficient for ignition in unfavorable circumstances. Attention must therefore also be paid to ignition sources in equipment and apparatus. Some materials (such as metals) have a tendency to spontaneously combust in fine dispersions. The reactivity can be very high and catalytic effects are also possible.

8.4.2 Inhalation Exposure

The use of standardized and tested laboratory fume cupboards is considered to be an effective protective measure for many types of nanomaterials. With more hazardous nanomaterials (e.g., including those created from toxic materials or for which there is already emerging evidence regarding toxicity or that are highly spontaneously flammable), glove boxes, glove bags or sealed equipment can also be used. When working in fume cupboards, it is vital that the front sash should be largely closed (as much as possible) to minimize the extent to which NOAA can escape. Other materials such as solvents or reagents may pose additional risks that require the proper use of a fumehood. Formation of dust should be prevented by using nanomaterials in a wet rather than dry form (suspensions, colloidal solutions, pastes) where possible. Integration in matrices (granules, compounds) is also helpful. However, if a situation requires dry, non-compound nanomaterials to be used, as little mechanical energy as possible should be applied to minimize the extent to which agglomerates break up and free nano-objects are thereby released into the air. Extraction systems with excessive flow rates can cause airborne particles (not just nanoparticles but also larger objects) to be carried out with the exhaust air stream. It is also therefore recommended to keep a distance of at least 15 cm to the front opening of fume cupboards. Open windows in the laboratory can generate air currents that can cause particles to escape into the laboratory because air velocities can be quite high near the gap of the front sash and so can cause nano-objects to be transported with the air stream. These nano-objects may be deposited on the inner surfaces of the fume cupboards or in the exhaust ducts, where they can form new and unexpected hazards. Leakage of nano-objects from fume cupboards has been reported, although tests on some fume cupboards as per EN 14175-1 [39] with a variety of nanomaterials showed the containment factor to be generally high.

Large equipment, such as tube furnaces for manufacturing, should ideally not be operated in the fume cupboard, as they can disturb the air flow. If such equipment is set up next to the fume cupboard, the reaction tube can be laid through the wall of and into the fume cupboard, thereby allowing extraction via the fume cupboard. If equipment of this nature needs to be set up within the fume cupboard, it is vital that sufficient distance from the inner surfaces of the fume cupboard be provided to ensure that the air currents are impaired as little as possible. A distance of at least 5 cm from the working area in particular should be maintained. Equipment should therefore be placed on spacers.

Where possible, nanomaterials that have a tendency to give off dust should be handled in closed systems if they cannot be handled safely in a fume cupboard. Working methods of this kind are similar to those commonly practiced when handling air-sensitive or moisture-sensitive compounds in the laboratory. Powdery nanomaterials can be moved from one vessel to another without leakage using glass transfer bows with ground glass joints (preferably not lubricated with grease but using Teflon sleeves or O-ring gaskets). Protection against air and moisture can be provided simultaneously by flushing the apparatus with argon. Apparatus can be filled (in portions) using glass vessels attached with flexible tubing (plastic or glass with ground glass joints) or — a particularly elegant solution — with powder metering funnels (also available with an electrical drive for improved dosing). Pre- and post-weighing can be taken in the corresponding parts of the equipment in the fume cupboard (the balance in the fume cupboard may have problems with the air draft), with vials or bottles closed in the fume cupboard and placed in a weighing chamber subsequently (the effectiveness of the containment has to be tested before) or in a glove box, precluding the risk of any exposure. The grinding beaker of a mill can be removed and transported unopened, with the contents only being removed later in a fume cupboard or in the glove box. Care needs to be taken when disassembling such systems to avoid leakage or contamination at that stage.

Weighing nanomaterials can be done in closed containers or using weighing cabinets that avoid drafts, typically for fume cupboards, disturbing the measurement. If electrostatic charging is a problem for powders, special electrostatic discharging devices are available for the balances.

Enclosed automatic balances are well suited for the dosing of nanomaterials, although at a comparatively higher cost. However, if a large number of weighings have to be carried out, this is a safe and efficient approach.

If the apparatus needs to be dismantled, for example for cleaning, it should be well rinsed before any of its contents can escape into the air in the lab. This is easy to arrange if connectors for the supply and draining of cleaning fluid, e.g., water, are included when the apparatus is assembled. The drained liquid can be collected as waste in a dedicated container. When setting up the apparatus, it is important to ensure that the rinsing fluid can reach all parts of the system in which nanomaterials may be located. To prevent a pressure buildup due to evaporation of the rinsing fluid and also stress fractures, the equipment should not be rinsed at excessive temperatures. The risk of dangerous reactions between the contents of the equipment and the rinsing fluid must also be eliminated. As such, it will not be possible to use water in all cases.

If devices are to be connected to an extraction system to reduce the risk of exposure, this can also take the form of an enclosure or extraction at the source of the emissions. Here, it is vital to ensure that extraction is as complete as possible, that the velocities of the air currents in the area of the extraction system are not high enough to cause vortexes to form and also that fractions of nanomaterials that move easily are carried along by the air draft. Expert design of the extraction apertures is necessary. Simply positioning an exhaust hose near the point of emission will generally be of little use or even have a potentially negative effect. For these two reasons, even in laboratory fume cupboards the air intake volume flow should not be set too high. The values set for the exhaust air volume flow (as a result of the type of approval test) should also not be increased at random for 'apparent safety reasons'. Nano-objects carried in the air stream can be deposited in the exhaust air systems (largely behind the extraction system's deflection wall or in exhaust channels, particularly in corners and bends). When performing maintenance work, appropriate consideration must be given to the health and protection of all persons working on the system, as well as to potential contamination of the area.

If nano-objects released in a room cannot be prevented, respiratory protection must be worn. Class P2 or P3 particle filters are effective here. When releasing such particles into the air, appropriate consideration must of course also be given to contamination of the room. In case of fibrous nano-objects, the use of a protective suit should be considered together with an air stream helmet. With respect to the hierarchy of control, we must stress here the point that release of fibrous nano-objects must be avoided and that the use of an air helmet can only be the last resort.

If nanomaterials in solution or suspension are spilled, they have to be removed before they dry as this would prevent a risk of them being airborne. If vacuum cleaners are used as part of a clean-up process, they have to be equipped with HEPA filters (explosion-proof systems may be required in some cases).

8.4.3 Dermal and Oral Exposure

Alongside protection from exposure by inhalation, protection from dermal and oral exposure is also necessary. Gloves should always remain in the fume cupboard (or any other potentially contaminated work areas). Contamination on the sleeves of lab coats can be avoided by using gloves with cuffs that can be pulled over the sleeves. These should then also be left in the fume cupboard. Gloves should not have any cuts or holes (due to poor quality or age). If solvents are used in production or use of nanomaterials (e.g., in suspension), the protective gloves must be qualified to withstand the solvents.

Airborne nano-objects are not only inhaled but can also deposit onto surfaces. This also applies to the agglomerates and aggregates that nano-objects form at varying rates with one another or with larger particles suspended in the air. From the skin on the face, these can then find their way into the digestive tract via the mouth. This can be caused by a lack of proper hygiene practice (e.g., scratching the forehead with contaminated gloves).

Personal protective equipment including lab coats should not be carried outside the laboratory to avoid the contamination of other areas. Thorough cleaning of hands and other possibly contaminated parts of the body before leaving the laboratory is mandatory.

8.4.4 Effectiveness of the Measures

Measurements can be helpful in assessing the situation, although issues such as the often high biogenic and anthropogenic background pollution with nano-objects and the lack of quantitative reference values can make the judgment of the measurement results difficult. When working in the lab, however, taking measurements while performing tasks can provide useful statements regarding the effectiveness of installed control measures. Portable or handheld devices are available for this (see Chapter 5).

8.4.5 Summary and Outlook on the Safe Handling of Nanomaterials in the Laboratory

Overall, no radically new protective measures are required from today's perspective. The key here is rather to ensure consistent application of existing measures that allow exposure against dangerous substances to be controlled in the laboratory. A risk assessment is used to determine whether additional measures might be necessary. However, future generations of further developed nanomaterials with new properties may require stricter protective measures. As more knowledge is gained, it is important to keep a close eye on findings to allow swift amendments to the level of protection as and when necessary.

8.5 HEALTH RISK MANAGEMENT OF ENGINEERED NANOMATERIALS

8.5.1 General

Responsible health risk management of engineered nanomaterials aims at securing and controlling, in a reliable and transparent way, safe use of these materials [6]. A balance has to be struck between a precautionary approach and a cost-effective and practical way of implementing control measures. While as indicated before in this chapter, in following the hierarchy of control most technological measures are available to control exposure of NOAA at the workplace, the real challenge in practice is the question of the appropriateness of the control measures [27] in the selection process of these measures. It is therefore important that the risk management implementation is carried out in a transparent and predictable fashion such that it may be adapted to emerging new information on a specific nanomaterial or changes in regulation. Health risk management of chemicals including engineered nanomaterials typically covers occupational environments during production, transport, storage and incorporation into products, the uses in consumer goods to maintain consumer exposure at a safe level and leaks into the environment to maintain the environmental burden at an acceptable and safe level.

This discussion focuses on health risk management in the occupational environment. Typical risk management means through which the use of chemicals can be managed include classification of materials into different safety categories, labeling of packages containing these materials and setting of different exposure limits such as occupational exposure limits values (OEL), recommended exposure limits values (REL) or other limits that restrict exposure. In case the health risks would otherwise be considered unacceptable, the use of the chemical can also be banned. One recent example of a chemical of a particulate nature banned in the European Union is asbestos [40].

8.5.2 Recommended Exposure Limits Based on Mass Concentration

The use of OEL is the most convenient and common tool for risk management in the workplace. However, binding OEL for nanomaterials are lacking at international level. The US NIOSH has carried out a health risk assessment for fine and nano-sized titanium dioxide by using carcinogenicity (lung cancer) as an endpoint for the compounds. By using the US approach of deriving OELs, NIOSH has proposed an in-house, non-binding REL in the occupational environment for fine titanium oxide of 2.4 mg/m^3 and for nano-sized titanium dioxide of 0.3 mg/m^3 for 8-hour daily exposure for lifetime, thus reducing the 'risks of lung cancer to below 1 in 1000' and clearly setting for the first time different OELs for particulate fractions in the nanoscale and larger particulate forms of the same substance [41]. In the US, a risk level for a severe disease (e.g., cancer) that is considered significant is 0.1%, or 1 case per 1000 workers exposed for a 45-year working lifetime [42]. NIOSH has carried out a similar health risk assessment for carbon nanotubes and nanofibers based on inflammatory effects (not fiber-specific effects) and proposed a nonbinding REL of 0.001 mg/m^3, based on the limit of quantification of the analytical method and stating that the remaining risk may be higher than in the case of titanium dioxide [29].

There have also been European science-based attempts to develop risk assessment of given engineered nanomaterials, including CNT and titanium dioxide (TiO$_2$). In the course of the European Framework project ENRHES, Stone et al. [43] proposed a derived no-effect level (DNEL) for multi-walled carbon nanotubes (MWCNT) of 0.2 mg/m^3 for short-term inhalation and a DNEL of 0.034 mg/m^3 for long-term inhalation, both for pulmonary effects. They have also proposed a DNEL for immunological effects for the same materials of 0.004 and 0.00067 mg/m^3 after short-term and long-term inhalation, respectively. For titanium dioxide (diameter of 21 nm), Stone et al. [43] have proposed a DNEL of 17 mg/m^3. One has to remember, though, that OEL values are derived from DNEL values, usually dividing them by a factor of 100, and hence the above values for both CNT and TiO$_2$ should be divided by this factor. Hence the probable OELs based on the DNEL values would be close to the proposals of NIOSH [41]. In addition to these studies, Pauluhn [44] reported on long-term inhalation studies of experimental animals exposed to Baytubes, a certain type of multi-walled carbon nanotubes (MWCNT) produced by the company Bayer Material Science. Based on these studies an OEL of 0.05 mg/m^3 for Baytubes was suggested. Ma Hock et al. [45] have shown that rats exposed for 90 days at even a low concentration of 0.1 mg/m^3 MWCNT − produced by the company Nanocyl − can cause inflammation and granuloma formation. Nanocyl then applied an overall assessment factor of 40 to the lowest observed effect level of 0.1 mg/m^3 to estimate an OEL of 0.0025 mg/m^3 [46].

Even though there are several European proposals for values to be used as the European OEL values or that may be used as a foundation for them, no European regulatory attempts have yet been made to utilize these values in managing of risks of engineered nanomaterials, and no expert bodies or institutes so far have proposed them to be used for setting of OEL values in the European Union. Hence, it seems that the currently available small amount of risk assessment data creates a major obstacle for the regulation of exposure to engineered nanomaterials.

8.5.3 Pragmatic Approach with Benchmark Levels Based on Particle Number Concentration

In an attempt to overcome the described obstacle, some European institutions have chosen an approach based on particle number concentration, rather than the classical metric of mass concentration, to provide a foundation for the safe management of engineered nanomaterials. One of the first efforts was by the British Standard Institution. In BSI PD 'Nanotechnologies — Part 2: Guide to safe handling and disposal of manufactured nanomaterials' [47], they adopted a pragmatic approach, proposing 'benchmark exposure levels' in order for a defensible safety level to be attained. The lower limit for the ubiquitous concentrations in contaminated areas of 20,000 particles/cm^3 was proposed as a benchmark. The authors probably had in mind a diameter range for which this maximum concentration should apply; such a range is not stated in the document, however. Thus the meaning of 20,000 particles/cm^3 in terms of mass concentration can be very diverse or one mass concentration, e.g., 0.1 mg/m^3 could equal very different particle number concentrations depending on the size and density of the material. For example it takes about 1.2 million 20-nm gold particles/cm^3 to carry the mass of 0.1 mg/m^3 whereas it takes only about 10,000 gold particles/cm^3 of the size of 100 nm.

In 2009 the German Institute for Occupational Safety and Health (DGUV-IFA) gave the British approach an additional context and proposed the following benchmark levels:

1. For metals, metal oxides and other biopersistent granular nanomaterials with a density of > 6000 kg/m^3, a particle number concentration of **20,000 particles/cm^3** above the background concentration in the range of measurement between 1 and 100 nm should not be exceeded.
2. For biopersistent granular nanomaterials with a density below 6000 kg/m^3, a particle number concentration of **40,000 particles/cm^3** above the background concentration in the measured range between 1 and 100 nm should not be exceeded.
3. In the view of DGUV-IFA, a substantial need for discussion remains with regard to evaluation of the effectiveness of protective measures against nanoscale particles or agglomerates/aggregates larger than 100 nm. For 500 nm titanium dioxide aggregates, 360 particles/cm^3 corresponds to a mass concentration of 0.1 mg/m^3. Measurement of this number concentration by means of the existing instruments would necessitate virtually clean-room conditions. In a typical industrial environment, this concentration would no longer be detectable, owing to the ubiquitous background level of 20,000 particles/cm^3 or more. The corresponding mass concentration of 0.1 mg/m^3 titanium dioxide can, however, be determined reliably by means of conventional analysis methods for documentation of the in-plant conditions. For 200-nm titanium dioxide aggregates, 20,000 particles/cm^3 corresponds to a mass concentration of 0.35 mg/m^3. This already exceeds the value of 0.3 mg/m^3 proposed by NIOSH for nanoscale titanium dioxide [41].

4. Owing to the mounting evidence that biopersistent carbon nanotubes (CNTs) that satisfy the WHO fiber definition or have similar dimensions may harbor effects similar to those of asbestos, it is urgently recommend that only CNTs be used that have been tested for this endpoint (according to the manufacturer's declaration) and that do not exhibit these properties. For carbon nanotubes for which no such manufacturer's declaration is available, a provisional fiber concentration of 10,000 fibers/m^3 is proposed for assessment, based upon the German exposure-risk ratio for asbestos [48,49]. In addition to use of state-of-the-art protective measures, the wearing of respiratory protection and protective clothing is advisable even if the recommended exposure limits are observed. The demarcation of contaminated and non-contaminated zones should be reviewed.

At present, however, monitoring of the above value in industrial plants is hampered by a lack of collection methods of verified suitability, corresponding analysis methods and criteria for counting of the fibers and determining of the fiber count concentration. An urgent need exists here for the development of analysis methods and conventions for interpretation.

For a transitional period, a particle number concentration of 20,000 particles/cm^3 should not be exceeded. In a worst-case scenario, however, this would correspond to a fiber concentration of 20 billion fibers/m^3, way higher than the 10,000 fibers/m^3 mentioned above, and this illustrates that the existing methods for determining the particle number concentration of CNTs at the workplace are unsatisfactory. Some companies have employed internal guide values as mass concentration values based upon the residual content of metallic catalysts in the CNTs. The following proposal was discussed by Schulte et al. [51] and on request of the Dutch parliament evaluated by a group of experts [52]:

5. For ultrafine liquid particles (such as fats, hydrocarbons, siloxanes), the applicable maximum workplace limit (MAK) or workplace limit (AGW) values should be employed owing to the absence of effects of solid particles.

6. The recommended benchmark levels stated above should not be applied to ultrafine particles. For some processes and technologies in which ultrafine particles are produced, proven protective measures and binding provisions exist for the handling of them. Welding fumes and diesel-engine emissions are examples. The body of rules and regulations that exists in this area and that has been drawn up with reference to the current state of knowledge should be applied until new findings become available [50].

This proposal was discussed by Schulte et al. [51] and on request of the Dutch parliament evaluated by a group of experts [52]. In the Netherlands a feasibility study proved that the further developed approach under the name of provisional 'nano reference values' (NRV), as an 8-h time-

TABLE 8.1 Provisional NRVs

Class	Description	Density	NRV (8-h TWA)	Examples
1	Rigid, biopersistent nanofibers for which effects similar to those of asbestos are not excluded	—	0.01 fibers cm^{-3}	SWCNT or MWCNT or metal oxide fibers for which asbestos-like effects are not excluded by manufacturer
2	Biopersistent granular nanomaterials in the range of 1–100 nm	>6000 kg m^{-3}	20,000 particles cm^{-3}	Ag, Au, CeO_2, CoO, Fe, Fe_xO_y, La, Pb, Sb_2O_5, SnO_2
3	Biopersistent granular and fiber form nanomaterials in the range of 1–100 nm	<6000 kg m^{-3}	40,000 particles cm^{-3}	Al_2O_3, SiO_2, TiN, TiO_2, ZnO, nanoclay, carbon black, C_{60}, dendrimers, polystyrene, nanofibers with excluded asbestos-like effects
4	Non-biopersistent granular nanomaterials in the range of 1–100 nm	—	Applicable OEL	e.g., fats, common salt (NaCl)

SWCNT, single-wall CNT; MWCNT, multi-wall CNT.
Source: The Social and Economic Council of the Netherlands (SER) Advisory Report (2012). 'Provisional nano reference values for engineered nanomaterials'. The Hague, The Netherlands, available at http://www.ser.nl/ ~/media/Files/Internet/Talen/Engels/2012/2012_01/2012_01.ashx.

weighted average, is a comprehensible and useful instrument for risk management' [53]. In 2012 the Dutch Social and Economic Council (SER), which plays a famous role in the so-called Dutch polder model, recommended these values to the Dutch parliament and government [54] as a tool in the precautionary approach in the handling of engineered nanomaterials. With reference to the Dutch Labour Conditions Act the Dutch government then stated that they regard the provisional NRVs (see Table 8.1) as the best available science and state-of-the art approach for risk assessment of nanomaterials [53].

It may be highlighted that in the approach described above, nanomaterials were categorized according to their biopersistence or as a proxy for this solubility and, in addition, according to their morphology, e.g., fiber-like properties. Interestingly no attempt was made to group nanomaterials according to quantitative structure-activity relationships (QSAR).

8.5.4 German Announcement on Hazardous Substances 527, 'Manufactured Nanomaterials'

Kuempel et al. [55] state that given the almost limitless variety of nanomaterials, it will be virtually impossible to assess the possible occupational health hazard of each nanomaterial individually. The development of science-based hazard and risk categories for nano-materials is needed for decision making about exposure control practices in the workplace. They propose as a possible strategy to select representative (benchmark) materials from various mode of action (MOA) classes, evaluate the hazard and develop risk estimates, and then apply a systematic comparison of new nanomaterials with the benchmark materials in the same MOA class. While this approach is pursued in the United States and the ISO TC 229 'Nanotechnology' accepted the new work item 'General framework for the development of occupational exposure limits for nano-objects and their aggregates and agglomerates' [56], the German Tripartite Committee on Hazardous Substances asked the Subcommittee 1 on Hazard Management to draft an announcement of manufactured nanomaterials. On May 7th 2013 the draft of the announcement was accepted in a final decision by the committee.

The Announcement on Hazardous Substances 527 'Manufactured Nanomaterials' from May 2013 was published in June/July 2013 by the Federal Ministry of Labour and Social Affairs [57]. In this announcement nanomaterials are grouped into four categories:

1. Soluble nanomaterials;
2. Biopersistent nanomaterials without specific toxicity (granular biopersistent nano-particles — nano GBP);
3. Biopersistent nanomaterials with specific toxicity;
4. Biopersistent fiber-like nanomaterials.

Interestingly, concerning the **1st category**, the terms soluble/insoluble or biopersistent are commonly used in particle toxicology but no exact definition or measurement method for these terms is at hand. Therefore as a proxy the solubility of nanomaterials in water is used, and substances with a solubility of less than 100 mg/L are coined 'practically insoluble' and nanomaterials with solubility in water greater of 100 mg/L belong to the category of soluble nanomaterials. Fully aware of the discussions on the possible enhanced proliferation or changed pathways of soluble nano-materials to different targets in the body, the advice has given to perform the risk assessment for nanomaterials in this category per default in treating them like bulk materials and neglecting possible nano-related properties.

For the **2nd category** (nano-GBP), reference was made to the study by Gebel [58] comparing the carcinogenicity of GBP micromaterials (micro-GBP) and nano-GBP in chronic rat inhalation studies. Gebel concluded that the difference in carcinogenic potency between GBP nanomaterials and GBP micromaterials is low and can be described by a factor of 2—2.5, referring to the dose metric mass concentration.

In July 2011, the Permanent Senate Commission for the Investigation of Health Hazards of Chemical Compounds in the Work Area of the German

Research Foundation proposed a reduced value of 0.3 mg/m^3 for a density of 1 kg/m^3 [59] for the respirable dust fraction (GBP) of the general dust limit value. This value is intended to prevent high concentrations of these dusts from having a carcinogenic effect. Aware of this proposal and the ongoing discussion, the recommendation is given in Announcement 527 to correct for the slightly higher potency of nano-GBP by applying a factor of 1/2 on the current occupational exposure limit for respirable dust in Germany, which is 3 mg/m^3 for materials with a density of 2.5 kg/m^3. For the time being this would result in a limit value of 1.5 mg/m^3. As this was regarded too high a value in terms of particle number concentration, it was stated that the REL for nano-GBP should not exceed 0.5 mg/m^3 for a density of 2.5 kg/m^3, measured as the respirable dust fraction. It must be highlighted that all other options to utilize REL for the risk management are still available for the companies. They may use the recommendations given by NIOSH or other organizations/companies, use the benchmark levels proposed by DGUV-IFA or set their own in-house standards.

Concerning the **3rd category** (biopersistent nanomaterials with a specific toxicity), reference is made to the handling of the bulk (non-nano) forms of these materials. With regard to existing OELs for most of these nanomaterials, it is stated that companies of course have to comply with the existing OELs. In Germany these OELs are usually below 0.1 mg/m^3. In fact, in the discussion on deriving exposure-risk relationships and the corresponding concentrations [48] for carcinogenic metals or metal compounds, like cobalt or nickel, mass concentrations in the range of 0.0001 mg/m^3 to 0.01 mg/m^3 are proposed. Announcement 527 states that in complying with OELs in this range of concentrations a strict regime of control measures with a high efficacy has to be employed, and in the case of handling the nanomaterial a further discrimination of control measures is not feasible.

For materials belonging into the **4th category** (biopersistent fiber-like nanomaterials), the distinction is made between biopersistent, rigid nanofibers adhering to the WHO fiber paradigm, for which one has to assume an asbestos-like effect, and biopersistent, entangled or spaghetti-like nanofibers. For the latter, asbestos-like effects can only be excluded in the risk assessment if the manufacturer or supplier of the given nanomaterial can give proof of evidence that the nanomaterial does not exhibit asbestos-like effects. Overall companies are strongly discouraged to use biopersistent, **rigid** nanofibers, and a very strict regime of control measures has to be followed if these materials are handled.

In conclusion one can summarize Announcement 527 on 'Manufactured Nanomaterials' in a way that, at least for the time being and for the handling of the first, passive generation of nanomaterials, these materials are mostly treated at the workplace like ordinary hazardous substances. The only exception is the evaluation of control measures for nano GBP, as in this case only half of the respirable dust limit values may be used. The Announcement will be adapted if new evidence on toxicological properties of manufactured nanomaterials emerges or if future generations of active nanomaterials find their way to the workplace and pose new hazards.

8.6 RISK GOVERNANCE, POLICY ASPECTS AND LEGISLATION IN THE EUROPEAN UNION AND THE UNITED STATES

8.6.1 General

Governance includes the processes, conventions and institutions that determine how power is exercised in managing resources and interests. It also includes how important decisions are made, how conflicts are resolved and how interactions among and between the key actors in the field are organized and structured. Also the development of resources, skills and capabilities, and how they are developed and mobilized for reaching desired outcomes, are an important part of governance, which also covers issues such as how participation of various stakeholders is accorded in the participation of these processes. In essence, governance sees the process of governing as a joint effort of industrial, public and civic society actors, usually within networks, with a goal of making the best use of available resources, skills and competencies for reaching specific goals or purposes [60].

Risk governance is focused on the institutional arrangement on how risk information is collected, analyzed and communicated and how risk management decisions are taken [61]. Risk governance includes all the risk-relevant decisions and actions, and it is of special importance on occasions where the nature of risks requires collaboration and coordination between various agencies and stakeholders and emphasizes the essence of consideration of contextual factors such as institutional arrangements and sociopolitical culture and perceptions.

These terminological issues are important because risk governance of engineered nanomaterials and nanotechnology is just one reflection of these rules in settings in which the issues related to nanosciences and technologies, engineered nanomaterials and especially governance of their potential environmental and health risks need to be dealt with. Thus, they need to be governed for the benefit of all stakeholders, and it is important to assure that the utilization of the potential and resources associated with engineered nanomaterials and nanotechnologies are not misused and that safety of these materials and technologies can be guaranteed. Risk governance of nanotechnologies and engineered nano-materials, therefore, includes much wider societal and policy and value-driven dimensions than just the management of these materials and technologies [62].

Societal and political governance of risks of nanotechnologies are a top priority for the acceptance of engineered nanomaterials and nanotech-nology by the public at large, key stakeholders such as social partners, civic society organizations, the regulatory community and downstream industry with a potential to use these materials in products. An open and reliable risk governance of these materials and technologies is an important prerequisite for the acceptance of nanotechnology-based ap-plications in the society, enabling benefits of nanotechnologies. In different societies, concerns related to nanotechnology are on an increase, and this puts a tremendous challenge to communication of the benefits and potential risks of these beneficial technologies. One has to be able to be open and convey reliable messages on nanotechnologies to prevent wrong perceptions of risks of these technologies.

To address those challenges four simultaneous characteristics of effective nanotechnology governance were proposed. Nanotechnology governance needs to be [10]:

- Transformative (including a results- or projects-oriented focus on advancing multidisciplinary and multisector innovation);
- Responsible (including EHS and equitable access and benefits);
- Inclusive (participation of all agencies and stakeholders);
- Visionary (including long-term planning and anticipatory, adaptive measures).

Concerning the regulation of nanotechnology, currently two approaches are being developed in parallel [10]:

- Probing the extendability of regulatory schemes like the Registration, Evaluation and Authorization Chemicals (REACH) Regulation Act in the European Union and the Toxic Substances Control Act (TSCA) in the United States (both following a 'developing the science' approach)
- Exploring (soft) regulatory and governance models that work despite insufficient knowledge for full risk assessment, including ethical, legal and social impacts (ELSI) research, voluntary codes, public engagement, observatories, public attitude surveys and other instruments.

However, one can observe that with the implementation of regulation of nanotechnology under the existing umbrella of regulatory frameworks like REACH and TSCA and the need for companies to immediately comply with these changes, the second approach is losing its attractive-ness, at least for the companies involved.

We therefore describe extensively the current activities to adapt the REACH regulation to nanomaterials and later on compare differences to the TSCA regulation.

8.6.2 The REACH Regulation (by Marita Luotamo)

The aim of REACH [63] is to improve the protection of human health and the environment from the risks that can be posed by chemicals through the better and earlier identification of the intrinsic properties of chemical substances while enhancing the innovation and competitiveness of the EU chemicals industry.

Industry has to ensure that chemical substances are used safely. This is achieved by using information on the properties of substances to assess their hazards both for classification and risk assessment and hence to develop appropriate risk management measures to protect human health and the environment. One of the main reasons for developing and adopting the REACH Regulation was to fill information gaps for the large number of substances already in use in the EU, as for many there is inadequate information on the hazards and the risks they pose. REACH prescribes that, in general, all substances manufactured or imported in quantities at 1 ton or more per year in the EU have to be registered. The European Chemicals Agency (ECHA) [64] in Helsinki acts as the central point in the REACH system.

If the risks cannot be adequately managed, authorities can restrict the use of substances in different ways. In the long run, the most hazardous substances should be substituted with less dangerous ones, i.e., the Regulation also calls for the progressive substitution when suitable alternatives have been identified.

There are no specific provisions in REACH for nanomaterial substances, but REACH deals with substances, in whatever size, shape or physical state. Substances at the nanoscale are therefore covered by REACH and its provisions apply. It thus follows that under REACH manufacturers, importers and downstream users have to ensure that their nanomaterials do not adversely affect human health or the environment (Document CA/59/2008_Rev1 [65]).

The European Commission published the 'Second Regulatory Review on Nanomaterials' (COM(2012) 572 final) in October 2012 [66] and the Commission Staff Working Paper on types and uses of nanomaterials, including safety aspects (SWD(2012) 288 final) [7] accompanying the communication from the Commission to the European Parliament, the Council and the European Economic and Social Committee on the Second Regulatory Review on Nanomaterials.

8.6.2.1 Definition of a Nanomaterial

According to the Commission Recommendation (adopted 18 October 2011) on the Definition of a Nanomaterial [67], a 'nanomaterial' means:

'A natural, incidental or manufactured material containing particles, in an unbound state or as an aggregate or as an agglomerate and where, for 50 % or more of the particles in the number size distribution, one or more external dimensions is in the size range 1 nm—100 nm.

In specific cases and where warranted by concerns for the environment, health, safety or competitiveness the number size distribution threshold of 50% may be replaced by a threshold between 1 and 50%.

By derogation from the above, fullerenes, graphene flakes and single wall carbon nanotubes with one or more external dimensions below 1 nm should be considered as nanomaterials.'

In order to understand the context of this definition, it is useful to refer to the definition in the DG ENV website [68]: 'the definition will be used primarily to identify materials for which special provisions might apply (e.g. for risk assessment or ingredient labelling). Those special provisions are not part of the definition but of specific legislation in which the definition will be used'.

8.6.2.2 REACH Registration of Nanomaterials

The registrant should collect all relevant available information on the substance in order to fulfill the information requirements: information on substance identity, physicochemical properties, toxicity, ecotoxicity, environmental fate, exposure and instructions for appropriate risk management. Data sharing with other registrants for the same substance is one of the major tools in REACH to avoid unnecessary testing. The higher the tonnage, the more information needs to be submitted. The submission includes a technical dossier including, for substances manufactured or imported in quantities of 10 tons p.a. or above, a Chemical Safety Report.

REACH deals with substances in whatever size, shape or physical state. If a substance exists in nanoform(s) as well as the bulk form, the differences between the nanoform(s) and bulk form will need to be clarified. It may be the case that the intrinsic properties of the substance differ between the nanoform(s) and the bulk form, hence the hazardous properties may differ. Indeed there may be different uses, and hence different exposures to humans and the environment. Hence the combination of different hazardous properties and/or different exposures would require a suitable coverage in the Chemical Safety Assessment.

8.6.2.3 REACH Guidance and Nanomaterials

REACH Guidance Documents [69] provided by ECHA apply to all chemical substances, including those in nano form. It was recognized that some specific guidance is necessary for nanomaterials, and in order to facilitate this process the Commission services initiated REACH Implementation Projects on Nanomaterials (RIP-oNs) in 2009, which have now reported [70], to evaluate the applicability of the existing guidance to nanomaterials regarding:

1. Substance Identification (RIP-oN 1);
2. Information Requirements including testing strategies (RIP-oN 2);
3. Chemical Safety Assessment (RIP-oN 3).

ECHA has published the Appendices for nanomaterials [71] to the Guidance on Information Requirements and Chemical Safety Assessment concerning Information Requirements (IR) and Chemical Safety Asessment (CSA) in line with the outcomes of RIP-oN 2 and 3 as the two projects gained broad support from the different stakeholders participating in the project.

Registrants may find it useful to refer to the RIP-oN 2 and 3 reports when preparing registration dossiers covering nanomaterials. However, registrants should ensure that safe use of their substance is demonstrated in the registration dossier and hence that the data of the submitted properties as well as related risk assessment and management information is applicable and appropriate for the nanoforms covered by the registration.

The RIP-oN 1 report on the substance identification of nanomaterials also contains useful information that can in due course be integrated by ECHA into the IUCLID 5 manuals. As the experience in addressing characterization of nanoforms is still developing, the current ECHA guidance on substance identity will at present not be revised. Furthermore, for other areas it is clear that further scientific development and research are still necessary before providing definitive guidance updates.

8.6.2.4 Technical Registration Dossier

The registrants provide the required information in a registration dossier consisting of two main components: a technical dossier and a Chemical Safety Report (required for substances at 10 tons or more per year) with information requirements based on the tonnage level (REACH Annexes VII to XI).

The registration dossier has to be prepared using the IUCLID 5 software application [72]. IUCLID 5 implements the Harmonised Templates developed by the OECD, and it is compatible with other chemical legislations around the world. There has been continuous development of the IUCLID software also towards the specific needs concerning registration of nanomaterials. The IUCLID 5.4 version has an additional field in all robust study summary templates to give the possibility to indicate the form of the test material via a selection in a picklist including 'nano-materials'. IUCLID is continuously improved and can be further developed also with regard to specific information needs on nanomaterials.

8.6.2.5 Evaluation

Evaluation work under REACH is divided between ECHA and Member States Competent Authority (MSCA). ECHA is responsible for dossier evaluation including the examination of all the testing proposals made by registrants and for undertaking compliance checks of least 5% of the registrations in each tonnage band. MSCAs are responsible for the substance evaluation where substances are examined to decide if further information is needed to clarify a suspected risk.

The substances are in the Community Rolling Action Plan (CoRAP) [73], which is a list of prioritized substances developed by the ECHA in cooperation with the Member States. The first CoRAP lists 90 substances for evaluation and was published on 29 February 2012, with substances distributed for evaluation in years 2012, 2013 and 2014 between the volunteering Member States. In the first CoRAP substances nanomaterials were also included: silicon dioxide to be evaluated by the Netherlands (NL) in 2012, silver also by NL in 2014 and titanium dioxide by France in 2014 [74].

In 2012 ECHA started to process dossiers related to nanomaterials covering both testing proposals and compliance. In order to share experiences with and generate consensus among MSCAs and Member State Committee members on safety information on nanomaterials in REACH registration dossiers, ECHA organized a nanomaterial work-shop in May 2012.

The workshop agreed on a way forward in further building capacity and forum of exchange of scientific expertise in support of the evaluation process under REACH. A pre-workshop on Group Assessing Already Registered Nanomaterials (GAARN) addressed the challenges faced on registering substances under REACH and on the information requirements regarding substance identification and physical chemical properties. Best practices for registrants of nanomaterials were derived following this pre-workshop and were published on the ECHA website [75], together with the proceedings of the workshop.

In October 2012, ECHA established a nanomaterials working group (ECHA-NMWG) to discuss scientific and technical questions relevant to REACH and Classification, Labelling and Packaging (CLP) processes and to provide recommendations on strategic issues. It is an informal advisory group consisting of experts from Member States, the European Commis-sion, ECHA and accredited stakeholders organizations, with the mandate to 'provide informal advice on any scientific and technical issues regarding implementation of REACH and CLP legislation in relation to nano-materials'. ECHA-NMWG also aims to facilitate discussions with industry regarding its experience gained in documenting intrinsic properties of the nano-forms of substances using recent methodologies and its obligations towards fulfilling REACH requirements.

ECHA together with the Joint Research Center (JRC) worked collaboratively on the project commissioned by DG Environment 'Scientific technical support on assessment of nanomaterials in REACH registration dossiers and adequacy of available information' starting January 2011, and the final report of task I is available [76]. It should be noted that there was no adopted European Commission Recommendation on the definition of nanomaterial at the time of the first registration deadline (December 2010). First part (task I) of the project entails identification of REACH registration dossiers that cover nanomaterials, scientific assessment of the information on nanomaterials contained in these dossiers and finally proposing how potential cross-cutting shortcomings can be addressed. The report provides summaries and options for adaptation of REACH for substance identification, characterization and physicochemical properties as well as human health and environment. The focus of task II is on assessing potential economic and environmental consequences of the technical proposals made in task I, and the final report is available [68].

8.6.2.6 *Authorization*

The aim of authorization is to ensure that the risks from Substances of Very High Concern (SVHC) are properly controlled and that these substances are progressively replaced by suitable alternative substances or technologies where possible, while ensuring the good functioning of the EU internal market. Authorization applies to SVHC regardless of their volumes, i.e., there is no tonnage trigger.

Substances having any one of the properties can be formally identified as SVHCs by an MSCA (or ECHA acting on instruction of the European Commission) submitting a dossier:

- Substances meeting criteria for classification as CMR substances (carcinogenic, mutagenic or toxic for reproduction) Category 1A and 1B in accordance with CLP Regulation;
- PBT substances that are persistent, bioaccumulating and toxic and vPvBs that are very persistent and very bioaccumulating according to Annex XIII of REACH;
- Substances identified on a case-by-case basis for which there is scientific evidence of probable serious effects to human health or the environment, which give rise to an equivalent level of concern as CMRs and PBTs/vPvBs.

Substances that have been identified as SVHCs are listed on the Candidate List [77]. In due course some substances are transferred onto the Authorisation List, with a view to being phased out of use. Manufacturers, importers or users can apply for authorization for continuing the specific uses of a substance on the Authorisation List before the substance-specific 'application date'.

A substance in nanoform could fulfill these criteria mentioned, so an Annex XV dossier identifying the nanomaterial as a SVHC could be submitted. If it is only the nanoform(s) of the substance that meets the SVHC criteria, this must be clarified in the dossier. Similarly, if an authorization application concerns a substance at the nanoscale/nanoform, this should also be clarified in the application [65].

8.6.2.7 *Restrictions*

REACH also contains provisions regarding restrictions that may limit or ban the manufacturing, placing on the market and use of a substance. A restriction applies to any substance on its own, in a mixture or in an article, including those that do not require registration, and it can also apply to imported articles. Hence, in principle such substances can be in nano form.

A Member State, or ECHA on request of the European Commission, can propose restrictions if they identify unacceptable risks and consider that these risks need to be addressed on a community-wide basis. ECHA Committees for Risk Assessment and for Socio-Economic Assessment provide scientific opinions on any proposed restriction that will help the European Commission, together with the Member States, to take the final decision.

8.6.2.8 *Conclusion*

Nanomaterials have many technical benefits, but it is possible that some may pose specific risks to human health and/or to the environment. These risks, and to what extent they can be tackled by the existing risk assessment methodologies used in the EU, have been the subject of several opinions of the Scientific Committee on Emerging and Newly Identified Health Risks [78]. The overall conclusion so far is that nanomaterials are similar to normal chemicals/substances in that some may be toxic and some may not. Possible risks are related to specific nanomaterials and specific uses. Therefore, there is still scientific uncertainty about the safety of nanomaterials in many aspects and the safety assessment of the substances must be done on a case-by-case basis using pertinent information. Current risk assessment methods are applicable, even if work on particular aspects of risk assessment is still required [66].

8.6.3 Comparing Risk Governance Approaches in the European Union and the US

In their discussion of policy considerations for responsible nanotechnology decisions Morris et al. [6] compare the US and European approaches in dealing with the regulation of risks of engineered nanomaterials and also discuss the environmental and health issues associated with them in a global perspective by using the OECD approach as an example. Hence, they take from a regulatory point of view a fresh regional and global look at dealing with risks of engineered nanomaterials and nanotechnologies. The background of this comparison is briefly discussed below.

The first widely used report to evaluate both the benefits and risks of nanotechnologies was published by the Royal Society and the Royal Academy of Engineering in the UK in 2004 [15]. The first European Union action plan on nanosciences and nanotechnologies was adopted in 2004. It covered the period of 2005–2009. It was followed by a report of the Scientific Committee on Emerging and Newly Identified Health Risks [79]. The first US report covering the environmental science implications of nanotechnology from a public policy perspective was the Nanotechnology White Paper published by the US Environmental Protection Agency [80]. Perhaps the two most important regulations on engineered nanomaterials and nanotechnologies are the REACH regulation [63] in the European Union and the Toxic Substances Control Act [81] in the US. Table 8.2 compares the approaches of the REACH regulation in the European Union and the Toxic Substance Control Act in the United States in the regulation, management and governance of engineered nanomaterials.

In the US, the main agency managing the safety of engineered nanomaterials is the US Environmental Protection Agency, but other agencies including the Food and Drug Administration and the National Institute for Occupational Safety and Health have issued framework documents or research strategies specific to their mandates. Since 2005, the US EPA has received more than 100 premanufacture notifications for specific engineered nanomaterials under the TSCA. Because not all engineered nanomaterials are new chemicals, the US EPA has implemented a voluntary stewardship program to gather information to better understand the potential risks of these materials already in use commercially or for research and development purposes. The US EPA issued a report on the program in 2008 and, because of the limited participation of various stakeholders, now follows up on it by issuing regulations for mandatory reporting and testing of engineered nanomaterials. One specifically important tool in

TABLE 8.2 Comparison of Regulations for New Chemicals and Existing Chemicals Under TSCA and REACH

TSCA	REACH
NEW CHEMICALS	
Baseline: TSCA Chemical Substance Inventory, 1979	Baseline: European Inventory of Existing Commercial Chemical Substances (EINECS, 1971–1981) and European List of Notified Chemical Substances (ELINCS, a cumulative listing of newly notified chemicals)
Requirement: Manufacturers must notify EPA 90 days before manufacturing a chemical substance not on the TSCA Chemical Substance Inventory. Information on chemical identity, production, use, exposure and hazard is required if available	Registration: Chemicals produced in quantities ≥ 1 t yr^{-1} per manufacturer must be registered. Registration is required 21 days before manufacturing. Scope and amount of data required depend on tonnage. A Chemical Safety Report for production ≥ 10 t yr^{-1} per manufacturer is compulsory. Companies must update registration on the basis of new risk-related information and change in status (for example, use and production volume)
Chemicals are added to inventory after a Notice of Commencement has been filed	Chemicals are added to inventory after a completeness check
Some exemptions are possible when submitter demonstrates that no unreasonable risk exists	All uses of the same chemical must be included in the registration dossier. Updates for new uses are compulsory
The EPA can request additional information, require testing and impose conditions of use through orders, and can regulate additional uses through a 'significant new use' rule	The European Chemicals Agency can request further information, and substances of very high concern (whether new or existing) are prioritized for evaluation. The EC maintains a candidate list of substances of very high concern that are considered for authorization
EXISTING CHEMICALS	
Routine reporting by Inventory Update Rule every 5 yrs for chemicals with production $\geq 25,000$ lb yr^{-1} per site. The EPA can, by rule, require exposure and health and safety data to be reported and chemicals to be tested	Registration required for production ≥ 1 t yr^{-1} per manufacturer. The timetable (2010–2018) is dictated by tonnage (≥ 1000 t; ≥ 100 t; ≥ 1 t). Registration is required by deadline to continue manufacturing. The scope and amount of required data depend on tonnage. A Chemical Safety Report for production ≥ 10 t yr^{-1} per manufacturer is compulsory
Existing chemicals are reviewed and prioritized on the basis of a number of factors, including hazard, persistence, bioaccumulation and exposure	Companies update registration on the basis of new risk-related information and change in status (for example, use and production volume)

(Continued)

TABLE 8.2 Comparison of Regulations for New Chemicals and Existing Chemicals Under TSCA and REACH—cont'd

TSCA	REACH
	Chemicals remain on inventory after a completeness check
Chemical management takes place through rule making to address possible significant new uses or unreasonable risks	Chemical management takes place through an Authorization List and a Restriction List
	Substances of very high concern are treated as described above, yet there is no tonnage trigger

Modified from [6].

governing existing chemicals (nanomaterials) is the 'Significant New Use Rules (SNUR)' issued by EPA for nanomaterials under Section 5(a) [2] of TSCA. Some controversy exists, however, over regarding what is considered as a 'new use'. US EPA has consistently maintained that nanomaterials already in use when the agency proposes a SNUR are not, in fact, 'new'. Given the fast pace of nanomaterials development, this position is likely to exempt many nanomaterials from this important TSCA section [82]. Some similar controversy exists under REACH on the information and characterization requirements issued by ECHA or Member States in the process of 'substance' evaluation, specifically to what extent further characterization of technical grades, of which there could be some hundreds, of one substance can be requested. Both controversies result from the fact that nanomaterials challenge our understanding and the definition of the term 'substance'.

Even though there are differences in the risk governance approaches within the European Union and the US, perhaps the similarities in risk governance of these materials and subsequent technologies exceed the differences. There is also a pressure for harmonization of the approaches due to increased trade of these materials and products incorporating engineered nanomaterials. Even though the Working Group on Manufactured Nanomaterials of the Organization for Economic Cooperation and Development (OECD) and the OECD in general is not a decision-making body, it provides an important forum for harmonization of approaches in global issues such as the safety and engineered nanomaterials and nanotechnologies [6].

8.6.4 Challenges for the Future of Risk Assessment and Risk Management of Nanomaterials

Compared to the United States Toxic Substance Control Act, some regard the European REACH regulation as a modern approach in regulating chemicals. Shifting the burden from government to prove harm post-marketing to manufacturers and importers to demonstrate safety pre-marketing is creating a very different world of risk assessment and risk management. On top of this challenge to companies and regulators, nanomaterials introduce even more challenges to cope with.

The **first challenge** is the blurring of the definitions of a 'substance' and 'material' by nanomaterials. So it is nowadays accepted that CNT are not just a form of graphite; it has been reported by the ENRHES project [43] that there are many different methods to produce CNT by different manufacturers who use different catalytic metals, carbon sources and processing conditions (such as temperature and pressure), resulting in batch-to-batch variation even for a given manufacturer for the 'pure' CNT. Further functionalization of the CNT just adds another layer of complexity. This highlights that CNT does not even describe 'one' substance being different from graphite. Instead CNTs are a 'heterogeneous population of materials, and there is potential that they are not all universally toxic' [43]. The question therefore arises when a material or a substance can be considered as identical or not and how this can be tested. This and other heterogeneous nanomaterials discussed by John Howard, the director of NIOSH, stating that the mid-20th century paradigm of a 'substance-by-substance' approach [83] might not be working anymore. In addition, the definition of a 'nano' material based solely on the metric of size has been questioned by Maynard [84], and instead it has been advocated to explore new risk assessment and risk management approaches for sophisticated materials [85] based on new properties and integrated within a life-cycle context [86].

The **second challenge** is presented by the sheer number of new substances or variations in a substance as highlighted by the example of CNTs before, because our testing resources do not allow adequate assessment of safety and risks of these materials [45,87]. Hartung [88], in his paper 'Toxicology of the 21st Century' published in *Nature*, called for departing from the existing experimental animal testing paradigm and moving to the use of high throughput testing approaches. This is a highly positive proposal supported by the recent proposal by Nel et al. [89], also emphasizing the use of *in vitro* approaches utilizing high-throughput technologies and -omics methodologies in the safety and toxicity testing of novel nanomaterials. From the regulatory point of view both proposals would be highly useful, but both lack a fundamental justification for their adoption. As long as there is not clear evidence of the predictive power of these approaches for *in vivo* systems, their acceptance as a part of the regulatory framework of nanomaterials or other chemicals is highly unlikely as they are associated with too many uncertainties.

Adding to the difficulties of the proposed approaches is the fact that even in conventional toxicology the applicability of test guidelines and measurement methods for a given nanomaterial are not always directly established [90], to the extent that the quality of papers in nanotoxicology has been questioned. 'One thing is at least clear for now: few studies offer consistent results that are of value, and it is difficult to compare studies because they are often carried out using poorly characterized nanomaterials and arbitrary experimental conditions' [91].

The third challenge may be the transition from the 'Stone Age' of nanomaterials we are currently in by using just passive nanomaterials to the Second—Fourth generation active nanomaterials. The reader is referred to Roco and Maynard [10,85] for a discussion of these aspects, as this indeed has the potential to be a game changer.

For today, still being in the 'Stone Age' of passive nanomaterials or even commodity nanomaterials like amorphous silica or titanium dioxide which we have already produced for six decades, all the tools to control exposure to these nanomaterials at the workplace are at hand by applying existing standards in OHS like the STOP principle together with known engineering control measures. We won't neglect the difficulties in actually exactly measuring exposure to NOAA and how to delineate the dose of workers, as this has been extensively discussed in Chapter 5. However, the challenge for tomorrow is what John Howard called the vigorous enforcement of existing standards in OHS [83].

DISCLAIMER

The views expressed in this paper are solely those of the authors and the content of the paper does not represent an official position of the European Chemicals Agency.

Acknowledgement

The work to develop this chapter was partly supported by the European FP7 project QualityNano.

References

[1] European Union (EU) Commission Communication s1 final. A European strategy for Key Enabling Technologies — A bridge to growth and jobs. Brussels. Available at: http://eur-lex.europa.eu/LexUriServ/LexUriServ.do?uri=COM:2012:0341:FIN:EN:PDF, 2012; [accessed 26.06.2012].

[2] European Parliament. European Parliament resolution of 24 April 2009 on regulatory aspects of nanomaterials (2008/2208(INI)). Brussels. Available at: http://www.europarl.europa.eu/sides/getDoc.do?type=TA&language=EN&reference=P6-TA-2009-328; 2009.

[3] ETC Group. No small matter II: The case for a global moratorium. Size matters! Occasional Paper Series vol. 7. No. 1. Canada: Winnipeg; 2003.

[4] Arnall AH. Future Technologies, todays choices. Nanotechnology, artificial intelligence and robotics: a technical, political and institutional map of emerging technologies. London: Greenpeace Environmental Trust; 2003. ISBN 1-903907-05-5.

[5] Hodge GA, Bowman DM, Maynard AD. International Handbook on Regulating Nanotechnologies. Cheltenham UK: Edward Elgar Publishing Limited; 2010. ISBN 978 1 848844 673 1.

[6] Morris J, Willis J, De Martinis D, et al. Science policy considerations for responsible nanotechnology decisions. Nat Nanotechnol 2011;6:784−7. http://dx.doi.org/10.1038/nnano.2010.191.

[7] European Commission. Commission Staff Working Paper: Types and uses of nanomaterials, including safety aspects. SWD (2012) 288 final, October 2012. http://eur-lex.europa.eu/Result.do?T1=V7&T2=2012&T3=288&RechType=RECH_naturel&Submit=Search; 2012.

[8] Roco MC. Nanoscale science and engineering: unifying and transforming tools. AIChE J 2004;50:890−7.

[9] Subramanian V, Youtie J, Porter AL, Shapira P. Is there a shift to "active nano-structures"? J Nanopart Res 2010;12:1−10.

[10] Roco MC, Harthorn B, Guston D, Shapira P. Innovative and responsible governance of nanotechnology for societal development. J Nanopart Res 2011. DOI 10.1007/s11051-011-0454-4.

[11] Council European. European Council Directive 89/391/EEC of 12 June 1989 on the introduction of measures to encourage improvements in the safety and health of workers at work. Official J Eur Communities L 1989;183:1−8.

[12] Council European. European Council Directive 98/24/EC of 7 April 1998 on the protection of the health and safety of workers from the risk related chemical agents at work. Official J Eur Communities L 1998;131:11−23.

[13] Health and Safety Executive. HSG 97: A step by step guide to COSHH assessment. Available at: http://www.hse.gov.uk/pubns/priced/hsg97.pdf; 2004.

[14] International Organization for Standardization. Nanotechnologies − Occupational risk management applied to engineered nanomaterials − Part 1: Principles and approaches. Switzerland: International Organization for Standardization, Geneva; 2012. ISO/TS 12901-1:2012.

[15] The Royal Society & The Royal Academy of Engineering. Nanoscience and Nano-technologies: opportunities and uncertainties. R Soc 2004. 6-9 Carlton House Terrace London SW1Y 5AG.

[16] Auffan M, Rose J, Bottero J-Y, et al. Towards a definition of inorganic nanoparticles from an environmental, health and safety perspective. Nat Nanotechnol 2009;4:634−41.

[17] Schinwald A, Murphy FA, Jones A, et al. Graphene-Based Nanoplatelets: A New Risk to the Respiratory System as a Consequence of Their Unusual Aerodynamic Properties. ACS NANO 2011;6(1):736−46.

[18] Van Tongeren M. NANEX Final publishable summary report. http://nanex-project.eu/mainpages/public-documents/doc_download/101-nanex-project-final-report-pdf; 2011.

[19] Australia SAFEWORK (SWA). Engineered Nanomaterials: Evidence on the Effectiveness of Workplace Controls to Prevent Exposure, Commonwealth of Australia. Available at: http://www.safeworkaustralia.gov.au/NR/rdonlyres/E3C113AC-4363-4533-A128-6D682FDE99E0/0/EffectivenessReport.pdf; 2009.

[20] International Organization for Standardization. Nanotechnologies — Nanomaterial risk evaluation. Switzerland: International Organization for Standardization, Geneva; 2011. ISO/TR 13121:2011.

[21] International Organization for Standardization. Nanotechnologies — Health and safety practices in occupational settings relevant to nanotechnologies. Switzerland: International Organization for Standardization, Geneva; 2008. ISO/TR 12885:2008.

[22] Hinds W. Aerosol technology: Properties and behavior and measurement of airborne particles. 2nd ed. NY: Wiley-Interscience, New York; 1999.

[23] Methner M, Hodson L, Dames A, Geraci C. Nanoparticle Emission Assessment Technique (NEAT) for the Identification and Measurement of Potential Inhalation Exposure to Engineered Nanomaterials- Part B: Results from 12 Field Studies. J Occup Environ Hyg 2010;7(3):163—76.

[24] Plitzko S. Workplace exposure to engineered nanoparticles. Inhalation Toxicol 2009; 21:25—9.

[25] Methner M, Crawford C, Geraci C. Evaluation of the Potential Airborne Release of Carbon Nanofibers During the Preparation, Grinding, and Cutting of Epoxy-Based Nanocomposite Material. J Occup Environ Hyg 2012;9:308—18.

[26] European Agency for Safety and Health at Work. Case Studies: Good Practice Handling Carbon Nano Tubes. Bilbao, Spain. Available at: http://ebookbrowse.com/nano-case-study-case-tno-nanocyl-pdf-d356770513; 2012.

[27] Geraci C. Efficiency of engineering controls, OECD Workshop on Exposure Assessment and Exposure Migitation. Germany: Frankfurt am Main; 2008.

[28] NIOSH. Interim Guidance for Medical Screening and Hazard Surveillance for Workers Potentially Exposed to Engineered Nanoparticles. Current Intelligence Bulletin 60, US Department of Health and Human Services, Centers for Disease Control, National Institute for Occupational safety and Health, Cincinnati, OH. DHHS (NIOSH), NIOSH Docket Number: NIOSH 116. Available at: www.cdc.gov/niosh/docs/2009-116; 2009.

[29] NIOSH. Occupational Exposure to Carbon Nanotubes and Nanofibers. Current Intelligence Bulletin 65, US Department of Health and Human Services, Centers for Disease Control, National Institute for Occupational safety and Health, Cincinnati, OH. DHHS (NIOSH), NIOSH Docket Number: NIOSH 145. Available at: www.cdc.gov/niosh/docs/2013-145; 2013.

[30] Balazy A, Toivola M, Reponen T, Podgorski A. Manikin-based performance evaluation of N95 filtering-facepiece respirators challenged with nanoparticles. Ann Occup Hyg 2006;50(3):259—69.

[31] Shaffer RE, Rengasamy S. Respiratory Protection against airborne nanoparticles: a review. J Nanopart Res 2009;11:1661—72.

[32] Health and Safety Executive. HSG 53: A guide to the selection and use of respirators. Available at: http://www.hse.gov.uk/Pubns/priced/hsg53.pdf; 2003.

[33] NIOSH. NIOSH Respirator Selection Logic. US Department of Health and Human Services, Centers for Disease Control, National Institute for Occupational Safety and Health, Cincinnati, OH. DHHS (NIOSH), NIOSH Docket Number: NIOSH 100. Available at: http://www.cdc.gov/niosh/docs/2005-100/pdfs/2005-100.pdf; 2004.

[34] Nanosafe2. Dissemination Report: Are conventional protective devices such as fibrous filter media, cartridge for respirators, protective clothing and gloves also efficient for nanoaerosols?. DR-325/326-200801-1. Available at: http://www.nanosafe.org/home/liblocal/docs/Dissemination%20report/DR1_s.pdf; 2008.

[35] Federal Institute for Occupational Safety and Health. Hazardous Substances Ordinance of 26 November 2010. I p. 1643. Federal Institute for Occupational Safety and Health (BAuA) Available at: http://www.baua.de/en/Topics-from-A-to-Z/Hazardous-Substances/TRGS/pdf/Hazardous-Substances-Ordinance.pdf?__blob=publication File&v=9; 2010.

[36] Deutsche Gesetzliche Unfallversicherung. Working Safely in Laboratories. Basic Principles and Guidelines. Jedermann-Verlag, Heidelberg. Available at: http://bgi850-0.vur.jedermann.de/index.jsp?isbn=bgi850-0&alias=bgc_bi850_0_bi850_0e_1_; 2012.

[37] NIOSH. General Safe Practices for Working with Engineered Nanomaterials in Research Laboratories. US Department of Health and Human Services Centers for Disease Control, National Institute for Occupational Safety and Health. Cincinnati, OH. DHHS (NIOSH), NIOSH Docket Number: NIOSH 147. Available at: http://www.cdc.gov/niosh/docs/2012-147/pdfs/2012-147.pdf; 2012.

[38] The UK NanoSafetey Partnership Group (UKNSPG). Working Safely with Nanomaterials in Research & Development. Available at: http://www.safenano.org/Portals/3/SN_Content/Documents/Working%20Safely%20with%20Nanomaterials%20-%20Release%201%200%20-%20Aug2012.pdf; 2012.

[39] The British Standards Institution. Fume Cupboards. Vocabulary. United Kingdom: The British Standards Institution, London; 2003. BS EN 14175−1.

[40] The European Union. Directive 2003/18/EC of the European Parliament and of the Council of 27 March 2003 amending Council Directive 83/477/EEC on the protection of workers from the risks related to exposure to asbestos at work. Available at: http://eur-lex.europa.eu/LexUriServ/LexUriServ.do?uri=OJ:L:2003:097:0048:0052:EN:PDF; 2003.

[41] NIOSH. Occupational Exposure to Titanium Dioxide. Current Intelligence Bulletin 63, US Department of Health and Human Services, Centers for Disease Control, National Institute for Occupational safety and Health. Cincinnati, OH. DHHS (NIOSH), NIOSH Docket Number: NIOSH 160. Available at: www.cdc.gov/niosh/docs/2011-160; 2011.

[42] U.S. Supreme Court. Industrial Union Department, AFLCIO v. American Petroleum Institute, et al., Case Nos. 78-911, 78-1036. Supreme Court Register 100; 1980. 2844−2905.

[43] ENRHES Project. Engineered Nanoparticles - Review of Health and Environmental Safety (ENRHES). Available at: http://ihcp.jrc.ec.europa.eu/whats-new/enhres-final-report; 2010.

[44] Pauluhn J. Subchronic 13-week inhalation exposure of rats to multiwalled carbon nanotubes: toxic effects are determined by density of agglomerate structures, not fibrillar structures. Toxicol Sci 2009;113:226−42.

[45] Ma-Hock L, Treumann S, Strauss V, et al. Inhalation toxicity of multiwall carbon nanotubes in rats exposed for 3 months. Toxicol Sci 2009;112:468−81.

[46] Nanocyl. Responsible Care and Nanomaterial Case Study Nanocyl. Prague. Available at: http://www.cefic.org/Documents/ResponsibleCare/04_Nanocyl.pdf; 2009.

[47] British Standards Institution. Nanotechnologies − Part 2: Guide to safe handling and disposal of manufactured nanomaterials. United Kingdom: The British Standards Institution, London; 2007. BSI PD 6690-2:2007.

[48] Federal Institute for Occupational Safety and Health. Risk figures and exposure-risk relationships in activities involving carcinogenic hazardous substances. Federal Institute for Occupational safety and Health (BAuA), Announcement on Hazardous Substances 910. Available at: http://www.baua.de/en/Topics-from-A-to-Z/Hazardous-Substances/TRGS/pdf/Announcement-910.pdf?__blob=publicationFile&v=2; 2008.

[49] German Committee for Hazardous Substances. The risk-based concept for carcinogenic substances developed by the Committee for Hazardous Substances, Federal Institute for Occupational safety and Health (BAuA). Available at: http://www.baua.de/de/Publikationen/Broschueren/A85.html;jsessionid=3B40A3D4BE97AEDC4FA14CC64142695A.1_cid389; 2013.

[50] Institute for Occupational Safety and Health of the German Social Accident Insurance. Available at: http://www.dguv.de/ifa/en/fac/nanopartikel/beurteilungsmasssstaebe/index.jsp.

[51] Schulte PA, Murashov V, Zumwalde R, et al. Occupational exposure limits for nano-materials: state of the art. J Nanopart Res 2010;12:1971–87. DOI 10.1007/s11051-010-0008-1.

[52] Dekkers S, de Heer C. Tijdelijke nano-referentiewaarden. Bruikbaarheid van het concept en van de gepubliceerde methode. Rapport 601044001/2010. Natl Inst Public Health Environ (RIVM). Available at: http://www.rivm.nl/bibliotheek/rapporten/601044001.pdf; 2010.

[53] Van Broekhuizen P, van Veelen W, Streekstra W-H, et al. Exposure Limits for Nanoparticles: Report of an International Workshop on Nano Reference Values. Ann Occup Hyg 2012;56(5):515–24. Available at DOI 10.1093/annhyg/mes043

[54] The Social and Economic Council of the Netherlands (SER). Provisional nano reference values for engineered nanomaterials. Advisory Report. The Hague, The Netherlands. Available at: http://www.ser.nl/~/media/Files/Internet/Talen/Engels/2012/2012_01/2012_01.ashx; 2012.

[55] Kuempel ED, Castranova V, Geraci CL, Schulte PA. Development of risk-based nanomaterial groups for occupational exposure control. J Nanopart Res 2012;14. 1029–10. DOI 10.1007/s11051-012-1029-8.

[56] International Organization for Standardization. General framework for the develop-ment of occupational exposure limits for nano-objects and their aggregates and ag-glomerates, International Organization for Standardization, Geneva, Switzerland ISO/AWI TR 18637. Available at: http://www.iso.org/iso/home/store/catalogue_tc/catalogue_detail.htm?csnumber=63096.

[57] Federal Institute for Occupational Safety and Health. Manufactured Nanomaterials. Announcement on Hazardous Substances 527. Federal Institute for Occupational safety and Health (BAuA). Available at: http://www.baua.de/en/Topics-from-A-to-Z/Hazardous-Substances/TRGS/pdf/Announcement-527.pdf?__blob=publicationFile&v=2; 2013

[58] Gebel T. Small difference in carcinogenic potency between GBP nanomaterials and GBP nanomaterials. Arch Toxicol 2012;86:995–1007. DOI 10.1007/s00204-012-0835-1.

[59] German Research Foundation (DFG). New Threshold Values for "Fine Dust" at the Workplace. Press Release No. 37 as of 19 July 2011 of German Research Foundation (DFG) - Commission for the Investigation of Health Hazards of Chemical Compounds in the Work Area. Available at: http://www.dfg.de/en/service/press/press_releases/2011/press_release_no_37/index.html; 2011.

[60] Lyall C, Tait J. Developing an Integrated Policy Approach to Science, Technology, Risk and the Environment. Ashgate Publishing Limited, Hants; 2005.

[61] International Risk Governance Council. White Paper on Risk Governance: Towards an Integrative Approach. Geneva. Available at: http://irgc.org/wp-content/uploads/2012/04/IRGC_WP_No_1_Risk_Governance__reprinted_version_3.pdf; 2005.

[62] International Risk Governance Council. White Paper on Nanotechnology Risk Governance. Geneva. Available at: http://irgc.org/wp-content/uploads/2012/04/IRGC_white_paper_2_PDF_final_version-2.pdf; 2006.

[63] The European Parliament and The Council of the European Union. Regulation (EC) No 1907/2006 of the European Parliament and of the Council of 18 December 2006 con-cerning the Registration, Evaluation, Authorisation and Restriction of Chemicals (REACH). Entered into force on 1 June 2007. Official J Eur Union; 2007.

[64] European Chemicals Agency. Helsinki, Finland. http://echa.europa.eu/

[65] Commission European. Nanomaterials in REACH. Commission document CA/59/2008 rev. 1, Brussels. Available at: http://ec.europa.eu/environment/chemicals/reach/pdf/nanomaterials.pdf; 2008.

[66] European Commission. Communication from the Commission to the European Parliament, the Council and the European Econimic and Social Committee – Second Regulatory Review on Nanomaterials (Text with EEA relevance) COM (2012)

572 final. Available at: http://eur-lex.europa.eu/Result.do?T1=V5&T2=2012&T3=572&RechType=RECH_naturel&Submit=Search; 2012.

[67] European Commission. Commission Recommendation of 18 October 2011 on the definition of nanomaterial (Text with EEA relevance) (2011/696/EU). Official J Eur Union L 275/38 of 20.10.2011, Brussels. Available at: http://eur-lex.europa.eu/LexUriServ/LexUriServ.do?uri=OJ: L:2011:275:0038:0040:EN: PDF; 2011.

[68] European Commission. Nanomaterials. http://ec.europa.eu/environment/chemicals/nanotech/

[69] European Chemicals Agency. "Guidance on REACH". Helsinki. Available at http://echa.europa.eu/web/guest/guidance-documents/guidance-on-reach

[70] European Commission. http://ec.europa.eu/environment/chemicals/nanotech/#ripon

[71] European Chemicals Agency. "Guidance on Information Requirements and Chemical Safety Assessment". Helsinki. Available at, http://echa.europa.eu/web/guest/guidance-documents/guidance-on-information-requirements-and-chemical-safety-assessment

[72] International Uniform Chemical Information Database 5. Available at: http://iuclid.eu/

[73] European Chemicals Agency. Community Rolling Action Plan (CoRAP) of 29 February 2012. Helsinki. Available at: http://echa.europa.eu/documents/10162/17221/corap_2012_en.pdf.

[74] European Chemicals Agency. Community Rolling Action Plan (CoRAP). Helsinki. Available at: http://echa.europa.eu/web/guest/regulations/reach/evaluation/substance-evaluation/community-rolling-action-plan

[75] European Chemicals Agency. Workshop on nanomaterials − proceedings. Helsinki. Available at: http://echa.europa.eu/documents/10162/5402174/2_workshop_on_nanomaterials_proceedings_en.pdf; 2012.

[76] European Commission. Nano Support Project − Scientific technical support on assessment of nanomaterials in REACH registration dossiers and adequacy of available information. Available at: http://ec.europa.eu/environment/chemicals/nanotech/pdf/jrc_report.pdf; 2012.

[77] European Chemicals Agency. Candidate List of Substances of Very High Concern for Authorization Helsinki. Available at: http://echa.europa.eu/web/guest/candidate-list-table.

[78] European Commission - Scientific Committee on Emerging and Newly Identified Health Risks (SCENIHR). http://ec.europa.eu/health/scientific_committees/emerging/opinions/scenihr_opinions_en.htm#nano.

[79] European Commission - Scientific Committee on Emerging and Newly Identified Health Risks (SCENIHR). "Modified Opinion (after public consultation) on: The appropriateness of existing methodologies to assess the potential risks associated with engineered and adventitious products of nanotechnologies". Adopted by the SCENIHR during the 10th plenary meeting of 10 March 2006 after public consultation. Available at: http://ec.europa.eu/health/archive/ph_risk/committees/04_scenihr/docs/scenihr_o_003b.pdf; 2006.

[80] United States Environmental Protection Agency. Nanotechnology White Paper. EPA 100/B-07/001. Available at: http://nepis.epa.gov/EPA/html/DLwait.htm?url=/Exe/ZyPDF.cgi?Dockey=60000EHU.PDF; 2007.

[81] United States Congress. The Toxic Substances Control Act (TSCA) (15 U.S.C. 2601-2692). TSCA, Public Law [Pub. L.] 94-469. Available at: http://www.epw.senate.gov/tsca.pdf; 1976.

[82] Nash J. The Massachusetts Toxic Use Reduction Act: a model for nanomaterials regulation? J Nanopart Res 2012;14(1070). DOI 10.1007/s11051-012-1070-7.

[83] Howard J. Commentary: Seven Challenges for the Future of Occupational Safety and Health. J Occup Environ Hyg 2010;7:D11−8.

[84] Maynard AD. Don't define nanomaterials. Nature 2011;475:31.

[85] Maynard AD, Warheit DB, Philbert MA. The New Toxicology of Sophisticated Materials: Nanotoxicology and Beyond. Toxicol Sci 2010;120(S1):S109—29.

[86] Som C, Berges M, Chaudhry Q, et al. The importance of life cycle concepts for the development of safe nanoproducts. Toxicology 2010;269:160—9.

[87] Meng H, Xia T, George S, Nel AE. A predictive toxicological paradigm for the safety assessment of nanomaterials. ACS Nano 2009;3:1620—7.

[88] Hartung T. Toxicology for the twenty-first century. Nature 2009;460:208—12. Available at: http://www.nature.com/nature/journal/v460/n7252/pdf/460208a.pdf.

[89] Nel A, Xia T, Meng H, et al. Nanomaterial Toxicity Testing in the 21st Century: Use of a Predictive Toxicological Approach and High-Throughput Screening. Acc Chem Res 2013;46(3):607—21. http://dx.doi.org/10.1021/ar300022h. Publication date: June 7, 2012.

[90] Roebben G, Rasmussen K, Kestens V, et al. Reference materials and representative test materials: the nanotechnology case. J Nanopart Res 2013;15:1455.

[91] Join the dialogue. Nat Nanotechnol 2012;7(545). Available at: http://www.nature.com/nnano/journal/v7/n9/full/nnano.2012.150.html.

9

Future Outlook of Engineered Nanomaterials and Nanotechnologies

Kai Savolainen

Nanosafety Research Centre, Finnish Institute of Occupational Health, Helsinki, Finland

9.1 INTRODUCTION

The use of engineered nanomaterials and applications of these materials in products incorporating them will increase rapidly by the end of this decade and further into the foreseeable future. The European Union has assessed that the nanotechnologies belong to the six key enabling technologies (KETs) that will be the key technological drivers of European competitiveness and play a key role in promoting well-being and contributing to the wealth of European citizens. Other KETs include, for example, biotechnology and the development of novel materials. Typical for all KETs is that they are closely interlinked and support each other [1]. Nanotechnological applications will become commonplace in European and global markets and the use of consumer products containing nano-enabled technologies will become routine [2].

Current trends suggest that the number of workers in companies using engineered nanomaterials and producing nano-enabled products worldwide will double every three years. This will represent a 3 trillion US dollar market with about 6 million jobs by 2020 [3]. For this reason, the importance of identification of those workplaces in which exposure to engineered nanomaterials will take place has become increasingly important [4]. Furthermore, the presence of ENM in addition to consumer and other products will make these materials ubiquitous in our societies

Handbook of Nanosafety
http://dx.doi.org/10.1016/B978-0-12-416604-2.00009-3

327

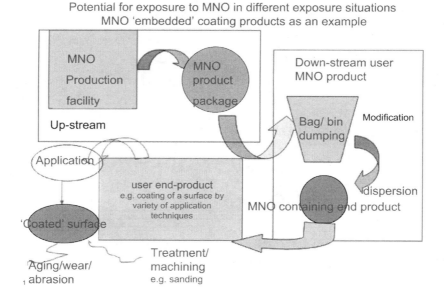

FIGURE 9.1 **Schematic diagram of the potential for exposure to MNO in different exposure situations [5].** *Reprinted from Brouwer, D. Exposure to manufactured nanoparticles in different workplaces. Toxicology 269: 120–7. Copyright (2010), with permission from Elsevier Ltd.*

in the future in such a way that contacts with these materials will become unavoidable. For this reason, the life-cycle and life-cycle assessment (LCA) of ENM and understanding the various steps of the ENM life-cycle will become increasingly important. Brouwer [5] depicts the life-cycle of the use of ENM which has been shown in Figure 9.1 and which also depicts the potential steps at which exposure to engineered nanomaterials can take place.

Systematic control of matter at the nanoscale by using direct measurement and simulations will lead in the near future to the creation of advanced materials, devices and nano-systems by design. However, this development also challenges our abilities to assure the safety of engineered nanomaterials and will require the creation of exposure-monitoring devices and technologies that allow assessment of exposure to engineered nanomaterials with easy-to-use, affordable, and even portable instruments which have on-line measurement capabilities allowing rapid actions to reduce levels of exposures when deemed necessary (www.nanodevice.org).

As indicated above, nanotechnology most likely will become widespread throughout societies globally by 2020. There is the potential to incorporate nanotechnology-based products and services to almost all consumer applications, industrial sectors and medical fields, e.g., medical imaging, non-invasive treatments and targeted drugs. This development will probably lead to mass production of nano-enabled products triggering remarkable changes in societies with much improved capabilities in terms of computing, energy production, production of clean water at a low price, and even environmental protection.

It is noteworthy that all these developments also carry the potential of inadvertent leakages and spillages leading to potential exposure of humans and the environment. Hence, at the same time as promoting the possibilities to exploit the benefits of nanotechnologies, one has to prepare for unexpected and unwanted effects of these materials and technologies. This latter goal can be achieved by developing technologies also for general monitoring purposes [3,6—8]. From the health and safety perspective, these new materials and technologies represent a remarkable challenge. An additional challenge for the monitoring of exposure and the assessment of safety of ENM will be the entry of 2nd-, 3rd- and 4th-generation complex materials with self-assembling properties into the market [3]. It is possible that the current risk and safety assessment tools and concepts have to be further developed and replaced with new ones as of yet unknown principles.

The new technology applications should not only be safe themselves but should also offer substantial improvements to human health and environment protection. Due to the rapid growth in the production and use of ENM and the utilization of nanotechnologies, safety aspects must be fully understood and addressed. This implies that safety assessment of the nanosized substances will be crucial. In fact, the safety of engineered nanomaterials and nanotechnologies has been emphasized, not only by the European Commission and several EU Member States, but also outside Europe, e.g., in the US 2011 National Nanotechnology Initiative's Environmental, Health, and Safety (EHS) Research Strategy [9] and in the recent report by the US National Research Council, 'A Research Strategy for Environmental, Health and Safety Aspects of Engineered Nanomaterials' [10]. The new EHS strategy of the US National Academy of Sciences [11] also emphasizes the importance of safety of ENM and nanotechnologies. It is also apparent that in addition to engineered nanomaterials and nanotechnologies, nanosafety has also become a generally global issue that has an impact on societies and economies at a global scale.

Indeed, it has been envisioned that ensuring the safety of processes as well as the technologies and products utilizing engineered nanomaterials will be crucial in guaranteeing the success of these materials, technologies and products. Successful promotion and growth of research on safety of engineered nanomaterials need to be able to promptly provide us with tools and methods that allow accurate and reliable prediction of hazards and risks associated with safe novel nanomaterials and nanotechnologies for the next 10–20 years.

EU-level research funding from the new framework program for research and innovation, Horizon 2020, and the recent National Nano-technology Initiative Environmental Safety and Health Strategy [9] have emphasized the need for a strategic vision of the research priorities associated with the safety of ENM. The safety of engineered nano-materials needs to extend through their entire life-cycle from production until their ultimate disposal or recycling. This will require the establish-ment of a new safety culture through the development and imple-mentation of a complete system of methods, techniques and equipment, a competent scientific and technical community and a portfolio of measures to promote a total safety paradigm and to inform the general public about the safety management and governance of the technologies utilizing ENM. This new safety governance approach will require the commitment of all important stakeholders, notably the regulators, the industry, the social partners, and the research community.

9.2 NEW PRINCIPLES FOR ENGINEERED NANOMATERIALS RISK AND SAFETY GOVERNANCE

It is essential that the assessment of risks and their management will become routine among regulators and industry will have adopted a way of working which will guarantee the incorporation of safety and health into the novel engineered nanomaterials and nano-based products and processes thus positively contributing to nanosafety in global terms. This will mean the wide acceptance of the safe-by-design principle in both the design of novel ENM and the shaping of the industrial processes utilizing these novel materials. The knowledge and awareness related to engineered nanomaterials and nanotechnologies will have markedly improved, i.e., citizens will be able to understand the fundamental issues surrounding nano-based products and ENM. This will have been due to neutral and reliable dissemination of information on ENM by regulators and academics as well as by the manufacturing industries.

One of the key drivers of this positive development will have been the creation of a positive industrial attitude towards nanosafety. It will have become self-evident to the different parties involved in nanosciences and nanosafety that knowledge and trust are the essential elements for building successful nanotechnologies. This positive development will allow regu-lators and decision-makers to interact with other parties to carry out effective regulation on the safe use of ENM, thus guaranteeing the safety of nanotechnologies and the products emerging from these processes.

9.3 KEY AREAS OF NANOSAFETY RESEARCH AND THE EXPECTED ACHIEVEMENTS IN THESE RESEARCH AREAS

Exploitation of the potential benefits of ENM and nanotechnologies will require, in addition to quantifying and characterization of the chemical and physical metrics of these materials, an understanding of those characteristics of engineered nanomaterials that are crucial in their interactions with biological molecules such as proteins, lipids, sugars and DNA, i.e., extending their impacts down to the organ and organism level [12] in mammalian and environmental species. New tools such as computational toxicology may clarify the crucial material characteristics, e.g., surface chemistry, porosity, or charge, shape or size of a given particle, or a combination of these elements which lead either to biological availability or consequences. These new techniques [13] will help pave the way for the knowledge-based development of safe nanomaterials by design. In addition to this knowledge, means to measure and quantify these features will be required. In fact, these can be viewed as new opportunities to develop technologies for the quantification and assessment of new, sustainable engineered nanomaterials. Once appreciated, these key features will then have been introduced into the development of novel nano-sized engineered materials, and this knowledge can be incorporated into the up-scaling of the production of these novel materials.

9.4 FUTURE HEALTH AND SAFETY REQUIREMENTS OF NANOMATERIALS AND NANOTECHNOLOGIES AND IDENTIFICATION OF THE KNOWLEDGE NEEDS AND GAPS

Governmental funding agencies globally, the European Union, and private funding agencies have recently invested an impressive amount of resources to support research into the safety of nanomaterials. Close to 200 million euros have been allocated to nanosafety research from the EU 7th Framework Programme for Research during 2007–2013 (http://cordis.europa.eu/fp7/home_en.html). This growth in funding has also been reflected in the rapid growth of published papers in peer-review journals on nanosafety [9,11]. In spite of this support, the results emerging from the research on the safety of ENM have contributed less than expected on systematic toxicity testing studies of ENM that could be directly used for regulations on ENM. The actual breakthroughs in the development of novel risk assessment tools and concepts have been slow due to complexity of the issues related to ENM and nanotechnologies.

This has prompted the funding agencies, governments and the academic community to try to find ways to improve the cost–benefit ratio of the research being undertaken, and to emphasize the social dimension and societal needs of this research. The data emerging from nanosafety research have been emphasized recently [14] and the importance of the research to serve risk assessment, management and governance of ENM has received much attention. Much of the academic research has focused on mechanistic issues which though they have significant merits do not meet the urgent needs of society. At the same time, it is important to convey the realization that mechanistic clarifications may hold major implications in the future also from the regulatory perspective.

Because of the above reasons, a number of strategy documents aiming at guiding the direction of research on the safety of ENM and nanotechnologies [9,15,16] seek to identify a set of priorities which would serve societal needs to assure the safety of these new materials and technologies. In these documents, it has been noted that duplication of research should be avoided. The following issues have been emphasized in the current research strategy documents [9,17]:

1. Description of the current state of knowledge, describing the existing research landscape and identification of the requirements of the research environments and infrastructures that are essential for the promotion of research on safe ENM and nanotechnologies;
2. Identification of the societal needs for the regulation of safety of these materials and technologies;
3. Identification of the necessary research goals for fulfilling of the societal needs and setting a time-line with milestones for the follow-up of the progress of the research endeavors;
4. Identification of the research priorities that will allow achievement of the set goals within the time-limit set by the strategy;
5. Identification of the means by which the results of the research can be disseminated, implemented and exploited to evoke a change in the ways that different industry sectors can promote the safe use of these materials and assure the safety of workers and consumers, and enabling regulators to make educated decisions;
6. Identification of the needs for further possible regulatory actions and possible further investments into infrastructures, educational and funding programs to be able to fully utilize the technological and economic benefits of these materials and technologies.

Many of these elements can be generalized to any given research strategy, but when dealing with new and emerging technologies such as nanotechnologies, public perception should always be at the heart of the strategic activities, and has to be carefully considered.

9.5 THE COMPLEXITY OF NANOMATERIALS AND THEIR CONTROL AND REGULATION

Engineered nanomaterials are characterized by their complexity in terms of their physico-chemical characteristics, behavior in different media and their complicated and unexpected interactions with living systems (see [12,18,19]). It is reasonable to consider these materials conceptually as a paradigm for the 21st century [20]. Indeed, nanotechnologies demand creativity because at the nano-scale of matter, new laws of chemistry and physics apply, e.g., at this scale matter has the potential to interact with living organisms at the cellular and molecular level — cells in essence are planets in a biological micro-cosmos, the elements of which are expressed on the nanometer scale, conferring a qualitatively novel dimension to the material-biology field of play [21].

Nanotechnology and engineered nanomaterials have already had a major impact on electronics, coatings, construction, food technology, the design of new materials such as nanocellulose applications, telecommunication, environmental technologies, medical technologies and drug development, and energy production as well as their potential use in agriculture, water purification systems and the utilization of solar energy. The possibilities associated with nanomaterials are seemingly limitless; at present this development seems to be in the direction towards greater complexity. This will also pose challenges for the regulation and societal control of unlimited exploitation of these possibilities. In addition to the societal considerations, the possible impacts of these materials and technologies on human health and the consequences of these materials and technologies on the environment will become increasingly important. It is not surprising that public concern about the potential hazards inherent in the applications of these materials and technologies is on the rise, and these legitimate concerns obviously need to be taken into account.

9.6 PROGRESS OF NANOTECHNOLOGIES ON SAFETY REQUIREMENTS OF NANOMATERIALS AND TECHNOLOGIES

Nanotechnologies and ENM incorporated into commercial products are emerging at a growing speed. There are hopes that this development will confer marked expected economic and technological benefits but there is only very limited knowledge on their possible hazard potential, or exposure to these materials in workplaces, and to the general public through consumer products, as well as their presence in food, water, and even the air. The urgent current challenge related to ENM and nanotechnologies is to acquire relevant knowledge that can be utilized for reliable risk and safety assessment, adequate risk management and governance of these materials and products [11,22]. Even though there are increasing amounts of information about the hazard potential of several engineered nanomaterials from a variety of *in vitro* systems, there is a serious lack of systematic and relevant information about the potential hazards of these materials which can be adopted to help decision-making regulators to have access to adequate risk assessment data.

In order to be able to respond to the needs of society for safe nano-materials and nanotechnologies it will be necessary to increase investments on nanosafety research with systematic short-term and long-term animal studies providing a reliable estimate of the possible hazards and risks of engineered nanomaterials.

Currently, the vast majority of the available studies which have investigated relatively small numbers of nanomaterials cannot be used widely for the assessment of risks and safety of ENM. In its assessment of the future needs for research on ENM and nanotechnologies, the EU NanoSafety Cluster (http://www.nanosafetycluster.eu) identified the following areas of nanosafety research of strategic importance as crucial for assuring the future safety and success of engineered nanomaterials and nanotechnologies. These four topics are: 1) material characteriza-tion, i.e., determining the bio-identity of the materials; 2) assessment of exposure, transformation and release of ENM with emphasis on novel measuring technologies of materials in different matrices during the whole life-cycle; 3) hazard mechanisms of the materials including studies on the biokinetics of the materials; and 4) risk assessment with an emphasis on risk prediction tools. Figure 9.1 also emphasizes the close association of the chosen research priorities with safety testing of engineered nanomaterials, and the importance of this testing in facili-tating innovations and breakthrough in the development of new engi-neered nanomaterials and nanotechnologies. In addition, the Cluster also identified a number of cross-cutting issues of high importance for the successful implementation of these types of strategic research un-dertakings. These issues will be discussed below. For clarity, Nano-Safety Cluster (NSC) is composed of all EU-Commission-funded nanosafety research projects from the Framework Programmes for Research and Innovation. The formation of the NSC in 2009 was encouraged by the EU Commission with which the NSC currently closely collaborates. More information on the cluster can be found at www.nanosafetycluster.eu.

The present situation is especially crucial for the future of safety and risk assessment of engineered nanomaterials because our testing re-sources do not allow an adequate assessment of safety and risks of these materials [19]. Hartung [20] in his paper 'Toxicology of the 21st Century' published in *Nature* emphasized the need to depart from the existing experimental animal testing paradigms and to adopt the use of high-throughput testing approaches. This is a highly positive proposal supported by the recent proposal by Nel et al. [23], also emphasizing the use of *in vitro* approaches applying high-throughput technologies and -omics methodologies in the safety and toxicity testing of novel nanomaterials. From the regulatory point of view, these proposals seem praiseworthy, but both lack the fundamental foundation on which to justify their adoption. As long as there is no clear evidence of the predictive power of these approaches for *in vivo* systems, it is highly unlikely that they can be accepted as a part of the regulatory framework of nanomaterials or other similar chemicals, i.e., these techniques are associated with too many uncertainties.

9.7 CONCLUSIONS

The challenges associated with the safety of engineered nanomaterials and nanotechnologies are global. The many benefits of engineered nanomaterials and nanotechnologies are threatened by the potentially harmful characteristics of these materials, since they are often intimately linked with the technological benefits. As these emerging materials and technologies have been identified as the key enabling technologies promoting the ability of technology-based industrial sectors to compete globally, their success has to be based on an assurance that those materials are safe, not toxic time-bombs. It has to be stated that today's tools used in the assessment of safety are often inappropriate, or so laborious that adequate assessment of safety of these materials and technologies remains highly problematic. The current resources or testing tools are not likely to provide a safety assessment of the novel nanomaterials that are flooding the markets. This means that new risk/safety assessment paradigms need to be promptly developed in order to solve this problem. At the same time, it is important to maintain the present level of resources which apply the currently available means of risk and safety assessment of existing engineered nanomaterials and to support the regulators and the nanotechnology industry. This is important since these enterprises may hold the keys to economic prosperity and well-being not only of EU citizens but also for all the world's inhabitants.

The current situation calls for the rapid identification of research priorities and the development of a roadmap to guide efforts in estimating the safety of nanomaterials and nanotechnologies in the future. This goal is not possible without a clear understanding of the current research landscape within Europe and beyond. It is important to create environments that encourage the effective use of the available resources.

The following multi-disciplinary issues have been identified by the NSC as being important drivers of successful execution of nanosafety research: 1) building key infrastructures around the European Union and worldwide supporting and encouraging this line of research; 2) setting up and developing of regulatory framework and funding that provides opportunities for nanosafety research; 3) promoting of dissemination results of the key findings of the research; 4) enabling active networking within the scientific community and between the international scientific communities, the industry and the regulators; and 5) removing obstacles that prevent translation of the results of the research towards their commercialization via innovations and new products. Due to the nature of research and the complexities inherent in interpreting the results of nanosafety research, this has to be a thorough process, and hence major breakthroughs and giant leaps in these areas cannot be expected overnight. A time frame of 10—15 years is more realistic.

Acknowledgements

The authors acknowledge of the support of the EU FP7-funded NANODEVICE project providing support for the preparation of the manuscript.

References

[1] Horizon 2020. http://ec.europa.eu/research/horizon2020/index_en.cfm; 2013.
[2] European Commission SWD. 288 final. (2012). Commission staff working paper — Types and uses of nanomaterials, including safety aspects. Available at http://eur-lex .europa.eu/LexUriServ/LexUriServ.do?uri=SWD:2012:0288:FIN:EN:PDF; 2012.
[3] Roco MC, Mirkin CA, and Hersam MC. 2010. Nanotechnology research directions for societal needs in 2020: retrospective and outlook summary. NSF/WTEC (National Science Foundation/World Technology Evaluation Center) report (summary of the full report published by Springer).
[4] Schulte P, Geraci C, Zumwalde R, et al. Sharpening the focus on occupational safety and health in nanotechnology. Scand J Work Environ Health 2008;34:471—8.
[5] Brouwer D. Exposure to manufactured nanoparticles in different workplaces. Toxicology 2010;269:120—7.
[6] Auffan M, Rose J, Bottero JY, et al. Towards a definition of inorganic nanoparticles from an environmental, health and safety perspective. Nat Nanotechnol 2009;4:634—41.
[7] van Broekhuizen PJC. Nano Matters: Building Blocks for a Precautionary Approach. PhD thesis. Available at http://www.ivam.uva.nl/?nanomatters; 2012. P.198.
[8] van Broekhuizen P, van Broekhuizen F, Cornelissen R, Reijnders L. Workplace exposure to nanoparticles and the application of provisional nano reference values in times of uncertain risks. J Nanopart Res 2012;14:770.
[9] National Nanotechnology Initiative (NNI). Environmental, Health, and Safety (EHS) Research Strategy. Available at http://www.nano.gov/sites/default/files/pub_ resource/nni_2011_ehs_research_strategy.pdf; 2011.
[10] National Research Council (NRC). A Research Strategy for Environmental, Health and Safety Aspects of Engineered Nanomaterials. Available at https://download.nap.edu/ catalog.php?record_id=13347#toc; 2012.

[11] National Academy of Sciences (NAS). A Research Strategy for Environmental, Health, and Safety Aspects of Engineered Nanomaterials. Natl Res Counc Natl Academies 2012:1–211.

[12] Fadeel B, Feliu N, Vogt C, et al. Bridge over troubled waters: understanding the synthetic and biological identities of engineered nanomaterials. WIREs Nanomed Nanobiotechnol 2013;5:111–29.

[13] Hartung T, Sabbioni E. Alternative in vitro assays in nanomaterial toxicology. Wiley Interdiscip Rev Nanomed Nanobiotechnol 2011 [Epub ahead of print]. doi: 10.1002/wnan.153.

[14] European Commission COM. 572 final. Second regulatory review on nanomaterials. Available at http://eur-lex.europa.eu/LexUriServ/LexUriServ.do?uri=COM:2012:0572:FIN:EN:PDF; 2012.

[15] European Commission COM. 543. (2005). Nanosciences and Nanotechnologies: An Action Plan for Europe 2006-2009. Available at http://ec.europa.eu/nanotechnology/pdf/nano_action_plan2005_en.pdf; 2005.

[16] National Science Foundation. Empowering the nation through Discovery and innovation NSF Strategic Plan for Fiscal Years 2011-2016. Available at https://www.nsf.gov/news/strategicplan/nsfstrategicplan_2011_2016.pdf; 2011.

[17] European Commission COM. 2020 final. (2010). Communication from the Commission: EUROPE 2020 — A strategy for smart, sustainable and inclusive growth. Available at http://eur-lex.europa.eu/LexUriServ/LexUriServ.do?uri=COM:2010:2020:FIN:EN:PDF; 2010.

[18] Nel A, Xia T, Mädler L, Li N. Toxic potential of materials at the nanolevel. Science 2006;311:622–7.

[19] Nel AE, Mädler L, Velegol D, et al. Understanding biophysicochemical interactions at the nano–bio interface. Nat Mater 2009;8:543–57.

[20] Hartung T. Toxicology for the twenty-first century. Nature 2009;460(7252):208–12. http://dx.doi.org/10.1038/460208a.

[21] Shvedova AA, Kagan VE, Fadeel B. Close encounters of the small kind: adverse effects of man-made materials interfacing with the nano-cosmos of biological systems. Annu Rev Pharmacol Toxicol 2010;50:63–88.

[22] International Risk Governance Council (IRGC). White paper on nanotechnology risk governance. Geneva: IRGC; 2006. Available at http://www.irgc.org/IMG/pdf/IRGC_white_paper_2_PDF_final_version-2.pdf.

[23] Nel A, Xia T, Meng H, et al. Nanomaterial Toxicity Testing in the 21st Century: Use of a Predictive Toxicological Approach and High-Throughput Screening. Acc Chem Res 2012. 2012 Jun 7. [Epub ahead of print].

Index

Tubing system cleaning in pilot plant
facility, 48—49
exposure estimation, 49
operational conditions and risk
management measures, 48—49
Turbulent diffusion, 139—140, 139f
Two-way coupling, 144—145

U

Ubiquitous background ultrafine
particles, 8
UHMA model, 148—149
Ultra clean areas of semiconductor
industry, equipment emissions
used in, 268—275
Ultrafine condensation particle counters
(UCPCs), 177—178, 232—233
Ultrafine metals, 18
Ultrafine particle counter, 42
Ultrafine water-based condensation
particle counter (UWCPC), 46—47
Ultraviolet lithography, 268—269
UNC Passive Aerosol Sampler, 190—191
Uncertainty
extended uncertainty, in measurements,
213
about safety of engineered
nanomaterials, 62—65
Unipolar diffusion charging, 178
United States, risk governance approaches
in, 316—317
US 2011 National Nanotechnology
Initiative
Environmental, Health, and Safety (EHS)
Research Strategy, 329—330
US Environmental Protection Agency
(EPA), 316—318
US National Academy of Sciences (NAS),
248, 254, 329
US National Institute for Occupational
Safety and Health, 112
US National Research Council, 61—62, 329
Users, responsibilities of, 213—216

V

Vacuum cleaning of reactor in multi-
walled carbon nanotubes pilot
plant facility, 50—51
exposure estimation, 51
operational conditions and risk
management measures, 50—51
Value chain of engineered nanomaterials,
3—4, 4f

van-der-Waals forces, 140
Vascular system, effects of engineered
nanomaterials on, 73—74
Ventilation, 290
LEV. *See* Local exhaust ventilation (LEV)
Victims of exposure, identification of, 285
Volatile organic compounds (VOCs),
226—227
Volume of inhalation, 161

W

Water condensation, 160
Weighing of nanomaterials, 298
of multi-walled carbon nanotubes in pilot
plant facility, 49—50
exposure estimation, 50
operational conditions and risk
management measures, 49—50
of silica nanospheres, 43—45
exposure estimation, 44—45
measurement data for, 44t
operational conditions and risk
management measures, 43—44
WHAT-IF function (MS-Excel), 260
Wide-range real-time classifier, 198
Workplaces
air, surveillance of, 191—192
exposure to engineered nanomaterials in,
6—7, 34, 118f
nanoparticles sources in, 135—136
paint-manufacturing, 225, 255—268
simulated, aerosols evolution modelling
in, 152—157, 153f, 154f, 155f, 155t,
156f

X

X-ray, 290—291
proton-induced X-ray emission, 188
soft X-ray bipolar chargers, 179—180, 186
total reflection X-ray fluorescence
spectometer, 224—225, 248, 250—253

Z

Zinc oxide nanoparticles (ZnONP), 9—10,
26—28
absorption of, 69
dermal absorption of, 91—92
dissolubility of, 67—68
effect on pulmonary inflammation, 85
genotoxicity of, 103
photo-genotoxic properties of, 100—101
Zirconia/zirconium dioxide (ZrO$_2$),
dissolubility of, 67—68

Edwards Brothers Malloy
Ann Arbor MI. USA
October 25, 2014